SOTHIS PRESS

Sothis Press aims to completely reimagine the public domain book with an emphasis on universal access to captivatingly designed print editions. We believe that technology will enable every single book ever published to be perpetually accessible in beautiful print form. Sothis is beauty in ubiquity.

www.sothis.press
1 Waterhouse Square, London, EC1N 2ST

Astronomy, Determination of Time, Longitude, Latitude, and Azimuth

William Bowie Coast and Geodetic Survey

SOTHIS PRESS

ISBN 978-1-0365-1040-4

Astronomy, Determination of Time, Longitude, Latitude, and Azimuth is in the public domain.

This edition is published by Sothis Press © 2023. The company is registered in England and Wales. The Company Number is 14900389.

DEPARTMENT OF COMMERCE

U. S. COAST AND GEODETIC SURVEY

O. H. TITTMANN

SUPERINTENDENT

ASTRONOMY

DETERMINATION OF TIME, LONGITUDE LATITUDE, AND AZIMUTH

FIFTH EDITION

BY

WILLIAM BOWIE

Inspector of Geodetic Work and Chief of the Computing Division

U. S. Coast and Geodetic Survey

SPECIAL PUBLICATION No. 14

WASHINGTON

GOVERNMENT PRINTING OFFICE

1913

CONTENTS.

TABLES.

ILLUSTRATIONS.

DETERMINATION OF TIME, LONGITUDE, LATITUDE, AND AZIMUTH.

By William Bowie,

Inspector of Geodetic Work and Chief of the Computing Division, U. S. Coast and Geodetic Survey.

INTRODUCTION.

From time to time during many years publications have been issued describing the instruments and methods used by the Coast and Geodetic Survey in the determination of time, longitude, latitude, and azimuth. The general aim has been to provide a working manual which would serve as a guide to the observer in the field and the computer in the office in carrying on the astronomic work of the Survey in a systematic manner. The exhaustion of previous editions and the introduction of new instruments and methods have made necessary the successive editions, in each of which much has been repeated from the preceding one.

The edition of the last publication is now exhausted, which gave in one volume descriptions of the instruments and methods, and was entitled "Determination of Time, Longitude, Latitude, and Azimuth." It was published as Appendix No. 7, Report for 1898. The needs of the members of this Survey for a similar manual, and requests for it by others, make it desirable to issue the present and fifth edition.

The subject matter includes most of that in the fourth edition, with a number of changes, however. Some of the most important additions to the previous edition are: The determination of time and longitude, using the transit micrometer; the description of the transit micrometer; determination of time with the vertical circle for use in connection with azimuth observations; a description of the method of observing azimuth coincidently with horizontal directions in primary triangulation; an example of the determination of an azimuth in Alaska with a transit equipped with a transit micrometer; examples of the records and computations in the different classes of work, as actually made at present by the Survey; and statements of the field cost of the different classes of work. A number of new illustrations have been added.

The writer takes pleasure in acknowledging here his indebtedness to Mr. H. C. Mitchell, Mr. C. R. Duvall, and several other members of the Computing Division who assisted in preparing this edition. The material is principally the work of former Assistant C. A. Schott, who prepared the first three editions, and of former Assistant John F. Hayford, who prepared the fourth edition.

It has not been deemed necessary to insert the derivation of formulæ, except in the few rare cases in which such derivation can not be found readily in textbooks on astronomy. For general developments the reader is therefore referred to Chauvenet's Astronomy, to Doolittle's Practical Astronomy, and to Hayford's Geodetic Astronomy. The last-mentioned book and the fourth edition of this publication appeared about the same time, and as they were by the same author it is natural that some of the text is identical in the two. Much of this publication was copied from the fourth edition without change, and some portions are necessarily identical with the corresponding parts of Prof. Hayford's textbook.

In addition to this manual on geodetic astronomy, the American Ephemeris and Nautical Almanac for the year of observation will be required in time and azimuth work, and the Boss Preliminary General Catalogue of 6188 stars, together with the Cape Tables, by Finlay, in latitude determinations.

William Bowie,
Inspector of Geodetic Work, Chief of the Computing Division.

PART I.

DETERMINATION OF TIME.

GENERAL REMARKS.

This part deals almost exclusively with the portable transit instrument in its several forms as used in the Coast and Geodetic Survey, and when mounted in the plane of the meridian for the purpose of determining local sidereal time from observations of transits of stars, in connection with an astronomic clock or chronometer regulated to sidereal time. The use of this instrument when mounted in the vertical plane of a close circumpolar star out of the meridian is not recommended on account of the greater complexity both in field and office work, as compared with the usual method herein discussed, especially when one considers the ease with which a transit may be placed approximately in the meridian. (See p. 16.) The observations are made either by the method of "eye and ear," or by chronographic registration. The latter method is used exclusively for all telegraphic longitude work and in making time observations for determining the periods of the pendulums in gravity determinations. In using the first method the observer will, of course, mark his own time; that is, he will pick up the beats of the chronometer and carry them forward mentally up to the time of transit of the star, which he will estimate to the nearest tenth of a second. In using the second method the chronograph record will be produced in one of two ways: First, when the observer sees the star bisected by a line of the diaphragm he will press an observing key (break-circuit) held in his hand and cause a record of that instant to appear on the chronograph sheet; or, second, he will follow the star across the field of the telescope with the movable wire of the transit micrometer, the star being continuously bisected as nearly as possible by the wire, and the record on the chronograph sheet will be made automatically by the make-circuit device of the micrometer.

DESCRIPTION OF LARGE PORTABLE TRANSIT.

Several sizes of portable transits are used in this Survey. The largest and oldest ones, made by Troughton & Simms, of London, were intended for use exclusively on the telegraphic determinations of longitude, but in 1888 a slightly smaller type of transit (described below) was made at the Survey office, and has been used very extensively since that time on the same class of work as the largest type. The smallest type of transit, known as the meridian telescope (described on p. 8), is used in the determination of the local time needed while observing astronomic azimuths and latitudes, and for other purposes. In the hands of skillful observers the instruments used for longitude determinations give results which compare favorably with the results obtained with the much larger transits usually employed at astronomic observatories, where special difficulties are encountered in consequence of strains or temporary instability of the instrument due to reversal of axis, and the more serious effect of flexure. In case of necessity, and when an approximate degree of accuracy suffices, any theodolite or altazimuth instrument may be converted temporarily into and used as an astronomic transit.

Illustration No. 1 shows Transit No. 18,[1] one of the second-sized portable transits made in the Survey office in 1888. It has a focal length of 94 cm. and a clear aperture of 76 mm. The magnifying power with the diagonal eyepiece ordinarily used is 104 diameters. It is provided with a convenient reversing apparatus, by means of which it can be reversed without lifting the

---

[1] For a full description of this instrument, see Appendix 9, Report for 1889, by Edwin Smith, Assistant.

telescope by hand. The value of one division (=2 mm.) of the striding level is 1''.35. The setting circles are 4 inches in diameter, are graduated to 20' spaces, and are read by verniers to single minutes.

Until about 1905 this, as well as the other transits of the Coast and Geodetic Survey, was supplied with a glass diaphragm, but, with the adoption of the transit-micrometer, the glass diaphragms were discarded. The glass diaphragm carries two horizontal lines which are simply to define the limits within which all observations should be made, and 13 vertical lines, 11 of which are used in making time observations with the chronograph and observing key and 5 of which (longer than the others) are used in making eye and ear observations. The shortest time interval between lines for chronographic observations is about 2½ seconds and for eye and ear observations about 10 seconds. The transit micrometer and its use are described below.

Transit No. 18 is provided with a sub-base which is firmly secured to the supporting pier. The transit proper is supported on this sub-base by three foot screws. At the left of the base in the illustration is shown a pair of opposing screws which serve to adjust the instrument in azimuth. One of these screws carries a graduated head which enables one to set the instrument very nearly in the meridian as soon as the azimuth error is known.

This instrument may serve as a typical illustration of the class of large portable transits.

The broken telescope transit, like that shown in illustration No. 2, has been used with marked success by other countries. This instrument may also be used in the determination of latitude by the Talcott method. This manual can be used with either type of instrument (broken or straight telescope).

DESCRIPTION OF MERIDIAN TELESCOPE.

Certain instruments are known in this Survey as meridian telescopes.[1] They are fitted both for time observations and for latitude observations by the Horrebow-Talcott method (see p. 103) and are provided with a frame which may be folded up for convenience in transportation. Illustration No. 3 shows Meridian Telescope No. 13, which may serve as an illustration of the type of smaller instruments used for time observations in this Survey.

This telescope has a focal length of 66 cm., a clear aperture of 5 cm., and a magnifying power of 72 diameters. The value of one division (=2 mm.) of the striding level is about 2¼''. During time observations the telescope is reversed by hand; during latitude observations it may be reversed by turning the upper half of the double base on the lower half. One of the two setting circles carries a delicate level for use in making latitude observations, and the eyepiece is fitted with a micrometer for measuring differences of zenith distance, in addition to the diaphragm carrying fixed vertical lines for use in making time observations. On one side of the base (the left-hand side in the illustration) is a slow-motion screw for accurate adjustment in azimuth.

THE TRANSIT MICROMETER.

The transit micrometer is a form of registering micrometer placed with its movable wire in the focal plane of an astronomic transit and at right angles to the direction of motion of the image of the star which is being observed at and near meridian transit. Certain contact points on the micrometer head serve to make an electric circuit as they pass a fixed contact spring, thus causing to be recorded upon the chronograph sheet each separate instant at which the micrometer wire reaches a position corresponding to a contact.

The transit micrometer in use on the transits of this Survey is hand driven and was designed by Mr. E. G. Fischer, Chief of the Instrument Division of the Survey, and made in that division. Much of the following description is copied from pages 458–460 of Appendix No. 8, Report for 1904, entitled "A test of the transit micrometer." The pages referred to were written by Mr. Fischer.

[1] See Appendix No. 7, Report for 1879, for a "Description of the Davidson Meridian Instrument."

LARGE PORTABLE TRANSIT (EQUIPPED WITH TRANSIT MICROMETER).

No. 2.

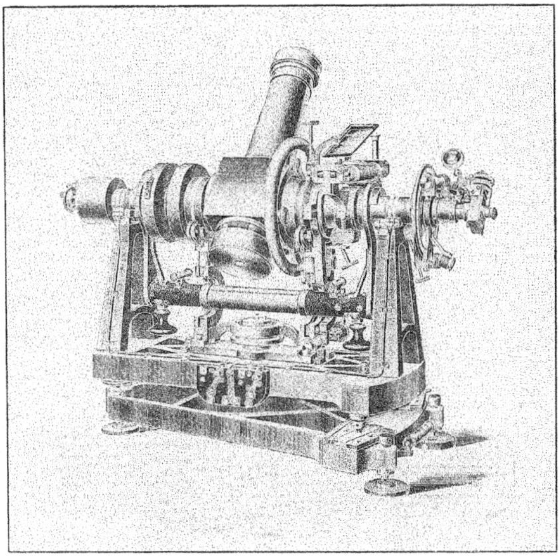

BROKEN TELESCOPE TRANSIT.

MERIDIAN TELESCOPE.

DESCRIPTION OF THE HAND-DRIVEN TRANSIT MICROMETER, MADE FOR COAST AND GEODETIC SURVEY TRANSIT NO. 2.

Before considering the details of this micrometer, three points were determined upon as being essential to insure accurate and decisive action, durability, and convenience in reading the chronograph record made by it.

First, it was decided that the mechanism of the slide carrying the wire should be of the form in which the screw is mounted in bearings at the extreme ends of the box or case holding the slide, the micrometer head being fast upon the end of the screw projecting from the box, because this insures greater stability under the side stress of the gears connecting the screw with the handwheel shaft than the form usually employed in theodolite and ocular micrometers, in which the screw is fastened to the slide and therefore takes part of whatever play there may be in the latter.

Second, it was decided that the electric recording device of the micrometer should be of the make-circuit form, transmitting its records to the chronograph, which is in the break-circuit of the chronometer, through a relay. This permits the use of a strong current through the contact points of the micrometer head, and therefore a minimum of pressure upon the latter by the contact spring.

Third, in order that the micrometer transmit no records except those made within an accepted space on either side of the line of collimation and forming the observations of the star transits proper, an automatic cut-out must be provided.

Illustrations 4 and 5 show the micrometer with draw tube and eye end of the telescope. The telescope has a focal length of 115 cm. and an aperture of 77 mm. It is of the straight type of the same general form as that shown in illustration No. 1 of Appendix 7 of the Report for 1898. (Illustration No. 1 of this publication.)

The micrometer box or case is 46 mm. in length and 31 mm. wide. Within it and near to one side is mounted the micrometer screw. Upon the latter fits, by a thread and cylindrical bearing, a rectangular frame forming the slide, which is 31 mm. long and 23 mm. wide. All play or lost motion, both of the slide upon the screw and the screw in its bearings, is taken up by means of a helical spring within the box, which, pressing from the inner end of the box against the slide and through it against the screw, holds the latter firmly against the point of an adjustable abutting screw, without impeding its free rotary motion. Upon the slide, at right angles to its line of motion, is mounted the single spider thread, which is used for bisecting the star during its passage across the field. Two threads, parallel to the line of motion, about four time seconds apart, and mounted against the inner surface of the box, define the space within which the observations should be made. A short comb of five teeth, with distances equal to one turn of the screw between them, is also provided and indicates the four whole turns of the screw within which the observations are to be made. The diameter of the field of view through the Airy diagonal eyepiece, which has an equivalent focal length of 12 mm., is something over 24 turns of the screw, thus giving a space of fully 10 turns of the screw on each side of the 4 turns in the center of the field.

That portion of the micrometer screw which projects through the box has the micrometer head fitted upon it and secured in position by a clamp nut. The cylindrical surface of this head, graduated at the edge nearest the box to 100 parts (*g*, illustration No. 4), also carries near its opposite edge a screw thread, *t*, of three turns with a pitch of 1 mm. and a diameter of 32 mm. Sunk into the outer face of the head and fitted concentrically with it is a thin metallic shell, which has fitted upon it a hollow cylinder, *e*, made of ebonite, 6 mm. long and 26 mm. in diameter. Five strips of platinum, each 0.4 mm. thick, and corresponding to the 12.5, 25.0, 50.0, 75.0, and 87.5 division points of the graduation, *g*, are slotted into the edge of the ebonite cylinder and secured in such manner as to make metallic contact with the micrometer head proper, and through it with the screw, micrometer box, telescope and telescope pivots, and the iron uprights of the transit. By releasing the clamp nut within the ebonite ring the graduated head, with its thread, *t*, can be adjusted, in a rotary sense, in relation to the thread of the screw, and therefore also to the spider thread upon the slide. At the same time the position of the platinum contact strips can be set to correspond to the zero of the graduation, *g*, which latter is read by the index, *i*, illustration No. 5.

A small ebonite plate, *p*, illustration No. 4, secured to the micrometer box, carries upon its outer end, mounted in a suitable metal block, the contact spring, *s*, which ends in a piece of platinum turned over so as to rest radially upon the ebonite cylinder. The width of this piece of platinum is 4 mm., and its thickness that of the contact strips, i. e., 0.4 mm. A small screw, *c*, illustration No. 5, serves to adjust the pressure of the spring upon the cylinder. Against one end of the micrometer box is fastened a small bracket, upon which is centered a small worm wheel, *w*, illustration No. 4, gearing into the screw thread, *t*, of the micrometer head. It has 40 teeth, and moves 1 tooth for each turn of the micrometer head. To this worm wheel is fastened a cup-shaped cylinder, *r*, which has cut into its rim a notch or depression with sloping ends not visible in the illustrations. A small steel pin in the end of the lever, *l*, rests upon the edge of this cup-shaped cylinder. The other end of the lever, *l*, fitted with a small ivory tip, presses upon the end of the contact spring, *b*, which is mounted upon an ebonite plate, and is therefore insulated electrically from the instrument. When the small steel pin rests upon the edge of the cup-shaped cylinder, the ivory tip presses the contact spring away from the platinum-tipped screw, *a*. When, however, the notch or depression comes below the steel pin, the contact spring, *b*, is free to press against the platinum-tipped screw, thus allowing the flow of an electric current through the coiled wires, *m* and *n*, and the contact spring, *s*. The length of the notch is chosen so as to allow the circuit to be closed during four revolutions of the micrometer head. As the ends of the notch are sloping, it will be seen that by raising or lowering the platinum-tipped screw, and consequently lowering or raising respectively the steel pin in the lever *l*, the time during which the current can flow can be made to correspond exactly to that of four revolutions of the micrometer head. But it is also important that the four revolutions during which the current can flow and record the contacts made on the ebonite cylinder, *e*, are those disposed symmetrically about the zero position of the micrometer, which indicates the meridian. This is accomplished for adjustments requiring corrections greater than one tooth of the worm wheel *w*, by removing the latter from its axis, turning and replacing it with the proper tooth engaging the screw thread, *t*. The adjustment for amounts less than that of one tooth, as the micrometer is now arranged, is made by loosening a capstan-headed screw (hidden in the illustration by the lever *l*), and turning to right or left the two screws *z*, thus moving the plate carrying the lever *l*, until the small steel pin at the end of lever *l* is in proper relation to the notch or depression in the cup-shaped cylinder *r*. It will be seen, therefore, that this arrangement permits of the motion of the spider thread across the entire field without transmitting records to the chronograph, except during the four revolutions symmetrically disposed about the line of collimation.

Against the inner face of the micrometer head is fastened a spur wheel, *k*, illustration No. 5, with 36 teeth of 48 diametral (inch) pitch, into which gears the wheel *f*, with 72 teeth, mounted on the handwheel shaft, *d*. This shaft is supported by arms from the micrometer box, as can readily be seen from illustration No. 5. The handwheels have a diameter of 33 mm., are 116 mm. apart, and equidistant from the middle of the telescope, allowing ample space for manipulating in either position of the eyepiece.

The pitch of the micrometer screw is about 48.4 threads per centimeter, or 123 per inch. In the telescope of Transit No. 2 the angular value of one revolution of the screw is 2.5 equatorial time seconds, nearly. As the gearing of the handwheel shaft to the micrometer screw is as 2 to 1 it follows that the hands must produce rotary motion of one revolution in about 5^s for an equatorial star.

The adjustment for collimation is made by means of two nuts, *x*, illustration No. 4, upon a small screw fastened to the micrometer box, which in turn is mounted by dovetail slides upon a short flanged cylinder, *y*. The latter is fixed in position by the screws, *h*, which, when loosened, also permit of a rotary motion for adjusting the transit wire into the vertical. Neither

TRANSIT MICROMETER.

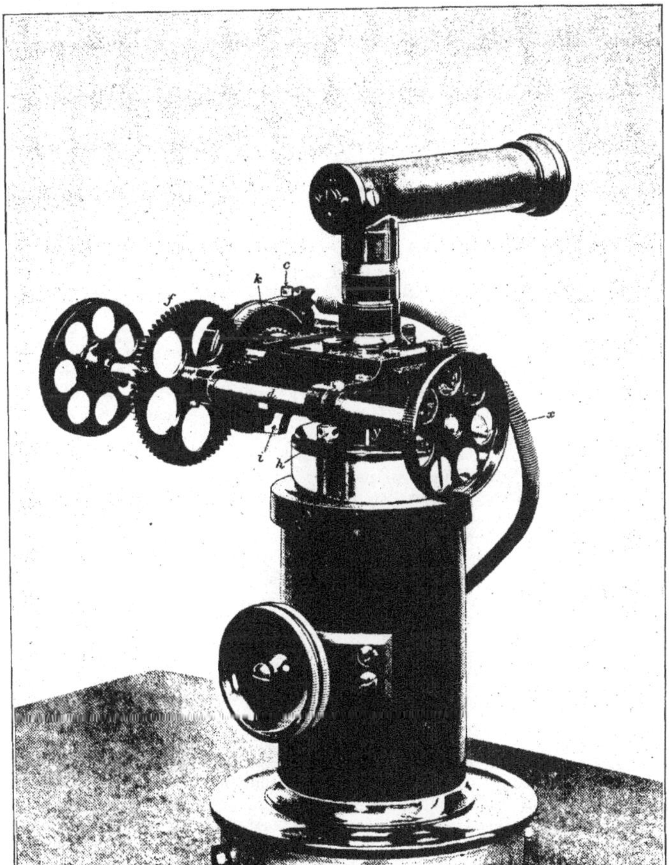

TRANSIT MICROMETER.

of these adjustments will disturb the rather delicate relations between the zero of the transit wire, the contact breaks upon the micrometer head, and the worm wheel with its electric cut-out attachment.

As indicated in the description of the ebonite head with its five platinum contact strips, the instrument itself is used as part of the electric conductor forming the transit circuit. The relay of 20 ohms resistance converts the makes of the transit circuit into breaks in the chronograph circuit. From the contact spring, *b*, through wire, *m*, connection is made with an insulated binding post at the eye end of the telescope tube, from which a wire leads along the telescope to and into the telescope axis and within the latter to an insulated metal cylinder projecting from the transit pivot. Each of the wye bearings of the transit has fastened to it an insulated contact spring, which, being connected with an insulated binding post at the foot of the instrument, establishes the circuit whether the telescope lies in either an east or west position. Another binding post, screwed directly into the iron foot of the transit, affords a ready means for making the necessary connection to begin observations.

It is necessary to use both hands in order to impart to the wire a steady motion. As explained above, the cut-out device allows only a limited portion of the field of observation to be registered, by automatically breaking the transit circuit while the wire is outside the limits. It requires four complete revolutions of the micrometer head to carry the wire across the field of record and as there are five contact strips on the micrometer head, the complete record of the observation of the transit of a given star consists of 20 breaks registered on the chronograph sheet. As the five contact strips are not equally spaced around the head of the micrometer wheel, it follows that the record is in four groups of five observations each. This facilitates the reading of the chronograph sheet. The transit of an equatorial star across the field of record occupies only about 10 seconds of time, a fact which makes it possible to observe stars which are quite close together in right ascension.

Adjustments of the transit micrometer.—Before using the transit micrometer it should be carefully examined to see that there is no loose play in any of its parts, that its contact strips and contact spring are clean and bright, and that the cut-out attachment permits the recording of 20 breaks which are symmetrical about the mean position of the micrometer wire. If a symmetrical record is not obtained, the adjustment must be made, as described on page 10.

The adjustment of the micrometer wire for collimation and verticality are described on page 15, under the heading "Adjustment of the transit instrument."

THE CHRONOGRAPH.

Illustration No. 6 shows the form of chronograph now in use in the Survey. The train of gears seen at the right is driven by a falling weight. It drives the speed governor (seen above the case containing the gears), the cylinder upon which the record sheet is wound, and the screw which gives the pen carriage a slow motion parallel to the axis of the record cylinder. When the speed governor is first released, the speed continually increases until the governor balls have moved far enough away from the axis of revolution to cause a small projection upon one of them to strike a small hook. This impact and the effect of the friction at the base of the weight attached to the hook causes the speed to decrease continually until the hook is released. The speed then increases again until the hook is engaged, decreases until it is released, and so on. The total range of variation in the speed is, however, surprisingly small, so small that in interpreting the record of the chronograph the speed is assumed to be uniform during the intervals between chronometer breaks. The speed may be regulated by screwing or unscrewing the movable weights which are above the governor balls and attached to the same arm. This moves them nearer to or farther from the axis, and thus decreases or increases the critical speed at which the hook is engaged. To get a convenient record it is desirable to adjust the speed so that the record cylinder makes just one revolution per minute with the ordinary arrangement of the train of gears. The gears may also be changed quickly to another combination which will run the record cylinder at double speed. This will require additional driving weights.

The chronograph circuit, passing through the coils of the pen magnet, is operated by a battery of two dry cells in series, so that a relatively strong spring may be used to draw the pen armature away from the pen magnet when the circuit is broken. This insures a sharp lateral movement of the recording pen, which is attached to the pen armature, on the breaking of the circuit, and a correspondingly sharp offset or break is secured in the helix which the pen traces on the drum.

When observations are made on the lines of a reticle, an observing key is placed in the chronograph circuit, which normally keeps the circuit closed, and breaks it only when the key is pressed by the observer as the star is bisected by each of the lines of the reticle.

When the transit micrometer is used, the transit circuit, passing through the transit, the micrometer head and the coils of the transit relay, and operated by two dry cells in series, is connected with the chronograph circuit through the points of the transit relay. The observing key and the transit circuit with its relay may be regarded as interchangeable, as either one may be joined into the chronograph circuit in the place of the other.

The chronometer circuit is operated by a single dry cell, and passes through the coils of a relay, through the points of which it is connected with the chronograph circuit. Breaks in the chronometer circuit are transmitted into breaks in the chronograph circuit by means of the chronometer relay. A condenser should be placed in the circuit across the terminals of the chronometer to prevent sparking and consequent injury to the contact points of the break circuit wheel in the chronometer.

The strength of the current, the tightness of the spring which draws back the pen armature, the distance of that armature from the magnet core, and the range of movement of the armature must all be adjusted relatively to each other so that the pen will furnish a neat and complete record of all the breaks in the circuit. The driving weight must be heavy enough to overcome all friction and cause the governor hook to be engaged frequently, but it must not be so heavy as to cause the hook to be carried forward continuously after it is once engaged. Where a transit micrometer is used and the chronograph circuit is broken by means of a relay placed in the transit circuit, this relay also must be adjusted to produce a short neat break of the chronograph circuit.

In operation the chronometer breaks the circuit automatically every second (or every two seconds) and the pen records the breaks upon the moving record sheet at equal or very nearly equal linear intervals. The chronometer is usually arranged to indicate the beginning of each minute by failing to make a break for the fifty-ninth second, or if it is a two-second chronometer, by making a break for the fifty-ninth second. The hours and minutes may be identified by writing upon some point of the record sheet the corresponding reading of the face of the chronometer. In longitude work it is not essential to have the hours and minutes on the chronograph sheet correspond to those shown on the face of the chronometer. It is customary to mark on the chronograph sheet such hours and minutes as will give the clock a correction of less than one minute, which is equivalent to setting the chronometer to produce that reading.

The record of the exact time of the transit of a star is obtained in the following manner: Where a transit micrometer is used the star is bisected with the wire of the micrometer soon after it enters the field of view of the telescope (see p. 18), and the observer endeavors to keep the star bisected as it crosses the field. As the wire passes the various positions corresponding to contacts on the micrometer head the transit circuit is automatically made, and through the action of a relay it automatically breaks the chronograph circuit and produces a record on the chronograph sheet. Where an observing key is used the observer breaks the chronograph circuit directly by pressing the key which he holds in his hand; this is done as the star transits each line of the reticle. In each case the position of the additional break or record on the chronograph sheet, with reference to the record made by the chronometer, indicates accurately the chronometer time at which it was made, the chronograph being assumed to run uniformly between adjacent chronometer breaks. (See illustration No. 7.) To read the fractions of seconds from the chronograph sheet one may use either a glass scale on which converging lines make it possible to divide varying lengths of seconds into 10 equal spaces, or a small linear

No. 6.

CHRONOGRAPH.

No. 7.

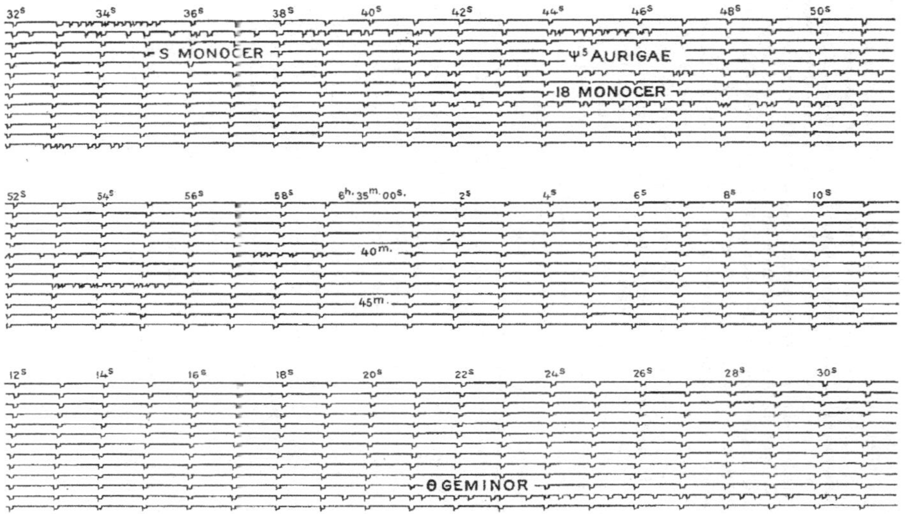

PORTION OF CHRONOGRAPH RECORD

rule, so divided that 10 of its spaces fit closely a second's interval of the chronograph, when the chronograph is making exactly one revolution per minute. Some of the chronographs now in use in the Survey are so constructed that when in perfect adjustment one second on the record will be exactly 1 cm. in length. Such a record may be easily read by using a meter scale. When the linear scale does not fit the chronograph record exactly a satisfactory reading is obtained by a slight shifting of the scale to fit the adjacent seconds marks as the transit records are successively read. This linear scale is much preferred to the glass scale, as it enables one to read the complete record for a star with one setting of the scale. Also by placing the 0 mark of the scale on an even 10-second mark (0, 10, 20, etc.) immediately preceding the star's record, not only the fractional part of the second may be read at once, but also the number of the second. The *beginning* of each break made by the observer and by the chronometer is the exact point to be used in reading the chronograph record, the *break* of the circuit being sharp and definite, while the *make* is indefinite. When an observing key is used and 11 breaks constitute a full record for a star, the star transits are usually read from the record sheet to the nearest half-tenths (0.05) of a second; when a transit micrometer is used and 20 observations constitute the full record of a transit, the readings are made to the nearest tenth (0.1) of a second only. In longitude work it is customary to read the time signals to the nearest hundredth (0.01) of a second, the chronograph then being run at double speed. There will occasionally be a slight interference between the chronometer and the star transit record caused by overlapping, but the time of the observation can usually be identified and closely estimated by comparing the distances between the successive breaks.

A correction, called the *contact correction*, is sometimes applied to the chronograph record of transits observed with a micrometer to account for the time required for the contact spring to cross the contact strip on the head of the micrometer. In order to insure a satisfactory record the contact strips on the micrometer are given material width, since if they were reduced too much there would be an occasional skipping of a record. The micrometer wire travels from a different side of the instrument for upper and lower culminating stars, and also before and after reversal of the telescope in its wyes, so that the contact spring produces a record sometimes from one edge of the contact strip and sometimes from the other. Theoretically, the proper reduction would be to correct all observations for one-half the movement of the micrometer wire from the beginning of the contact to its end. This may be measured on the micrometer head. The micrometer is turned very slowly until the armature of a relay in the transit circuit is heard to make the circuit; the micrometer head is then read. The motion is continued until the armature sounds the breaking of the circuit, and the micrometer is read again. The difference between the two readings is the movement of the wire in terms of divisions on the micrometer head. This may be reduced to time when the equatorial value of the micrometer division is known. This correction is always plus, since the middle of the strip must always come under the contact spring later than does its near edge. But being very small and having nearly the same effect on all time determinations with similar instruments it is without appreciable effect on the observed differences of longitude. Nor is this correction necessary in time determinations for gravity observations with pendulums. If we designate the contact correction on an equatorial star for any transit micrometer as n, then the contact correction for any star is $n \sec \delta$ or $n\ C$, where C, the collimation factor, is obtained directly from the table on pages 62–77, or graphically as shown in illustration No. 8. The equatorial contact correction on transit No. 18 is 0.008 second.

THEORY OF THE TRANSIT INSTRUMENT.

The meaning of the phrase *line of collimation* used in the preceding edition of this publication (Appendix No. 7, of 1898) is adhered to in the present publication. The *line of collimation* may be defined as the line through the optical center of the objective and the middle point of the mean vertical line of the diaphragm or the micrometer wire in its mean position. It may be considered synonymous with the pointing line, sight line, or line of sight. The term *collimation axis* as used in this publication may be defined as the line through the optical center of the objective, and perpendicular to the horizontal axis (axis of rotation) of the telescope. The *line of collimation* and *collimation axis* of a telescope coincide only when there is no error of collimation in the instrument.

If a transit instrument were in perfect adjustment the line of collimation of the telescope would be at right angles to the transverse axis upon which the telescope rotates, and that transverse axis would be horizontal and in the prime vertical. Under these circumstances the line of collimation would always lie in the meridian plane, and local sidereal time at the instant when a given star crossed the line of collimation would necessarily be the same as the right ascension of that star. The difference then between the *chronometer* time of transit of a given star across the line of collimation and the *right ascension* of that star would be the error of the chronometer on local sidereal time. Before observing meridian transits for the determination of time, the conditions stated in the first sentence of this paragraph are fulfilled as nearly as possible by careful adjustment of the instrument. The time observations themselves and certain auxiliary observations are then made in such a manner that the small remaining errors of adjustment may be determined, and the observed times of transit are corrected as nearly as may be to what they would have been had the observations been made with a perfectly adjusted instrument. The observed chronometer time of transit of any star across the line of collimation *as thus corrected* being subtracted from the right ascension of that star gives the correction (on local sidereal time) of the chronometer used during the observations.

ADJUSTMENTS OF THE TRANSIT INSTRUMENT.

Let it be supposed that observations are about to be commenced at a new station at which the pier and shelter for the transit have been prepared. (See p. 105.) By daylight make the preparations described below for the work of the night.

By whatever means are available determine the approximate direction of the meridian and mark it on the top of the pier or by an outside natural or artificial signal. Place the sub-base or footplates of the instrument in such position that the telescope will swing closely in the meridian. It is well to fix the sub-base or footplates firmly in place by cementing them to the pier with plaster of Paris when a stone, concrete, or brick pier is used, and by screws or bolts when a wooden pier is used. The meridian may be determined with sufficient accuracy for this purpose by means of a compass needle, the magnetic declination being known and allowed for. A known direction from triangulation or from previous azimuth observations may be utilized. All that is required is that the telescope shall be so nearly in the meridian that the final adjustment will come within the scope of the screws provided upon the instrument for the azimuth adjustment.

Set up the instrument and inspect it. The pivots and wyes of both instrument and level should be cleaned with watch oil, which must be wiped off to prevent its accumulating dust. They should be carefully inspected to insure that there is no dirt gummed to them. The lens should be examined occasionally to see that it is tight in its cell. It may be dusted off with a camel's-hair brush, and when necessary may be cleaned by rubbing gently with soft, clean tissue paper, first moistening the glass slightly by breathing on it.

Focus the eyepiece by turning the telescope up to the sky and moving the eyepiece in and out until that position is found in which the most distinct vision is obtained of the micrometer wire. If any external objects are visible through the eyepiece in addition to the micrometer wire seen projected against a uniform background (the sky, for example) the eye will attempt, in spite of its owner, to focus upon those objects as well as upon the micrometer wire and the object of the adjustment, namely, to secure a focus corresponding to a minimum strain upon the eye, will be defeated to a certain extent.

Focus the objective by directing the telescope to some well-defined object, not less than a mile away, and changing the distance of the objective from the plane in which the micrometer wire moves until there is no apparent change of relative position (or parallax) of the micrometer wire and the image of the object when the eye is shifted about the front of the eyepiece. The object of the adjustment, namely, to bring the image formed by the objective into coincidence with the micrometer wire is then accomplished. If the eyepiece has been properly focused this position of the objective will also be the position of most distinct vision. The focus of the objective will need to be inspected at night, using a star as the object, and corrected if necessary. Unless the focus is made nearly right by daylight none but the brightest stars will be seen at all at night and the observer may lose time trying to learn the cause of the trouble. If the objective is focused at night a preliminary adjustment should be made on a bright star and the final adjustment on a faint star, as it is almost impossible to get a very sharp image of a large star. A planet or the moon is an ideal object on which to focus the objective. A scratch upon the draw-tube to indicate its approximate position for sidereal focus will be found a convenience. After a satisfactory focus has been found the drawtube is clamped in position with screws provided for that purpose.

Methods exactly similar to those described in the two preceding paragraphs are employed in focusing the eyepiece and objective when a diaphragm is used instead of the micrometer.

If unusual difficulty is had with the illumination at night, it is advisable to remove the eyepiece and look directly at the reflecting mirror in the telescope tube. The whole surface of the mirror should be uniformly illuminated. If this is not the case, the mirror should be rotated until a satisfactory illumination is obtained. Occasionally the mirror must be removed from the telescope and its supporting arm bent in order to make the reflected rays of light approximately parallel with the tube of the telescope.

Adjust the striding level in the ordinary manner, placing it on the pivots direct and reversed. If the level is already in perfect adjustment the difference of the two east (or west) end readings will be zero for a level numbered in both directions from the middle, or the sum of the two east (or west) end readings will be double the reading of the middle of the tube for a level numbered continuously from one end to the other. The level must also be adjusted for wind. In other words, if the axis of the level tube is not parallel to the line joining the wyes, the bubble will move longitudinally when the level is rocked back and forth on the pivots. The adjustment for wind is made by means of the side adjusting screws at one end of the level. To adjust for wind, move the level forward and then back and note the total movement of the bubble. The wind will be eliminated by moving the bubble back one-half of the total displacement by means of the side adjusting screws. Then test again for wind, and repeat adjustment if necessary. In placing the level upon the pivots it should always be rocked slightly to insure its being in a central position and in good contact.

Level the horizontal axis of the telescope.—This adjustment may, of course, be combined with that of the striding level.

Test the verticality of the micrometer wire (or of the lines of the diaphragm) by pointing on some well-defined distant object, using the apparent upper part of the wire (or of the middle line of the diaphragm). Rotate the telescope slightly about its horizontal axis until the object is seen upon the apparent lower part of the line. If the pointing is no longer perfect, the micrometer box (or reticle) must be rotated about the axis of figure of the telescope until the wire (or line) is in such a position that this test fails to discover any error.

To adjust the collimation proceed in the following manner: If a transit micrometer is used, place the micrometer wire in its mean position, as indicated by the middle point of the rack or comb in the apparent upper (or lower) edge of the field, the graduated head reading zero. Point on some well-defined distant object by means of the azimuth screws, keeping the wire in the position indicated above. Reverse the telescope in its wyes and again observe the distant object. If the wire again bisects the object, the instrument has no error of collimation. If upon reversal the wire does not again bisect the object, then the adjustment is made by bringing the wire halfway back to the object with the screw x, illustration No. 5. Set on the object again, using the azimuth screws, and test the adjustment by a second reversal of the telescope.

If the transit has a diaphragm instead of a transit micrometer, the process is very similar to that described above, though simpler. Point on some well-defined distant object, using the middle vertical line of the diaphragm. Reverse the instrument in its wyes and again observe the same distant object. If after reversal the wire covers the object no adjustment is needed. If an adjustment is necessary it is made by moving the diaphragm halfway back to the object by means of the adjusting screws which hold it in place. A second test should be made to show whether the desired condition has been obtained.

Wherever practicable, the adjustment for collimation should be made at sidereal focus on a terrestrial object at least 1 mile distant, or on the cross wires of a theodolite or collimator which has previously been adjusted to sidereal focus, set up just in front of the telescope of the transit. If necessary the lines of the theodolite are artificially illuminated. Occasionally, if neither a distant object nor a theodolite is available for making the collimation adjustment, a near object may be used for the purpose. In this case, however, collimation error may exist when the telescope is in sidereal focus. If such error is not large, the method of computations of the observations will eliminate its effect from the results. A rapid and careful observer may sometimes be able to make this collimation adjustment on a slow-moving close circumpolar star. In so doing he will have to estimate the amount the star moves while he is reversing his instrument and securing the second pointing. No attempt should be made to adjust the collimation error to zero. If it is already less than say 0.2 second of time it should not be changed, for experience has shown that frequent adjustment of an instrument causes looseness in the screws and the movable parts.

To test a finder circle which is supposed to read zenith distances, point upon some object, placing the image of the object midway between the two horizontal lines (guide lines); bring the bubble of the finder circle level to the center and read the circle. Next reverse the telescope and point again on the same object; bring the bubble to the center and read the same finder circle as before. The mean of the two readings is the true zenith distance of the object, and their half difference is the index error of the circle. The index error may be made zero by setting the circle to read the true zenith distance, pointing on the object, and bringing the vernier bubble to the center with the level adjusting screw. At night this adjustment may be made by keeping a known star between the horizontal lines as it transits the meridian. While the telescope remains clamped in this position set the finder circle to read the known zenith distance of the star and bring the bubble to the middle position of the tube as before. A quick test when there are two finder circles is to set them at the same angle and see if the bubbles come to the center for the same position of the telescope.

Adjust the transit micrometer so that it will give 20 records which are symmetrical about the mean position of the micrometer wire. For a description of this adjustment see page 10.

The preceding adjustments can not always be made in the order named, as, for instance, when a distant mark cannot be seen in the meridian, nor need they all be made at every station. The observer must examine and correct them often enough to make certain that the errors are always within allowable limits.

The azimuth adjustment.—In the evening, before the regular observations are commenced, it will be necessary to put the telescope more accurately in the meridian. If the chronometer correction is only known approximately, say within one or two minutes, set the telescope for some bright star which is about to transit within 10°, say, of the zenith. Observe the chronometer time of transit of the star. This star being nearly in the zenith, its time of transit will be but little affected by the azimuth error of the instrument.[1] The collimation and level errors having previously been made small by adjustment, the right ascension of this star minus its chronometer time of transit will be a close approximation to the chronometer correction. Now set the telescope for some star of large declination (slow-moving) which is about to transit well to the northward of the zenith. Compute its chronometer time of transit, using the chronometer correction just found. As that time approaches bisect the star with the micrometer

[1] To avoid waiting for stars close to the zenith the chronometer correction may also be estimated closely by comparing observations of two stars not very distant from the zenith, one north and one south, and these at the same time will give some idea of the amount and direction of the azimuth error.

wire in its mean position or with the middle vertical line of the diaphragm and keep it bisected, following the motion of the star in azimuth by the slow-motion screws provided for that purpose, until the chronometer indicates that the star is on the meridian.

The adjustment may be tested by repeating the process; that is, by obtaining a closer approximation to the chronometer error by observing another star near the zenith and then comparing the computed chronometer time of transit of a slow-moving northern star with the observed chronometer time of transit. If the star transits apparently too late, the objective is too far west (if the star is above the pole), and vice versa. The slow-motion azimuth screw may then be used to reduce the azimuth error. This process of reducing the azimuth error will be much more rapid and certain if, instead of simply guessing at the movement which must be given the azimuth screw, one computes roughly what fraction of a turn must be given to it. This may be done by computing the azimuth error of the instrument roughly by the method indicated on page 35, having previously determined the value of one turn of the screw.[1]

If from previous observations the chronometer correction is known within, say, five seconds, the above process of approximation may be commenced by using a northern star at once, instead of first observing a zenith star as indicated above.

Or, the chronometer correction being known approximately, and the instrument being furnished with a screw or graduated arc with which a small horizontal angle may be measured, the first approximation to the meridian may be made by observing upon Polaris, computing the azimuth approximately by use of tables of azimuth of Polaris at different hour angles then by means of the screw or graduated arc swinging the instrument into the meridian. The tables referred to are given in Appendix No. 10 of the Report for 1895, in "Principal Facts of the Earth's Magnetism, etc.," (a publication of the Coast and Geodetic Survey), or in the American Ephemeris and Nautical Almanac. Where saving of time is an important consideration, the latter method has the advantage that Polaris may be found in daylight, when the sun is not too high, by setting the telescope at the computed altitude and moving it slowly in azimuth near the meridian. It is advisable to use a hack chronometer and the eye and ear method in making the azimuth adjustments, the chronograph being unnecessary for this purpose, even when available.

OBSERVING LIST.

The following is an example of the list of stars selected for time observations at stations of a lower latitude than 50°. The second time set shown in this list is computed on page 26, and enters into the longitude determination shown on page 84. Each set consists of two half sets of six stars each, selected in accordance with the instructions shown on page 80. Such a list, prepared in easily legible figures, should be posted in the observatory.

[1] Some of the meridian telescopes carry a small graduated arc on the double base of the frame, which may be used for measuring the small angle here required.

Star list for Key West, Fla.

Form 256.*

$\phi = 24° 33'$

Catalogue	Star	Magnitude	Right ascension α			Declination δ		Zenith distance ζ		Star factors			Diurnal aberration κ
										A	C	B	
			h	m	s	°	′	°	′				
B†	β Tauri	1.8	5	20	25	+28	32	N 3	59	−0.08	1.14	1.14	−0.02
A‡	χ Aurigae	5.0		26	40	+32	07	N 7	34	−0.15	1.18	1.17	−0.02
B	ι Orionis	2.8		30	53	− 5	58	S 30	31	+0.51	1.01	0.87	−0.02
B	ο Aurigae	5.7		38	42	+49	47	N 25	14	−0.66	1.55	1.40	−0.03
B	ζ Leporis	3.5		42	44	−14	51	S 39	24	+0.65	1.04	0.80	−0.02
A	ν Aurigae	3.9		45	03	+39	07	N 14	34	−0.32	1.29	1.25	−0.02
B	δ Aurigae	3.8	5	51	52	+54	17	N 29	44	−0.85	1.71	1.48	−0.03
B	θ Aurigae	2.7		53	23	+37	12	N 12	39	−0.28	1.26	1.22	−0.02
B	ν Orionis	4.4	6	02	16	+14	47	S 9	46	+0.18	1.04	1.02	−0.02
B	η Geminor.	3.3		09	16	+22	32	S 2	01	+0.04	1.08	1.08	−0.02
B	*8* Monocer.	4.5		18	50	+ 4	38	S 19	55	+0.34	1.01	0.94	−0.02
B	*10* Monocer.	5.0		23	22	− 4	42	S 29	15	+0.49	1.01	0.88	−0.02
B	*S* Monocer.	4.4	6	35	51	+ 9	59	S 14	34	+0.26	1.02	0.98	−0.02
A	ψ^5 Aurigae	5.5		40	02	+43	40	N 19	07	−0.45	1.38	1.31	−0.03
B	*18* Monocer.	4.7		43	01	+ 2	31	S 22	02	+0.37	1.01	0.93	−0.02
B	θ Geminor.	3.4		46	40	+34	04	N 9	31	−0.20	1.21	1.19	−0.02
B	ζ Geminor.	3.8		58	36	+20	42	S 3	51	+0.07	1.07	1.07	−0.02
B	*63* Aurigae	5.0	7	05	16	+39	28	N 14	55	−0.34	1.30	1.25	−0.02
B	ε Geminor.	3.8	7	19	57	+27	59	N 3	26	−0.07	1.13	1.13	−0.02
B	β Canis Min.	2.9		22	06	+ 8	29	S 16	04	+0.28	1.02	0.97	−0.02
B	α Canis Min.	0.5		34	26	+ 5	28	S 19	05	+0.33	1.01	0.95	−0.02
B	β Geminor.	1.1		39	38	+28	15	N 3	42	−0.08	1.13	1.13	−0.02
B	π Geminor.	5.5		41	31	+33	39	N 9	06	−0.19	1.21	1.18	−0.02
A	φ Geminor.	5.0		47	48	+27	00	N 2	27	−0.05	1.12	1.12	−0.02

* Form 256, known as "Coast and Geodetic Survey, Longitude Record and Computation," is a book containing all the different forms used in observing and computing, time and longitude, except form 34 shown on p. 20.
† Berliner Astronomisches Jahrbuch.
‡ American Ephemeris and Nautical Almanac.

DIRECTIONS FOR OBSERVING.

Everything being in readiness and the instrument completely adjusted set the telescope for the first star. It is not advisable to use the horizontal axis clamp during observations, for its action may have a slight tendency to raise one end of the axis. See to it, loading one end if necessary, that the center of gravity of the telescope is at its horizontal axis, and then depend upon the friction at the pivots to keep the telescope in whatever position it is placed. Watch the chronometer[1] so as to know when to expect the star to appear in the field of view of the telescope. When the star enters the field, bring it between the horizontal lines of the diaphragm, if it is not already there, by tapping the telescope lightly.

If a transit micrometer is used the process of observing consists simply in bisecting the star's image with the micrometer wire soon after it appears and in keeping it bisected as it moves across the field of the telescope. The record is made automatically by the contact of a spring with certain metal strips on the micrometer head. A cut-out device allows only 10 such contacts on either side of the mean position of the micrometer wire to register on the chronograph. The observer learns by experience at what part of the field the wire begins to register and he should endeavor to keep the star bisected several seconds before it reaches that point. Similarly, he knows when the record is complete and he can cease observing a particular star, and set for the next one on his observing list.

If an instrument with a diaphragm is being used in connection with a chronograph, the process of observing the transit of a star across a line of the diaphragm consists in waiting, observing key in hand, until the instant when the star is apparently bisected by the line and then pressing the key as soon as possible thereafter. The time record thus made on the chrono-

[1] When a chronograph is being used, it is customary to keep the chronometer which is connected with the chronograph protected as carefully as possible from rapid changes of temperature and from jars. During the observations it is not usually removed from its protecting box, but instead an extra chronometer (sometimes called a hack chronometer) is used at the instrument.

graph will always *follow* the event by a time interval, known as personal equation, which depends mainly on the rapidity of the action of the nerves and brain of the observer.

It may occur to a new observer to attempt to make this time interval zero by anticipating the bisection of the star's image, and this he may succeed in doing. He may even make the personal equation negative. The accumulated experience of many observers, however, is that it is better to observe in the manner first indicated and have a large and constant personal equation, rather than to reduce this personal equation to a small but at the same time rather variable quantity. The method of observing with a transit micrometer practically eliminates the personal equation from the time observations. In other methods it may be eliminated from the results by special observations, or by programs of observing especially devised for that purpose. (See p. 91.)

At about the middle of the observations which are to constitute a set the telescope should be reversed, so that the effects of the error of collimation and inequality of pivots upon the apparent times of transit may be reversed in sign. Three or four readings of the striding level, in each of its positions (direct and reversed) should be taken during each half set. To eliminate, in part at least, the effects of irregularities in the figure of the pivots upon the determination of the inclination of the axis, it is desirable to take the level readings with the telescope inclined at the various practicable angles at which stars are observed, and to make half of them with the objective to the northward and half with the objective southward. Great care should be taken to avoid unequal heating of the two ends of the striding level. The level readings may be checked and possible errors often detected by the fact that the bubble length should be constant except for the effect of change of temperature (the bubble shortens with rise of temperature) and in observing and computing this should be kept in mind. A very short length of bubble should not be used on account of increased tendency to stick, and extreme length should be avoided because of danger of running off the graduation. In using the striding level it is important that the bubble be given time to come to rest before reading.

The only difference between the eye and ear method of observing time and the chronograph and key method just described is in the process of observing and recording the times of transit of the star image across the separate lines of the diaphragm.

Before using the eye and ear method the observer must first learn to pick up the beat of a chronometer and to carry it even while paying attention to other matters. To pick up the beat of a chronometer, first look at some second's mark two or more seconds ahead of the second hand. Fix the number of that second in mind as the second hand approaches it. Name it exactly with the tick at which the second hand reaches it. Then, keeping the rhythm of the chronometer beat, count the seconds and half seconds (aloud, in a whisper, or mentally), always keeping the count exactly with the tick of the chronometer. In counting it will be found easier to keep the rhythm if the names of the numerals are elided in such a way as to leave but a single staccato syllable in each. The half-second beat should be marked by the word "half," thus—one, half, two, half, three . . . *twenty*, half, twenty-*one*, half, twenty-*two* . . . and so on.[1] With practice, an observer can carry the count of the beat for an indefinite period without looking at the chronometer face if he can hear the tick. If he becomes expert, he will even be able to carry the count for a half minute or more during which he has not even heard the tick. The chronometer should, of course, be placed where it can be seen and heard by the observer with as little effort as possible.

To observe the time of transit of a star across a given line the observer first picks up the beat of the chronometer as the star approaches the line. At the last tick of the chronometer occurring before the transit he notes mentally the number of the tick, and also carefully observes the apparent distance of the star from the line. At the next tick the star is on the other side of the line and the observer notes again the apparent distance of the star from the line. By a mental comparison of these two *distances* he estimates fifths of the *time interval* between the two ticks of the chronometer and obtains his estimate of the time of transit to the nearest tenth of a second. Though the mental processes involved may seem difficult at first, practice soon makes them easy. An experienced observer using this process is able to estimate the time of transit

[1] Another method often used is to count only to 10 (thus using only words of one syllable) and to glance at the chronometer after the observation to show the position in the minute.

of a star's image across a line of the diaphragm with a probable error of about $\pm 0^s.1$. It is conducive to accuracy for the observer to acquire the habit of deciding definitely, without hesitation, upon the second and tenth as soon as the event is complete. Hesitation in this matter is likely to cause inaccuracy.

EXAMPLE OF RECORD AND PART OF THE COMPUTATIONS.

There are shown on pages 18, 20–22 examples of the list of stars and the original transit level readings made in the observatory at the time of the observations, a set of time observations as read from the chronograph sheet, and the computation of $\alpha - t$ (right ascension minus the chronometer time of transit) for each star. The computation of ΔT (the mean correction to the chronometer) is shown on page 26. These computations are for the second set of stars given on page 18.

These observations were made under the General Instructions for Longitude Determinations with the Transit-Micrometer, which are given on page 79 of this publication.

Form 34.

Longitude record.

[Station, Key West. Date, Feb. 14, 1907. Instrument, Transit No. 2. Observer, J. S. Hill.]

Set I				Set II			
Stars	Levels W		Levels E	Stars	Levels W		Levels E
	d	N	*d*		*d*	N	*d*
Clamp or band, W	17.7		58.8	Clamp or band, W	62.0		20.0
β Tauri	60.1		19.0	S Monocer.	17.7		59.5
χ Aurigae				ψ^5 Aurigae			
ι Orionis		S		18 Monocer.		S	
o Aurigae	17.7		58.8	θ Geminor.	61.2		19.4
ν Aurigae	61.2		20.0	ζ Geminor.	17.7		59.6
				63 Aurigae			
		N				N	
	17.5		58.9		61.5		19.5
	60.7		19.3		17.7		59.7
		S					
	17.6		59.0				
	61.7		20.2				
		N				N	
Clamp or band, E	17.0		58.7	Clamp or band, E	16.8		58.9
δ Aurigae	61.3		19.7	ι Geminor.	61.6		19.5
θ Aurigae				β Canis Min.			
η Geminor.		S		α Canis Min.		S	
8 Monocer.	17.2		59.0	β Geminor.	17.4		59.7
10 Monocer.	61.9		20.0	π Geminor.	62.1		19.7
				ϕ Geminor.			
		N				N	
	16.8		58.7		17.0		59.4
	61.3		19.4		62.0		19.5
						S	
					16.9		59.4
					62.3		19.9

1 div. of level scale = 2''.322. Chronometer 1824.
Pivot inequality = 0.000.
Remarks: Cable was used direct, without repeaters, between Miama and Key West.

While the following method of computing was devised for observations with the transit micrometer, it is not limited in its use to such observations. The star list for which observations and computations are shown on the following pages could have been observed with a key and the computation made in the same manner as the one which follows. The only difference is that had the observations been made with a key not so many records would have been obtained and the observations would have been subject to a large observation error, calle l personal equation. (See p. 90.)

Explanation of the formulæ and methods used in this computation follows the examples of the record and computation.

Form 256.*

[Station, Key West. Date, Feb. 14, 1907. Instrument, transit No. 2, with transit micrometer. Observer, J. S. Hill. Recorder, J. S. Hill. Chronometer, Sidereal 1824.]

Star: S. Monocer.	ψ^5 Aurigae	18 Monocer.	ζ Geminor.	ζ Geminor.	63 Aurigae
Clamp: W	W	W	W	W	W
Level: W *d* N 62.0, 17.7 — +44.3 E *d* 20.0, 59.5 — −39.5 +4.8		W *d* S 61.2, 17.7 — +43.5 E *d* 19.4, 59.6 — −40.2 +3.3		W *d* N 61.5, 17.7 — +43.8 E *d* 19.5, 59.7 — −40.2 +3.6	

Computation of level constant: Mean N +4.20 *d*, S +8.30; +3.75 × 0.039 = +0.146 *s* = b_W

h m 6 35			h m 6 39			h m 6 42			h m 6 46			h m 6 58			h m 7 04		
s	*s*	Sums	*s*	*s*	Sums	*s*	*s*	Sums	*s*	*s*	Sums	*s*	*s*	Sums	*s*	*s*	Sums
32.0	41.4	73.4	41.3	54.0	95.3	41.5	50.5	92.0	19.5	30.4	49.9	16.2	26.0	42.2	55.3	67.0	122.3
32.4	41.1	.5	41.8	53.5	.3	41.9	50.2	.1	20.0	30.1	50.1	16.5	25.5	2.0	55.6	66.5	.1
33.1	40.4	.5	42.8	52.6	.4	42.5	49.7	.2	20.6	29.4	.0	17.2	24.8	2.0	56.4	65.8	.2
33.6	39.8	.4	43.5	51.9	.4	43.1	49.1	.2	21.3	28.7	.0	17.7	24.3	2.0	57.1	65.1	.2
33.9	39.5	.4	43.9	51.4	.3	43.3	48.8	.1	21.7	28.3	.0	18.0	23.9	1.9	57.5	64.6	.1
34.6	38.8	.4	44.7	50.6	.3	44.0	48.1	.1	22.3	27.6	49.9	18.8	23.1	1.9	58.4	63.9	.3
35.0	38.5	.5	45.3	50.3	.6	44.3	47.9	.2	22.8	27.1	9.9	19.1	22.9	2.0	58.8	63.4	.2
35.6	37.9	.5	46.0	49.3	.3	44.8	47.3	.1	23.6	26.4	50.0	19.8	22.3	2.1	59.5	62.6	.1
36.1	37.4	.5	46.9	48.5	.4	45.4	46.6	.0	24.3	25.7	.0	20.5	21.6	2.1	60.3	61.9	.2
36.4	37.1	.5	47.2	48.1	.3	45.7	46.3	.0	24.6	25.4	.0	20.7	21.4	2.1	60.7	61.5	.2

	6 35	6 39	6 42	6 46	6 58	7 04
Sum	734.6	953.6	921.0	499.8	420.3	1221.9
Mean	36.73	47.68	46.05	24.99	21.02	01.10
R†						
κ	− .02	− .03	− .02	− .02	− .02	− .02
Bb	+ .14	+ .19	+ .14	+ .17	+ .16	+ .18
t	6 35 36.85	6 39 47.84	6 42 46.17	6 46 25.14	6 58 21.16	7 05 01.26
α	6 35 51.85	6 40 02.92	6 43 01.21	6 46 40.17	6 58 36.16	7 05 16.28
(α−*t*)	+15.00	+15.08	+15.04	+15.03	+15.00	+15.02

* See note below table on p. 18.

† R, correction for rate, is negligible in this time set.

Form 256.*

[Station, Key West. Date, Feb. 14, 1907. Instrument, transit No. 2, with transit micrometer. Observer, J. S. Hill. Recorder, J. S. Hill. Chronometer, Sidereal 1824.]

Star: ι Geminor.			β Canis Min.			α Canis Min.			β Geminor.			π Geminor.			φ Geminor		
Clamp: E			E			E			E			E			E		
Level:																	
W	E		W	E					W	E		W	E				
d	d		d	d					d	d		d	d				
N 16.8	58.9		S 17.4	59.7					N 17.0	59.4		S 16.9	59.4				
61.6	19.5		62.1	19.7					62.0	19.5		62.3	19.9				
+44.8	−39.4		+44.7	−40.0					+45.0	−39.9		+45.4	−39.5				
+5.4			+4.7						+5.1			+5.9					
Computation of level constant: Mean N+5.25 (d); S+5.30; +5.28×0.039 = +0.206 = b_E (s)																	
h m 7 19			h m 7 21			h m 7 34			h m 7 39			h m 7 41			h m 7 47		
s	s	Sums	s	s	Sums	s	s	Sums	s	s	Sums	s	s	Sums	s	s	Sums
37.8	48.3	86.1	47.9	57.1	105.0	07.5	16.7	24.2	18.5	28.8	47.3	11.3	22.3	33.6	29.5	39.6	69.1
38.3	47.9	.2	48.2	56.8	5.0	07.8	16.4	.2	18.8	28.5	.3	11.6	21.9	.5	29.8	39.4	.2
38.9	47.3	.2	48.7	56.1	4.8	08.4	15.7	.1	19.5	27.7	.2	12.5	21.1	.6	30.3	38.5	68.8
39.6	46.5	.1	49.3	55.5	4.8	09.0	15.1	.1	20.1	27.0	.1	13.2	20.4	.6	31.0	37.8	.8
39.9	46.3	.2	49.7	55.2	4.9	09.2	14.8	.0	20.5	26.8	.3	13.6	20.1	.7	31.3	37.5	.8
40.7	45.6	.3	50.2	54.6	4.8	09.9	14.2	.1	21.2	26.1	.3	14.3	19.4	.7	32.0	36.8	.8
41.0	45.1	.1	50.6	54.4	5.0	10.2	13.9	.1	21.6	25.7	.3	14.7	19.0	.7	32.3	36.5	.8
41.7	44.6	.3	51.1	53.7	4.8	10.8	13.3	.1	22.3	25.0	.3	15.4	18.3	.7	33.1	35.9	69.0
42.5	43.8	.3	51.8	53.0	4.8	11.4	12.6	.0	23.1	24.3	.4	16.1	17.5	.6	33.8	35.1	68.9
42.8	43.4	.2	52.1	52.7	4.8	11.7	12.3	.0	23.3	24.1	.4	16.3	17.2	.5	34.1	34.8	.9
	Sum	862.0		Sum	1048.7		Sum	240.9		Sum	472.9		Sum	336.2		Sum	689.1
Mean		43.10.			52.44			12.04			23.64			16.81			34.46
R†																	
κ		− .02			− .02			− .02			− .02			− .02			− .02
Bb		+ .23			+ .20			+ .20			+ .23			+ .24			+ .23
t	7 19	43.31		7 21	52.63		7 34	12.22		7 39	23.85		7 41	17.03		7 47	34.67
α	7 19	57.74		7 22	07.08		7 34	26.67		7 39	38.26		7 41	31.45		7 47	49.14
(α−t)		+14.43			+14.45			+14.45			+14.41			+14.42			+14.47

* See note below table on p. 18.

† R, correction for rate, is negligible in this time set.

CORRECTION FOR INCLINATION OF AXIS.

If the horizontal axis of the telescope is slightly inclined to the horizon and the telescope is otherwise in perfect adjustment, the line of collimation will, when the telescope is rotated about its horizontal axis, describe a plane which passes through the north and south points of the horizon and makes an angle with the meridian plane equal to the inclination of the axis to the horizon. If the eastern end of the axis is too high, the transits of all the stars above the pole (apparently moving westward) will be observed too late, and the transits of all subpolars will be observed too early, and it is therefore necessary to correct the observed times of transit by means of the readings of the striding level, taking into account the inequality of the pivots, if appreciable.

Let w and e be the readings of the west and east ends, respectively, of the bubble of the striding level for a given position of the telescope axis. Let w' and e' be the corresponding west and east readings after the level is reversed, the telescope axis remaining as it was. Let d be the value of a division of the level in seconds of arc. Then for β, the apparent inclination of the telescope axis expressed in seconds of time, we may write, if the level divisions are numbered in both directions from the middle:

$$\beta = \tfrac{1}{4}\left\{(w+w')-(e+e')\right\}\frac{d}{15} = \left\{(w+w')-(e+e')\right\}\frac{d}{60}$$

in which $\frac{d}{60}$ is a constant for the level, $\frac{d}{15}$ being the value of one division of the level in seconds of time.

If the level divisions are numbered continuously from one end of the level to the other the above formula takes the form

$$\beta = \left\{(w-w')+(e-e')\right\}\frac{d}{60}$$

in which the primed letters refer to that position of the level in which the zero end of the tube is to the west.[1]

Inequality of pivots.—The level readings give a determination of the inclination of the line joining the points of the two pivots, which are midway between the lines of contact of the pivots and the wyes of the level, but do not give the required inclination of the axis of rotation of the telescope (which is the line joining the centers of the two pivots) unless the pivots are of the same size. Let p, the pivot inequality, be the angle, expressed in seconds of time, between the line joining the centers of the pivots and the line whose inclination is determined by the level readings, and let this angle be called positive if the pivot nearest the designating mark (band, clamp, or illumination) is the smaller.

Then

$$b_W = \beta_w + p \text{ and } b_E = \beta_e - p \quad ^2$$

in which b is the required inclination of the axis of rotation of the telescope. The subscripts indicate the position, to the westward or to the eastward, of the bright band, the clamp, or the illumination, or whatever mark is used to distinguish between the two positions of the telescope axis. The pivot inequality, p, is ordinarily derived from a special series of observations taken for that purpose. For an example of such a series, with the corresponding formula and computation, see page 44.

The correction to the observed time of transit of any star for inclination is

$$b \cos \zeta \sec \delta = bB,$$

in which δ is the declination of the star and ζ is its zenith distance ($=\phi-\delta$ for all stars above the pole, and $=\phi+\delta-180°$ for subpolar stars). The factor $B = \cos \zeta \sec \delta$ is tabulated on pages 62–77, but is much more easily obtained with the graphical device shown in illustration No. 9 and explained on page 61. It is positive for stars above the pole and negative for subpolars.

It is the present practice in this Survey to assume that b, the inclination, is constant for each half set, and it is computed in the following manner: Within each half set the mean of the observed values of β with objective northward is first derived, then the corresponding mean with objective southward, and finally the mean of these two means is taken as the β for the half set.

The value of B for each star, as taken from either the table on pages 62–77 or the graphical device shown in illustration No. 9, is given in the observing list on page 18.

---

[1] As w is always greater than w' and e is always less than e', the sign of the west difference is always + and of the east difference is always −, so that when the differences are taken vertically, the resulting sign of the level correction will at once be apparent, as shown in the following example:

West	East
d	d
62.0	20.0
17.7	59.5
+44.3	−39.5
+4.8	

[2] These formulæ are exact only in case the angle of the level wyes is the same as the angle of the supporting wyes.

INCOMPLETE TRANSITS WITH TRANSIT-MICROMETER.

If the transit of a star observed with the transit-micrometer is incomplete, only the observations which are symmetrical with regard to the mean position of the micrometer wire are used and those for which the symmetrical observations are lacking are rejected. (See General Instructions for Longitude Determinations, p. 79.) Incomplete transits by other methods of observing are utilized by a method of reduction shown on page 32.

CORRECTION FOR RATE.

If the chronometer rate is not zero, the chronometer correction changes during the progress of the time set. To reduce each observed time of transit across the mean line to what it would have been had the rate been zero (and the correction equal to that which actually existed at the mean epoch of the set) apply the following correction:

$$R = (t - T_o)\, r_h$$

in which t is the chronometer time of transit of a star, T_o is the mean epoch of the time set, that is, the mean of all the chronometer times of transit, and r_h is the hourly rate of the chronometer on sidereal time, + when losing and − when gaining. The quantity $(t - T_o)$ is expressed in hours. The above is the correction as applied to the observed time of transit of the star; applied to $\alpha - t$, the sign is reversed.

The correction for rate may be looked upon as a refinement which is not always essential. If a time set has perfect symmetry of arrangement, the effect of introducing a rate correction into the computation will be shown only in the residuals, as it will have no effect on the computed clock correction. If the daily rate of the chronometer is less than five seconds, it can be ignored in the computation of all time sets except those in which one of the half sets contains many more or less stars than the other, or in which one of the half sets extends over a very much longer period of time than the other. In all cases where the rate is greater than five seconds per day it should be considered, and it should be omitted only after a preliminary test shows its effect on the chronometer correction to be negligible.

CORRECTION FOR DIURNAL ABERRATION.

The effect of the annual aberration due to the motion of the earth in its orbit is taken into account in computing apparent star places and need not be considered here.

The correction for diurnal aberration to be applied to an observed time of transit across the meridian is

$$\kappa = 0^s.021 \cos\phi \sec\delta$$

This correction may be obtained easily by the graphical device shown and described on page —, but it is also given in the following table. It is *minus* for all stars observed at upper culmination and *plus* for stars observed at lower culmination.

Table of diurnal aberration (κ).

Latitude $=\phi$	Declination $=\delta$										
	0°	10°	20°	30°	40°	50°	60°	70°	75°	80°	85°
	s	s	s	s	s	s	s	s	s	s	s
0°	0.02	0.02	0.02	0.02	0.03	0.03	0.04	0.06	0.08	0.12	0.24
10°	.02	.02	.02	.02	.03	.03	.04	.06	.08	.12	.24
20°	.02	.02	.02	.02	.03	.03	.04	.06	.08	.11	.23
30°	.02	.02	.02	.02	.02	.03	.04	.05	.07	.10	.21
40°	.02	.02	.02	.02	.02	.03	.03	.05	.06	.09	.18
50°	.01	.01	.01	.02	.02	.02	.03	.04	.05	.08	.15
60°	.01	.01	.01	.01	.01	.02	.02	.03	.04	.06	.12
70°	.01	.01	.01	.01	.01	.01	.01	.02	.03	.04	.08
80°	.00	.00	.00	.00	.00	.01	.01	.01	.01	.02	.04

DERIVATION OF $(\alpha - t)$.

The correction for diurnal aberration, inclination of axis, and rate (if considered) being applied to the observed time of transit across the mean position of the micrometer wire (or mean line of the diaphragm) as shown in the computation on pages 21–22, the result is t, an approximate time of transit across the meridian. The apparent right ascension at the time of observation is taken from some star catalogue, giving apparent places, such as the American Ephemeris and Nautical Almanac or the Berliner Astronomisches Jahrbuch (preferably the former). The difference between t and the right ascension, α, of the star at the time of observation, is $(\alpha - t)$, an approximate correction to the chronometer time.

In taking right ascensions from the star catalogue it is necessary to interpolate for the longitude of the observer, and to consider second differences when they affect the result by as much as a hundredth of a second.

THE COLLIMATION CORRECTION.

If the instrument is otherwise in perfect adjustment, but has a small error in collimation, the micrometer wire in its mean position (or the mean line of the diaphragm) will describe a small circle parallel to the meridian and at an angular distance, the error of collimation, from it, when the telescope is rotated about its horizontal axis.

$$\text{The collimation correction} = c \sec \delta = Cc,$$

in which c is the angle, expressed in seconds of time, between the line of sight defined by the micrometer wire when in its mean position (or by the mean line of the diaphragm) and a plane perpendicular to the horizontal axis of the telescope. In other words, c is the angle between the *line of collimation* and the *collimation axis*. (See p. 13.) It is considered positive for a given telescope if the line of sight is too far east (and stars at upper culmination are therefore observed too soon) when the illumination (or bright band) is to the westward. This convention of sign is purely arbitrary, however. c is derived from the time computations by one of the processes shown on pages 26, 34, and 42.

The factor C is written for sec δ and is tabulated on pages 62–77. It is more easily obtained from the graphical device shown in illustration No. 9 and described on page 61. For observations made with illumination (or band) to the westward C is to be considered *positive* for stars at upper culmination and *negative* for stars at lower culmination. The signs are reversed with illumination (or band) east.

THE AZIMUTH CORRECTION.

If the instrument is otherwise in adjustment, but has a small error in azimuth, the micrometer wire in its mean position (or the mean line of the diaphragm) will describe a vertical circle on the celestial sphere at an angle with the meridian. The correction in seconds to an observed time of transit for this azimuth error is,

$$\text{Azimuth correction} = a \sin \zeta \sec \delta = Aa,$$

in which a is the angle expressed in seconds of time between the meridian and the vertical circle described by the mean position of the micrometer wire.[1] It is considered positive when the collimation axis is too far to the east with the telescope pointed south.

For convenience A is written for sin ζ sec δ and will be found tabulated on pages 62–77. It can be more easily obtained with the graphical device shown in illustration No. 9 and described on page 61. The factor A is considered positive for all stars except those between the zenith and the pole.

[1] In practice there always exists an error of collimation, so in general a is the angle between the meridian and the axis of collimation.

a is derived from the observations by one of the processes shown on pages 26, 34, 39, and 42, attention being paid to sign as indicated above.

COMPUTATION OF ΔT, c, AND a WITHOUT LEAST SQUARES.

The following method of computation was devised shortly after the time (1905) the transit-micrometer was adopted by this survey for use on longitude work and it is used both in the field and in the office for the final computation of all time observations made with the transit micrometer at stations in latitude less than 50°. In all latitudes greater than 50° the least-square solution is used in obtaining the final results. There is also a somewhat different method of computation (shown on p. 34) used when the stars of a time set consist of four time stars and one azimuth star. This method was used in the field for a number of years.

Computation of time set.

Form 256.*

[Station, Key West, Florida. Date, Feb. 14, 1907. Set, 2. Observer, J. S. Hill. Computor, J. S. Hill.]

Star	Clamp	$\alpha - t$	δt	C	A	Cc	Aa	$\Delta T = (\alpha - t) - Cc - Aa$	v
		s	s			s	s	s	s
1. S Monocer.	W	+15.00	0.00	+1.02	+0.26	+0.27	+0.02	+14.71	+.02
2. ψ^5 Aurigae	W	+15.08	+0.08	+1.38	−0.45	+0.36	−0.03	+14.75	−.02
3. 18 Monocer.	W	+15.04	+0.04	+1.01	+0.37	+0.26	+0.03	+14.75	−.02
4. θ Geminor.	W	+15.03	+0.03	+1.21	−0.20	+0.32	−0.01	+14.72	+.01
5. ζ Geminor.	W	+15.00	0.00	+1.07	+0.07	+0.28	0.00	+14.72	+.01
6. 63 Aurigae	W	+15.02	+0.02	+1.30	−0.34	+0.34	−0.02	+14.70	+.03
7. ι Geminor.	E	+14.43	−0.57	−1.13	−0.07	−0.30	0.00	+14.73	.00
8. β Can. Min.	E	+14.45	−0.55	−1.02	+0.28	−0.27	+0.01	+14.71	+.02
9. α Can. Min.	E	+14.45	−0.55	−1.01	+0.33	−0.26	+0.01	+14.70	+.03
10. β Geminor.	E	+14.41	−0.59	−1.13	−0.08	−0.30	0.00	+14.71	+.02
11. π Geminor.	E	+14.42	−0.58	−1.21	−0.19	−0.32	−0.01	+14.75	−.02
12. ϕ Geminor.	E	+14.47	−0.53	−1.12	−0.05	−0.29	0.00	+14.76	−.03

Mean $\Delta T = +14.727$

1.	$3.00\ \delta t + 3.10\ c + 0.70\ a_W$	$-0.04 = 0$					
2.	$3.00\ \delta t + 3.89\ c - 0.99\ a_W$	$-0.13 = 0$					
5.	$2.12\ \delta t + 2.75\ c - 0.70\ a_W$	$-0.09 = 0$	(2)×.707				
6.	$5.12\ \delta t + 5.85\ c$	$-0.13 = 0$					
9.	$4.71\ \delta t + 5.38\ c$	$-0.12 = 0$	(6)×.920				
10.	$9.53\ \delta t$	$+2.61 = 0$		11.	$\delta t = -0.274^s$		

$$\Delta T = +15.00 - 0.274 = +14.726^s$$

3.	$3.00\ \delta t - 3.15\ c + 0.56\ a_E$	$+1.63 = 0$			
4.	$3.00\ \delta t - 3.47\ c - 0.34\ a_E$	$+1.74 = 0$			
7.	$1.82\ \delta t - 1.91\ c + 0.34\ a_E$	$+0.99 = 0$	(3)×.607		
8.	$4.82\ \delta t - 5.38\ c$	$+2.73 = 0$			
12.	$-1.32\ \ -5.38\ c$	$+2.73 = 0$		13.	$c = +0.262^s$
14.	$-0.82\ \ +1.02\ \ -0.99\ a_W$	$-0.13 = 0$		15.	$a_W = +0.071$
16.	$-0.82\ \ -0.83\ \ +0.56\ a_E$	$+1.63 = 0$		17.	$a_E = +0.036$

* See note below table on p. 18.

EXPLANATION OF ABOVE COMPUTATION.

The serial numbers indicate the order of the various steps of the computation. Each equation, for a star, is of the form:

$$\Delta T + Cc + Aa - (\alpha - t) = 0$$

Equation 1 is obtained by adding corresponding terms of the three such observation equations for the three south stars (1, 3, and 5). Equations 2, 3, and 4 are obtained in a similar manner, there being two equations in each half set, one involving the three stars farthest south, the other the remaining stars of the half set, in this case three in number. There are then four equations, involving four unknowns, which can be solved by simple algebraic elimination. In the above computation this has been reduced to systematic mechanical operations. The azimuth constants are first eliminated, next c is eliminated, and then δt is obtained. The computation is so arranged that the multipliers are always less than unity, which are used to reduce coefficients in certain equations to equality with corresponding coefficients in other equations. This makes it possible to carry through the entire computation with the aid of Crelle's (or other similar) tables. In making substitutions in equations, such as 14 and 16, where there is a choice between two equations, it is always well to select the equation having the larger coefficient for the unknown sought. If the computation is followed in these respects and a sufficient number of whole seconds are dropped from the $(\alpha - t)$ to insure that δt will be less than one second, there is no necessity, in any given case, of carrying the computation to a greater number of decimal places than are shown above.

The checks which must be satisfied, if the computation is correct, are: (1) The algebraic sum of all the residuals must not in hundredths of seconds be more than one-half the number of stars in the complete set; (2) the sum of the two, three, or four residuals corresponding to each of the four equations designated above as 1, 2, 3, and 4 must seldom be as large as, and never exceed, $0^s.02$.

If these checks are not satisfied, the following principle may be found useful in detecting whether the error was made during the process of solution of the four equations. If the work of solution is correct, the derived values of the unknowns substituted in any one of the equations should give a residual not greater than $0^s.01$ (the substitution being carried to thousandths of seconds), but if any equation shows a residual greater than this, the error in the solution was made in deriving an equation of a higher serial number, the serial numbers having been assigned in the order in which the computation was made.

The chronometer correction ΔT is then equal to δt plus the number of whole seconds which were dropped from $(\alpha - t)$ in order to lighten the work involved in making the computation. In this case it is equal to $-0^s.274 + 15^s.00 = +14^s.726$. The chronometer epoch for which this correction applies is the mean of the chronometer times of the observed transits; that is, the mean of the t's. It is not the mean of the right ascensions—unless, of course, the chronometer correction happens to be zero.

While it is advisable to have the instrumental constants c, a_W, and a_E small, it is not desirable to strive to have them close to zero. For the azimuth constant one second is a good limit to keep within, while if the collimation constant is less than $0^s.2$ it is well not to attempt further adjustment with a view of reducing it.

The computations are somewhat simpler when the transit is reversed on each star and one-half the observations on a star are made in each of the positions—band west and band east—for the collimation is eliminated by the method of observing and the only unknowns are one azimuth constant and the clock correction, ΔT.

A SECOND EXAMPLE OF RECORD AND COMPUTATION.

On page 26 reference is made to a second method of solution for ΔT, a, and c, without the use of least squares. This second method is used when a different selection of stars is made from that shown on page 18. The difference between the two star sets is that in the example of computation shown on page 26 the instrumental constants c and a are determined from all the stars, each star being given unit weight, while in the method which follows there is observed in each half set a slow-moving star, called the azimuth star, from which the azimuth constant for that half set is principally determined. Besides this azimuth star there are four time stars in each half set, and it is from the eight time stars in the entire set that the collimation constant is mainly derived. It seems that the method of having all time stars in a set is preferable to the other method, in which both time and azimuth stars are used. In the former, the clock correction depends on all 12 stars instead of being derived mainly from 8 stars only, and the collimation correction is more accurately determined. The azimuth constants, however, are not so accurately determined by the first as by the second method, but this is immaterial if the plus and minus azimuth factors in each half set are about equally balanced.

While this second method has been superseded in the longitude work of the Coast and Geodetic Survey, it is considered desirable to continue it in this publication.

Using this second method, time acceptable for latitude or azimuth work can be easily obtained with a meridian telescope, a zenith telescope, or even with an engineer's transit or theodolite. In its usual form the star set consists of four time stars and an azimuth star with the instrument in each position, band west and band east. If greater accuracy is desired the number of time stars in a half set may be increased, or if less accuracy is needed the number may be decreased. In the work of the Survey up to the time of the adoption of the transit micrometer and the method of computation shown on pages 20–27, the standard time set consisted of two half sets, in each of which was one azimuth star and four time stars.

The following set of observations was made with a small portable transit, using an observing key to record the observations chronographically. With the record of observations there are given the readings of the level, the correction for inclination of the horizontal axis of the telescope (which in this case includes a correction for inequality of pivots), and the computation of $(\alpha-t)$. A correction for rate has been introduced. The correction for diurnal aberration and the correction for rate are obtained in the same manner as shown on page 24. The form on which the level readings are recorded is shown on page 20.

Star list for Washington, D. C.—Latitude 38 ° 54′ N.

Star	Catalogue	Magnitude	Right ascension α	Declination δ	Zenith distance ζ	Diurnal aberration κ	Star factors		
							Inclination B	Collimation C	Azimuth A
			h m s	° ′	° ′	s			
17 H. Can. Ven.	B	5.5	13 30 12	+37 43	+ 1 11	−.02	1.26	1.26	+ .02
η Ursæ Maj.	B	2.0	43 30	+49 50	−10 56	−.02	1.53	1.55	− .30
η Bootis	B	3.0	49 47	+18 55	+19 59	−.02	0.99	1.06	+ .36
11 Bootis	B	6.0	56 31	+27 53	+11 01	−.02	1.11	1.13	+ .22
α Draconis	B	3.3	14 01 39	+64 52	−25 58	−.04	2.12	2.36	−1.03
d Bootis	B	5.0	05 42	+25 35	+13 19	−.02	1.08	1.11	+ .25
α Bootis	B	1.0	10 58	+19 43	+19 11	−.02	1.01	1.06	+ .35
λ Bootis	B	4.0	12 29	+46 34	− 7 40	−.02	1.44	1.46	− .19
θ Bootis	B	3.8	21 43	+52 20	−13 26	−.03	1.59	1.64	− .38
5 Ursæ Min.	A	4.5	27 51	+76 09	−37 15	−.06	3.33	4.18	−2.53

B=Berliner Astronomisches Jahrbuch. A=American Ephemeris.

Following the computation are given any explanations needed to supplement or qualify the explanations of computations given on pages 22–27.

[Station, Washington, D. C. Date, May 17, 1896. Observer, G. R. P.

Star	17 H. Can. Ven.		η Urs. Maj.		η Bootis		π Bootis		α Draco.	
Position of band	West		West		West		West		West	
Direction of objective for level reading	S				S				N	
	W	E			W	E			W	E
	d	d			d	d			d	d
	22.7	24.1			27.8	20.0			28.0	19.9
Level readings	27.1	19.9			22.9	24.8			22.9	24.9
ΣW and ΣE	49.8	44.0			50.7	44.8			50.9	44.8
ΣW−ΣE	+5.8				+5.9				+6.1	
Remarks and computation of b			Means of levels d N +6.10 S +5.85 ——— +5.98×.0279=		s + .167=β_W − .010=pivot inequality ——— + .157=b_W					
Observed transit	h m s		h m s		h m s		h m s		h m s	
Line 1	13 29 56.90		13 43 10.60		13 49 34.45		13 56 17.20		14 01 07.55	
2	13 30 00.10	Mean	14.35		37.05		20.00		13.30	
3	03.30	s	18.30		39.70		22.80		19.35	
4	09.70	14.20	26.15		45.00		28.40		31.25	
5	12.90	Correc-	30.15	s	47.75	s	31.50	s	37.30	s
6	16.00	tion	33.80	33.80	50.25	50.25	34.30	34.30	43.00	43.00
7	19.30	15.15	37.95	68.10	52.95	100.70	36.90	68.40	49.00	86.30
8	22.60	=———	41.70	7.85	55.70	0.70	40.05	8.45	55.20	6.45
9	29.00	10	49.70	8.00	13 50 00.90	0.60	45.65	8.45	14 02 06.90	6.25
10	32.20	× 1.26	53.60	7.95	03.75	0.80	48.60	8.60	12.90	6.20
11	—	= + 1.92	57.55	8.15	06.50	0.95	51.50	8.70	18.85	6.40
Mean	16.12		33.99	10.85	50.36	4.00	34.26	2.90	43.15	1.60
R	+ .03		+ .02		+ .01		+ .01		.00	
κ	− .02		− .02		− .02		− .02		− .04	
Bb	+ .20		+ .24		+ .16		+ .17		+ .33	
t	13 30 16.33		13 43 34.23		13 49 50.51		13 56 34.42		14 01 43.44	
α	13 30 12.26		13 43 30.14		13 49 46.62		13 56 30.53		14 01 38.92	
α−t	−4.07		−4.09		−3.69		−3.89		−4.52	

Instrument, transit No. 18. Chronometer, Negus, 1836 (daily rate, 1s.51 gaining).]

d Bootis		α Bootis		λ Bootis		θ Bootis		5 Urs. Min.	
East		East		East		East		East	
S				N				N	
W	E			W	E			W	E
d	d			d	d			d	d
27.1	20.9			27.2	20.9			22.2	26.0
22.7	25.2			22.9	25.3			27.2	21.0
49.8	46.1			50.1	46.2			49.4	47.0
+3.7				+3.9				+2.4	
		Means of levels d N. +3.15 S. +3.70 s +3.42×.0279=		s +.095= β_E +.010= pivot inequality +.105= b_E				Thin clouds and hazy Temperature 76° F	
h m s		h m s		h m s		h m s		h m s	
14 05 29.40	Mean	14 10 45.50		14 12 11.15		14 21 22.30		14 26 53.15	
32.20	s	48.20		14.80		26.40		27 03.15	
34.85	44.76	50.90		18.60		30.60		14.25	
40.60	Correc-	56.20		25.95		38.90		35.30	
43.35	tion	58.90	s	29.50	s	42.90	s	45.85	s
46.20	12.69	14 11 01.65	01.65	33.45	33.45	47.35	47.35	57.15	57.15
48.90	= ——	04.30	03.20	37.00	66.50	51.35	94.25	14 28 07.00	112.85
51.90	10	07.10	3.30	40.70	6.65	55.40	4.30	18.00	3.30
57.30	× 1.11	12.30	3.20	48.00	6.60	14 22 03.60	4.20	38.70	2.95
—	= 1.41	15.20	3.40	51.80	6.60	07.80	4.20	49.50	2.65
14 06 02.90		17.70	3.20	55.25	6.40	11.95	4.25	14 29 00.05	3.20
46.17		01.63	6.95	33.29	3.20	47.14	1.55	56.55	72.10
.00		− .01		− .01		− .02		− .03	
− .02		− .02		− .02		− .03		− .06	
+ .11		+ .11		+ .15		+ .17		+ .35	
14 05 46.26		14 11 01.71		14 12 33.41		14 21 47.26		14 27 56.81	
14 05 42.32		14 10 57.90		14 12 29.18		14 21 42.97		14 27 51.37	
−3.94		−3.81		−4.23		−4.29		−5.44	

REDUCTION OF INCOMPLETE TRANSITS.

If the transit of a star across every line of the diaphragm is observed, the mean of the times is the required time of transit across the mean line. In obtaining the sum of the several observed times any gross error in any one of the times may be detected by using the auxiliary sums, shown in the example on pages 30–31, in the little column just after the observed times, namely, the sum of the first and last times, of the second and last but one, third and last but two, etc. These auxiliary sums should be nearly the same and nearly equal to double the time on the middle line. This is also a convenient method of taking means, as it is in general only necessary to sum the decimal columns.

When the star was observed on some of the lines but missed upon the others, the time of transit over the mean of all the lines may be found as follows:

$$t_m = \text{mean of observed times} - \frac{(\text{sum of equatorial intervals of observed lines}) \cdot (\sec \delta)^1}{\text{number of observed lines.}}$$

$$\text{or } t_m = \text{mean of observed times} + \frac{(\text{sum of equatorial intervals of missed lines}) \ (\sec \delta)}{\text{number of observed lines.}}$$

The first of these formulæ is the more convenient if but few lines were observed and the second the more convenient if but few lines were missed. The two incomplete transits shown in the example on pages 30–31 were reduced by the second formula.

t_m is the time of transit across the mean of all the lines of the diaphragm. The equatorial interval of a given line is the time which would elapse between the transit of an equatorial star over the mean line of the diaphragm and the transit over the line in question. It is, in seconds of time, $\frac{1}{15}$ the angular interval between the lines expressed in seconds of arc. An equatorial interval is called positive when the transit across the line in question occurs later than the transit across the mean line. The signs of all the equatorial intervals are therefore reversed when the horizontal axis of the telescope is reversed.

For an example of the method of computing the equatorial intervals see page 44.

The above formulæ for reduction to the mean line are approximate, and the maximum possible error of the approximation increases with an increase in the declination of the star and with an increase in the equatorial intervals of the extreme lines. If the extreme equatorial interval is 60^s, the maximum error is less than $0^s.01$ for a star of which $\delta = 70°$, and is only $0^s.3$ if $\delta = 85°$. If the extreme interval is 15^s, the maximum error is less than $0^s.01$ if $\delta = 85°$.

The more exact formula for use with circumpolar stars is the same as that given above, except that for each equatorial interval, i, must be substituted $i \sqrt[3]{\sec \tau}$, in which τ is the hour angle of the star at transit across the line, or with sufficient accuracy $\tau = i \sec \delta =$ the actual time interval from the mean line.

The following table will be found useful in connection with this formula.

τ	$\log \sqrt[3]{\cos \tau}$	$\log \sqrt[3]{\sec \tau}$	τ	$\log \sqrt[3]{\cos \tau}$	$\log \sqrt[3]{\sec \tau}$	τ	$\log \sqrt[3]{\cos \tau}$	$\log \sqrt[3]{\sec \tau}$
m			*m*			*m*		
1	9.99999	0.00000	16	9.99965	0.00035	31	9.99867	0.00133
2	99	01	17	960	040	32	858	142
3	99	01	18	955	045	33	849	151
4	98	02	19	950	050	34	840	160
5	97	03	20	945	055	35	831	169
6	95	05	21	939	061	36	821	179
7	93	07	22	933	067	37	811	189
8	91	09	23	927	073	38	800	200
9	89	11	24	921	079	39	789	211
10	86	14	25	914	086	40	778	222
11	83	17	26	907	093	41	767	233
12	80	20	27	899	101	42	756	244
13	77	23	28	892	108	43	744	256
14	73	27	29	884	116	44	732	268
15	9.99969	0.00031	30	9.99876	0.00124	45	9.99719	0.00281

[1] The collimation factor C (as given in the star list on p. 29) is the $\sec \delta$.

If the chronometer rate exceeds 15^s per day it will be desirable to take it into account in making the reduction of incomplete transits to the mean line.

Another method of reducing incomplete transits is to construct from the known equatorial intervals a table similar to that of which a portion is printed below showing the interval of each line from the mean line corresponding to various declinations. The correction of each observed line to the mean line is then taken out directly from the table and the mean of the various corrected transits taken.

Intervals of lines of Transit No. 18 from mean line.

[The numbering of the lines is for band west.]

δ	Line I	Line II	Line III	Line IV	Line V	Line VI	Line VII	Line VIII	Line IX	Line X	Line XI
°	s	s	s	s	s	s	s	s	s	s	s
0	+15.20	+12.69	+10.15	+5.06	+2.52	−0.09	−2.52	−5.11	−10.09	−12.65	−15.15
10	15.43	12.89	10.31	5.14	2.56	0.09	2.56	5.19	10.25	12.84	15.38
15	15.74	13.14	10.51	5.24	2.61	0.09	2.61	5.29	10.45	13.10	15.68
......	......	......	......	......	......	......	......	......	......	......	......
36	18.79	15.69	12.55	6.25	3.11	0.11	3.11	6.32	12.47	15.64	18.73
38	19.29	16.10	12.88	6.42	3.20	0.11	3.20	6.48	12.80	16.05	19.23
40	19.84	16.57	13.25	6.61	3.29	0.12	3.29	6.67	13.17	16.51	19.78
......	......	......	......	......	......	......	......	......	......	......	......
51	24.15	20.17	16.13	8.04	4.00	0.14	4.00	8.12	16.03	20.10	24.07
52	24.69	20.61	16.49	8.22	4.09	0.15	4.09	8.30	16.39	20.55	24.61
53	25.26	21.09	16.87	8.41	4.19	0.15	4.19	8.49	16.77	21.02	25.17
......	......	......	......	......	......	......	......	......	......	......	......

Transit No. 18 was the instrument used for the observations shown on pages 30–31. The incomplete transit of the star 17 H. Can. Ven., of which the declination is 37° 43′, may be computed as indicated below:

Line	Correction	Corrected transit
I	+19.22	16.12
II	+16.04	16.14
III	+12.83	16.13
IV	+ 6.40	16.10
V	+ 3.19	16.09
VI	− 0.11	15.89
VII	− 3.19	16.11
VIII	− 6.46	16.14
IX	−12.75	16.25
X	−15.99	16.21

Mean=16.12, agreeing with the result shown in the example on page 30.

The special advantage of this method of reducing incomplete transits is that a wild observation upon any one line is at once detected. Such wild observations are apt to occur under the conditions which produce incomplete transits, viz., clouds, haste, or difficulty with illumination.

CORRECTION FOR RATE.

The method of computing this correction is shown on page 24.

CORRECTIONS FOR DIURNAL ABERRATION, COLLIMATION, AND AZIMUTH.

The correction for diurnal aberration and general expressions, for the collimation and azimuth corrections are shown on pages 24–25.

COMPUTATION OF ΔT, a AND c, USING AZIMUTH STARS AND METHOD OF APPROXIMATIONS.

The method of computation shown below was in use in the field by parties of this Survey for many years.[1] It is now replaced by the method shown on page 26.

[Station, Washington, D. C. Date, May 17, 1896.]

Star	Band	$\alpha-t$	C	A	Cc	Aa	ΔT† $\alpha-t-Cc-Aa$	Δ	
		s			s	s	s	s	
17 H. Can. Ven.	W	−4.07	+1.26	+ .02	+.04	+ .01	−4.12	+.10	
η Urs. Maj.	W	−4.09	+1.55	− .30	+.05	− .17	−3.97	−.05	
η Bootis	W	−3.69	+1.06	+ .36	+.03	+ .20	−3.92	−.10	
11 Bootis	W	−3.89	+1.13	+ .22	+.04	+ .12	−4.05	+.03	at $14^h\ 02^m.00$
α Draconis	W	−4.52	+2.36	−1.03	+.08	− .58	−4.02	.00	$\Delta T = -04^s.024$
d Bootis	E	−3.94	−1.11	+ .25	−.04	+ .13	−4.03	.00	
α Bootis	E	−3.81	−1.06	+ .35	−.03	+ .18	−3.96	−.07	
λ Bootis	E	−4.23	−1.46	− .19	−.05	− .10	−4.08	+.05	
θ Bootis	E	−4.29	−1.64	− .38	−.05	− .19	−4.05	+.02	
5 Urs. Min.	E	−5.44	−4.18	−2.53	−.13	−1.28	−4.03	.00	
		$\alpha-t$	C	A	Cc	$\alpha-t-Cc$	Aa	$\alpha-t-Cc-Aa$	
First approximation:									s
Mean of time stars	W	−3.94	+1.25	+ .08	+.06	−4.00	+ .05	−4.05	$c = +.051$
Azimuth star	W	−4.52	+2.36	−1.03	+.12	−4.64	− .59	−4.05	$a_W = +.577$
Mean of time stars	E	−4.07	−1.32	+ .01	−.07	−4.00	.00	−4.00	$a_E = +.484$
Azimuth star	E	−5.44	−4.18	−2.53	−.21	−5.23	−1.23	−4.00	
Second approximation:									
Mean of time stars	W				+.04	−3.98	+ .04	−4.02	$c = +.032$
Azimuth star	W				+.08	−4.60	− .58	−4.02	$a_W = +.559$
Mean of time stars	E				−.04	−4.03	.00	−4.03	$a_E = +.504$
Azimuth star	E				−.13	−5.31	−1.28	−4.03	

† The complete formula for the chronometer correction is $\Delta T = \alpha - (t_m + R + \kappa + Bb + Cc + Aa)$. Let $t = t_m + R + \kappa + Bb$, then $\Delta T = (\alpha - t) - Cc - Aa$, so that it will be seen that the corrections Cc and Aa are to be subtracted algebraically from $\alpha - t$.

EXPLANATION OF THE COMPUTATION.

The first five columns of the upper portion of the computation were compiled from the record and computation shown on pages 30–31 and from the observing list shown on page 29. The remaining columns were filled out after the computation of a and c, shown in the lower portion of the form, was completed.

It should be noted that the five stars of each group, observed in one position of the instrument, have been so selected that one is a slowly moving northern star at a considerable distance from the zenith, while the other four are all comparatively near the zenith, some transiting to the northward of it and some to the southward, and at such distances from it that their mean azimuth factor, A, is nearly zero. These four stars of each group may be for convenience called *time stars,* since the determination of time falls mainly upon them, while the slowly moving northern star serves to determine the azimuth error of the instrument, and may be called the *azimuth star.*

In this computation to derive c and a the time stars in each position of the instrument are combined and treated as one star by taking the means of their $(\alpha - t)$'s, and of their star factors C and A, respectively, these means being written below the separate stars in the form, together with the azimuth stars. On the assumption that the means of the time stars in the two positions of the instrument are equally affected by the azimuth correction, the first approxi-

[1] It was devised in the seventies by Assistant Edwin Smith, then an aid in this Survey. See p. 280, Appendix 4 of the Report for 1904.

mation to c is found by dividing the difference between the mean $(\alpha-t)$'s by the difference between the C's. In the example,

$$c \text{ (first approximation)} = \frac{(\alpha-t)_W-(\alpha-t)_E}{C_W-C_E} = \frac{-3.94-(-4.07)}{+1.25-(-1.32)} = \frac{+0.13}{+2.57} = +0^s.051.$$

Using this approximation to c, the correction Cc is then subtracted from the $\alpha-t$ of each mean of the time stars and of each azimuth star, and the values of $\alpha-t-Cc$, in the seventh column on the fifth to eighth lines from the bottom of the form, are obtained.

Separate values for the azimuth error of the instrument are then derived for each position of the instrument as follows:

$$a_W = \frac{(\alpha-t-Cc)_{\text{time stars}}-(\alpha-t-Cc)_{\text{azimuth star}}}{A_{\text{time stars}}-A_{\text{azimuth star.}}} = \frac{-4.00-(-4.64)}{+0.08-(-1.03)} = \frac{+0.64}{+1.11} = +0^s.577.$$

$$a_E = \frac{-4.00-(-5.23)}{+0.01-(-2.53)} = \frac{+1.23}{+2.54} = +0^s.484.$$

With these values of a_W and a_E the corrections Aa are applied, giving the values $\alpha-t-Cc-Aa$ in the last column but one. If these do not agree for the stars east and west it indicates that the mean values $\alpha-t$, used in deriving c, were not equally affected by the azimuth error, so that their difference was not entirely due to c, as was assumed. An improved value of c may now be obtained by treating the difference in the last column as still an error of collimation, and thus obtaining a correction to the first approximate value of c. Thus, in the example,

$$\frac{-4.05-(-4.00)}{+1.25-(-1.32)} = \frac{-0.05}{+2.57} = -0^s.019.$$

Applying this correction to the first approximate value of $c=+0.051$, we have for a second approximation $c=+0.032$. Proceeding as before, improved values for a_W and a_E are found. If the star sets are well chosen and the instrumental errors small, the first approximation will generally suffice. If the values of $\alpha-t-Cc-Aa$ differ by but a few hundredths, east and west, there is little gained by making a closer adjustment. The chronometer correction will probably not be changed at all, but the instrumental errors and star residuals will be slightly altered, as is apparent from the example, where the closer adjustment is made for the purpose of illustrating the method.

In the first approximation the value of c may at once be derived more closely when there is much difference between the mean A's for the time stars, by estimating the effect of this difference in A on the ΔT, and allowing for this effect when deriving c in the first place. The formula for c then becomes

$$c = \frac{(\alpha-t)_W-(\alpha-t)_E-(A_W-A_E)\times a}{C_W-C_E}.$$

It is here necessary to estimate the azimuth of the instrument, a, roughly in advance, and this may be done by inspection. Thus, in the example, assuming $a=+0^s.5$, we have

$$c = \frac{-3.94+4.07-(+.07)\times(+0.5)}{+1.25+1.32} = \frac{+.09}{+2.57} = +0^s.037$$

agreeing closely with the value given by the second approximation.

When satisfactory values of c, a_W, and a_E have been obtained, the corrections Cc and Aa are applied separately to each star, as shown in the upper part, and the values of the chronometer correction $(\triangle T)$ derived separately. The residuals are taken for each group from the mean of that group, and thus furnish a convenient check on the computation, as their sums for each group should approximate zero. Unusual residuals also point to possible errors in $\alpha-t$. The mean of the ΔT's from the separate stars gives the final chronometer correction at the epoch of the mean of the chronometer times of transit of the stars observed.

This whole computation may be made with rapidity by the use of Crelle's multiplication tables.

The field computation having been made as outlined above,[1] the more refined office computation may be made as indicated on pages 39–41. It is desirable in this office computation to introduce weights dependent upon the declination of the star and the number of lines of the reticle upon which the star was observed.

The four equations, solved by successive approximations above, may be solved by direct elimination, in case the coefficients of a_W and a_E do not become relatively small in the two equations gotten by taking the mean of the time stars in the two half sets.

RELATIVE WEIGHTS FOR INCOMPLETE TRANSITS.

Sometimes the transit of a star is observed over some of the lines of the diaphragm and missed over the others. Obviously the deduced time of transit over the mean line from such an incomplete transit should be given less weight than that from a complete transit.

For observations made by the *eye and ear method* the relative weights given by Chauvenet may be used, viz:

$$p=\frac{n\ (N+3)}{N\ (n+3)}$$

in which p is the weight to be assigned to the computed time of transit over the mean line, N is the total number of lines in the diaphragm, and n is the number of lines upon which observations were made.[2] This formula is based upon the assumption that $(\epsilon)^2=3(\epsilon_1)^2$, in which $(\epsilon)=$ the probable error of an observed transit of an equatorial star over a *single* line and $(\epsilon_1)=$the probable culmination error referred to the equator, a constant for all the lines of the diaphragm for any one star, but variable from star to star, and supposed to be due mainly to atmospheric displacement, to outstanding instrumental errors, to irregularities in clock rate, and to changes in personal equation.

The following table shows the values of p and $\sqrt{p}$ for the two cases of 5 and 7 lines in the diaphragm:

Table of weights for incomplete transits for use with eye and ear observations.

n	$N=5$		$N=7$	
	p	$\sqrt{p}$	p	$\sqrt{p}$
1	0.40	0.63	0.36	0.60
2	0.64	0.80	0.57	0.75
3	0.80	0.89	0.71	0.84
4	0.92	0.96	0.82	0.91
5	1.00	1.00	0.90	0.95
6			0.95	0.97
7			1.00	1.00

[1] For a more complete account of this method of computation, see Appendix No. 9, Report for 1896. The above account is largely taken from that appendix.

[2] See Chauvenet's Astronomy, Vol. II, p. 198. The derivation of this formula follows the same lines as that given on the following pages for weights to be assigned to incomplete transits taken by the chronographic method.

The relative weights to be assigned to incomplete transits observed by the *chronograph method* may be derived as follows:

$$r^2 = (\epsilon)^2 + \frac{(\epsilon)^2}{n}$$

in which $r =$ the probable error of the time of transit over the mean line, arising from the combined effect of the culmination error referred to the equator (ϵ_1) and of the probable error of the transit of an equatorial star over a single line (ϵ).

To find r, individual determinations of right ascensions of stars, all referred to the same epoch (mean place), may be compared with their respective average values; thus, from 558 results of 36 stars observed at the United States Naval Observatory with the transit circle (using a magnifying power of 186) in 1870 and 1871, it was found that $r = \pm 0^s.034$. To apply this value to our instruments it must be somewhat increased, though not in proportion to the respective magnifying powers, since some of the errors involved approach the character of constants; multiplying it by 1.5 and 1.75 for our larger and smaller transits, respectively, there is obtained $r = \pm 0^s.051$ and $r = \pm 0^s.060$. For the larger transits $(\epsilon) = \pm 0^s.063$ and for the smaller $(\epsilon) = \pm 0^s.080$. (See p. 39.) Substituting these values in the above formula, together with the values 25 and 15 for n as actually used in the observations cited on page 38, there is obtained

$$(0.051)^2 = (\epsilon_1)^2 + \frac{(0.063)^2}{25} \text{ and } (0.060)^2 = (\epsilon_1)^2 + \frac{(0.080)^2}{15}$$

which give

$$(\epsilon_1) = \pm 0^s.049 \text{ and } (\epsilon_1) = \pm 0^s.056$$

for the larger and smaller instruments, respectively.

If the weight for a complete transit is unity, the weight for an incomplete transit is

$$p = \frac{(\epsilon_1)^2 + \frac{(\epsilon)^2}{N}}{(\epsilon_1)^2 + \frac{(\epsilon)^2}{n}}.$$

Hence, for the larger instruments, using the above values for (ϵ_1) and (ϵ),

$$p = \frac{1 + \frac{1.6}{N}}{1 + \frac{1.6}{n}}$$

and for the smaller instruments

$$p = \frac{1 + \frac{2.0}{N}}{1 + \frac{2.0}{n}}$$

very nearly. From these expressions the relative weights have been computed for total number of threads $N = 25$, 17, 13, and 11 for the larger instruments and for $N = 15$, 13, 11, and 9 for the smaller ones, and are shown in the following table.

Table of weights for incomplete transits for use with chronographic observations.

Number of lines n	For large portable transits								For small portable transits							
	$N=25$		$N=17$		$N=13$		$N=11$		$N=15$		$N=13$		$N=11$		$N=9$	
	p	$\sqrt{p}$	p	$\sqrt{p}$	p	$\sqrt{p}$	p	$\sqrt{p}$	p	$\sqrt{p}$	p	$\sqrt{p}$	p	$\sqrt{p}$	p	$\sqrt{p}$
1	.41	.64	.42	.65	.43	.66	.44	.66	.38	.62	.38	.62	.39	.63	.41	.64
2	.59	.77	.61	.78	.62	.79	.64	.80	.56	.75	.58	.76	.59	.77	.61	.78
3	.69	.83	.71	.84	.73	.86	.75	.86	.68	.83	.69	.83	.71	.84	.73	.86
4	.76	.87	.78	.88	.80	.90	.82	.90	.75	.87	.77	.88	.79	.89	.82	.90
5	.81	.90	.83	.91	.85	.92	.87	.93	.81	.90	.82	.91	.84	.92	.87	.93
6	.84	.91	.86	.93	.89	.94	.90	.95	.85	.92	.87	.93	.89	.94	.92	.96
7	.87	.93	.89	.94	.91	.96	.93	.97	.88	.94	.90	.95	.92	.96	.95	.97
8	.89	.94	.91	.95	.94	.97	.96	.98	.91	.95	.92	.96	.95	.97	.98	.99
9	.90	.95	.93	.96	.95	.98	.97	.99	.92	.96	.94	.97	.97	.98	1.00	1.00
10	.92	.96	.94	.97	.97	.98	.99	.99	.94	.97	.96	.98	.99	.99		
11	.93	.96	.95	.98	.98	.99	1.00	1.00	.96	.98	.98	.99	1.00	1.00		
12	.94	.97	.96	.98	.99	1.00			.97	.99	.99	1.00				
13	.95	.97	.97	.99	1.00	1.00			.98	.99	1.00	1.00				
14	.96	.97	.98	.99					.99	1.00						
15	.96	.98	.99	1.00					1.00	1.00						
16	.97	.98	1.00	1.00												
17	.97	.98	1.00	1.00												
18	.98	.99														
19	.98	.99														
20	.98	.99														
21	.99	.99														
22	.99	1.00														
23	.99	1.00														
24	1.00	1.00														
25	1.00	1.00														

RELATIVE WEIGHTS TO TRANSITS DEPENDING ON THE STAR'S DECLINATION.

The following tables of the probable error (ε) of an observation of a transit of a star over a *single* line have been derived from a discussion of 1047 transits taken in February and March, 1869, at San Francisco, by Assistant G. Davidson, with the large transit C. S. No. 3 (aperture 2¾ inches, magnifying power 85); and 875 transits taken about the same time at Cambridge by Assistant A. T. Mosman, including some observations by Subassistant F. Blake, with the large transit C. S. No. 5 (aperture 2¾ inches, magnifying power 100). For the discussion of observations with a smaller instrument, 330 transits were used, taken in September, October, and November, 1871, at Cleveland, Ohio; and 585 transits, taken in December and January, 1871–72, at Falmouth, Ky., by Assistant E. Goodfellow, with a meridian telescope C. S. No. 13 (aperture 1¾ inches, magnifying power about 70).

Transit No. 3		Transit No. 5		Meridian telescope No. 13		Meridian telescope No. 13	
δ	(ε)	δ	(ε)	δ	(ε)	δ	(ε)
°	s	°	s	°	s	°	s
87.2	±0.74	86.9	±0.66	81.9	±0.62	76.3	±0.20
86.6	0.49	80.0	0.20	76.9	0.18	68.2	0.16
83.0	0.38	76.3	0.19	67.4	0.11	55.8	0.13
81.0	0.31	72.6	0.12	62.0	0.14	48.4	0.15
68.4	0.12	68.8	0.11	55.8	0.09	23.2	0.102
62.9	0.088	3.2	0.066	44.8	0.088	20.4	0.089
48.6	0.075	....	....	29.7	0.067	17.3	0.110
28.5	0.058	....	....	0.7	0.071	6.1	0.080
7.8	0.060	....	....	....	....	....	....

These tabular values are fairly represented by the expressions

Transit, No. 3 $(\varepsilon)=\sqrt{(0.060)^2+(0.036)^2 \tan^2\delta}$

Transit, No. 5 $(\varepsilon)=\sqrt{(0.066)^2+(0.036)^2 \tan^2\delta}$

Meridian telescope, No. 13 $(\varepsilon)=\sqrt{(0.069)^2+(0.078)^2 \tan^2\delta}$

Meridian telescope, No. 13 $(\varepsilon)=\sqrt{(0.087)^2+(0.055)^2 \tan^2\delta}$

Combining these expressions for the larger and smaller instruments, we obtain

$$(\varepsilon)=\sqrt{(0.063)^2+(0.036)^2 \tan^2\delta} \text{ and } (\varepsilon)=\sqrt{(0.080)^2+(0.063)^2 \tan^2\delta}$$

respectively,[1] from which the following tables of probable errors (ε), of relative weights p, and of the multipliers $\sqrt{p}$ for the conditional equations, have been computed:

Table of weights to transits depending on the star's declination.

	δ	For large portable transits			For small portable transits		
		(ε)	p	$\sqrt{p}$	(ε)	p	$\sqrt{p}$
	° ′	s			s		
	0	±0.06	1	1	±0.08	1	1
	10	.06	1	1	.08	0.98	1
	20	.06	0.98	1	.08	.92	0.96
	30	.07	.91	0.95	.09	.83	.91
	40	.07	.82	.90	.10	.70	.83
	45	.07	.76	.87	.10	.62	.79
	50	.08	.69	.83	.11	.53	.73
	55	.08	.61	.78	.12	.44	.66
	60	.09	.51	.71	.14	.34	.59
	65	.10	.40	.63	.16	.26	.51
	70	.12	.29	.54	.19	.18	.42
	75	.15	.18	.43	.25	.10	.32
	80	.21	.09	.30	.37	.05	.22
	85	.42	.02	.15	.72	.01	.11
δ Ursæ Minoris	86 37	0.61	0.011	0.103	1.1	0.006	0.075
51 Cephei	87 12	0.74	0.007	0.085	1.3	0.004	0.062
α Ursæ Minoris	88 46	1.7	0.001	0.037	2.9	0.001	0.027
λ Ursæ Minoris	88 59	2.0	0.001	0.031	3.5	0.001	0.023

COMPUTATION OF ΔT AND a BY LEAST SQUARES.

A field computation made by the approximate method indicated on page 34 gives values for ΔT, a, and c, which are of a high degree of accuracy. It should be noted that the derived values of a and c depend upon *all* the observations and not simply upon observations on a few stars only of the set, as is frequently the case with other approximate methods. Experience shows that the value of c especially, as thus derived in the field computation, is so accurate that a value derived from a subsequent rigid least square adjustment will in general be substantially identical with it, provided the stars of the set are chosen as indicated on pages 34 and 43. Accordingly, in the final computations by this method, only the unknowns a_W, a_E, and ΔT are to be determined by least squares, while c is taken from the field computations, revised and corrected if necessary. This method of computation is shown below.

Let $\Delta t_c = (\alpha - t) - Cc$ in which t is the chronometer time of transit across the mean line of the diaphragm corrected for rate, diurnal aberration and inclination and $\alpha - t$ is therefore the

[1] The following formula has been published by Dr. Albrecht on p. 23 of his Formeln und Hülfstafeln, etc., Leipzig, 1894, viz:

$$(\varepsilon)=\sqrt{(0.05)^2+\left(\frac{3.18}{v}\right)^2 \sec^2\delta}$$

Putting $v=85$ for the magnifying power and changing sec into tan, this expression is equivalent to

$$(\varepsilon)=\sqrt{(0.062)^2+(0.037)^2 \tan^2\delta}$$

quantity on the last line of the field record and computation as shown on pages 30–31. Let Δt be an assumed value of the chronometer correction and δt a correction to Δt to be derived from the computation. The final value of the chronometer correction will then be $\Delta T = \Delta t + \delta t$. Let d, for each star $= \Delta t_c - \Delta t$.

Then for each star observed an observation equation of the form

$$\sqrt{p}\,\delta t + \sqrt{p} A a = \sqrt{p}\, d,$$

may be written, in which the weights p are assigned according to the tables on pages 38–39.

In forming the normal equations each half set, made with the horizontal axis in one position, is treated independently of the other half set.

The normal equations corresponding to the half set made with illumination (or bright band) to the westward are

$$\Sigma p\,\delta t + \Sigma p A a_W = \Sigma p d$$
$$\Sigma p A \delta t + \Sigma p A^2 a_W = \Sigma p A d$$

and similarly for the other half set.

The most convenient arrangement of this computation is shown below, this example being a computation of the time set treated on pages 29–31 and 34.

WASHINGTON, D. C., *May 17, 1896.*

$c = +.032$ $\Delta t = -4^s.01$

Star	Band	$\alpha - t$	C	Cc	Δt_c	d	A	p*	pA	pA^2	pd	pAd	Aa	ΔT	Δ	$p\Delta$	$p\Delta^2$
17 H.Can.Ven.	W	−4.07	+1.26	+.04	−4.11	− .10	+ .02	.83	+.02	.00	−.08	.00	+ .01	−4.12	+.10	+.08	.0083
η Urs. Maj.	W	−4.09	+1.55	+.05	−4.14	− .13	− .30	.69	−.21	.06	−.09	+.03	− .18	−3.96	−.06	−.04	25
η Bootis	W	−3.69	+1.06	+.03	−3.72	+ .29	+ .36	.98	+.35	.13	+.28	+.10	+ .22	−3.94	−.08	−.08	63
Π Bootis	W	−3.89	+1.13	+.04	−3.93	+ .08	+ .22	.93	+.20	.04	+.07	+.02	+ .13	−4.06	+.04	+.04	15
α Draconis	W	−4.52	+2.36	+.08	−4.60	− .59	−1.03	.40	−.41	.42	−.24	+.24	− .62	−3.98	−.04	−.02	06
								3.83	−.05	.65	−.06	+.39					
d Bootis	E	−3.94	−1.11	−.04	−3.90	+ .11	+ .25	.93	+.23	.06	+.10	+.03	+ .14	−4.04	+.02	+.02	04
α Bootis	E	−3.81	−1.06	−.03	−3.78	+ .23	+ .35	.98	+.34	.12	+.23	+.08	+ .19	−3.97	−.05	−.05	25
λ Bootis	E	−4.23	−1.46	−.05	−4.18	− .17	− .19	.74	−.14	.03	−.13	+.02	− .10	−4.08	+.06	+.04	27
θ Bootis	E	−4.29	−1.64	−.05	−4.24	− .23	− .38	.65	−.25	.09	−.15	+.06	− .21	−4.03	+.01	+.01	01
5 Urs. Min.	E	−5.44	−4.18	−.13	−5.31	−1.30	−2.53	.16	−.40	1.02	−.21	+.53	−1.37	−3.94	−.08	−.01	10
								[illegible]	[illegible]	[illegible]	[illegible]	[illegible]					[illegible]

* These weights are taken from the column headed "For large portable transits" in the table on p. 39.

Normal equations:

$$+3.83\,\delta t - .05\,a_W = -.06 \qquad +3.46\,\delta t - .22\,a_E = -.16$$
$$-.05\,\delta t + .65\,a_W = +.39 \qquad -.22\,\delta t + 1.32\,a_E = +.72$$
$$a_W = +^s.601 \qquad a_E = +^s.543$$
$$\delta t = -^s.008 \qquad \delta t = -^s.012$$

At $14^h\,02^m$ $\Delta T = -4^s.020$

$$+7.29\,Q - .27\,q = 1 \qquad Q = 0.138 \qquad \epsilon_1 = \pm^s.044$$
$$-.27\,Q + 1.97\,q = 0 \qquad \epsilon = \pm^s.016$$

In the above computation a check on the correctness of the assumed value of c is furnished by the nearness of agreement of the two values of δt resulting from the two groups of stars. The normal equations are solved most conveniently by successive approximations, as, for instance, in the second equation the value of a_W can be closely derived at once on the assumption that δt is small. The residuals (Δ) are taken for each group separately, using its own δt[1] to derive a $\triangle T$ for this purpose, and the sums of the $p\Delta$'s should of course nearly equal zero for each set. The probable error of a single observation of unit weight is

$$\varepsilon_1 = 0.674\sqrt{\frac{\Sigma p\Delta^2}{n_o - n_e}}$$

where $\Sigma p\Delta^2$ is the sum of the weighted squares of the residuals (last column in form), n_o is the number of stars and n_e is the number of unknown quantities or number of normal equations, remembering in this example that there are four unknowns, δt, a_W, a_E, and c, the latter being taken from the field computation. To obtain the probable error ε of the computed ΔT, add the corresponding normal equations of the two sets, put Q in place of δt, q in place of a, 1 in place of Σpd, and 0 in place of ΣpAd, as shown. Then $\varepsilon = \varepsilon_1\sqrt{Q}$.

THE COMPLETE LEAST SQUARE COMPUTATION.

When time observations are taken in Alaska unusual conditions are encountered, arising from the high latitude of the station—from 55° to 65° for the regions in which the Survey observers are called upon to observe most frequently. Zenith stars are there slow-moving stars (and consequently have small weights); for stars between the zenith and the pole pA is comparatively small; the rapidly moving stars are far to the southward of the zenith, and it is easy to observe subpolars, as the northern horizon is far below the pole. Moreover the very prevalent cloudy weather is apt to break in upon any previously arranged program. The combined result of these conditions is in general that the sets of stars actually observed are poorly balanced; that is, the algebraic sum of the A factors for each half set and of the C factors for the whole set will differ considerably from zero. In extreme cases it is sometimes desirable to resort to the complete least square computation in which c, a_W, a_E, and ΔT are all derived by the principle of least squares.

We here start with $\alpha - t$ (as shown on pp. 30–31), and the remaining notation stands as on page 40, except that we must here distinguish by the subscripts $_W$ and $_E$ between A factors belonging to the two half sets.

An observation equation of one of the following forms may be written for each star observed:

$$\sqrt{p}\delta t + \sqrt{p}A_E a_E + \sqrt{p}Cc = \sqrt{p}d$$
$$\sqrt{p}\delta t + \sqrt{p}A_W a_W + \sqrt{p}Cc = \sqrt{p}d$$

The normal equations will be—

$$\Sigma p\delta t + \Sigma pA_E a_E + \Sigma pA_W a_W + \Sigma pCc = \Sigma pd$$
$$\Sigma pA_E\delta t + \Sigma pA^2_E a_E + \Sigma pA_E Cc = \Sigma pA_E d$$
$$\Sigma pA_W\delta t + \Sigma pA^2_W a_W + \Sigma pA_W Cc = \Sigma pA_W d$$
$$\Sigma pC\delta t + \Sigma pA_E Ca_E + \Sigma pAC_W a_W + \Sigma pC^2 c = \Sigma pCd$$

The following will serve as a concrete illustration of this method of computation. The only preliminary assumption in this computation is an approximate value of the chronometer correction, Δt.

Owing to the high latitude of St. Michael, 63° 29′, the time stars are all south of the zenith, and the average value of A is far from zero.

[1] The two δt's here happen to be so nearly equal that Δ's are the same as if taken by using the ΔT for the whole group.

ST. MICHAEL, ALASKA, *March 19, 1891.*

$\Delta t = -20.^s10.$

Star	Clamp	$\alpha - t$	d	A	C	p	pA	pC	pA^2	pAC	pC^2	pd	pAd	pCd	Aa	Cc	ΔT	Δ	$p\Delta$	$p\Delta^2$
		s	s																	
1	E	−21.27	−1.17	+ .66	−1.13	0.9	+ .59	−1.02	.39	− .67	1.15	−1.05	− .69	+1.19	− .89	− .21	−20.17	+.05	+.04	.0022
2	E	−21.22	−1.12	+ .72	−1.08	0.9	+ .65	− .97	.47	− .70	1.05	−1.01	− .73	+1.09	− .97	− .20	−20.05	−.07	−.06	44
3	E	−21.40	−1.30	+ .76	−1.05	0.9	+ .68	− .94	.52	− .72	.99	−1.17	− .89	+1.23	−1.02	− .19	−20.19	+.07	+.06	44
4	E	−23.09	−2.99	+2.89	+4.58	0.08	+ .23	+ .37	.66	+1.06	1.68	− .24	− .69	−1.10	−3.88	+ .84	−20.05	−.07	−.01	04
5	E	−21.23	−1.13	+ .73	−1.07	0.9	+ .66	− .96	.48	− .70	1.03	−1.02	− .74	+1.09	− .98	− .20	−20.05	−.07	−.06	44
							+2.81		2.52	−1.73			−3.74							
6	W	−20.98	−0.88	+ .85	+1.01	1.0	+ .85	+1.01	.72	+ .86	1.02	− .88	− .75	− .89	−1.05	+ .18	−20.11	−.01	−.01	01
7	W	−20.86	−0.76	+ .72	+1.08	0.9	+ .65	+ .97	.47	+ .70	1.05	− .68	− .49	− .73	− .89	+ .20	−20.17	+.05	+.04	22
8	W	−20.70	−0.60	+ .64	+1.14	0.9	+ .58	+1.03	.37	+ .66	1.17	− .54	− .35	− .62	− .79	+ .21	−20.12	.00	.00	00
9	W	−20.95	−0.85	+ .85	+1.01	1.0	+ .85	+1.01	.72	+ .86	1.02	− .85	− .72	− .86	−1.05	+ .18	−20.08	−.04	−.04	16
10	W	−25.39	−5.29	+3.46	−5.83	0.05	+ .17	− .29	.60	−1.01	1.70	− .26	− .92	+1.54	−4.27	−1.07	−20.05	−.07	.00	02
						7.53	+3.10	+0.21	2.88	+2.07	11.86	−7.70	−3.23	+1.94						.0199

Normal equations:

$$+7.53\ \delta t + 2.81\ a_E + 3.10\ a_W + 0.21\ c = -7.70$$
$$+2.81\ \delta t + 2.52\ a_E - 1.73\ c = -3.74$$
$$+3.10\ \delta t + 2.88\ a_W + 2.07\ c = -3.23$$
$$+0.21\ \delta t - 1.73\ a_E + 2.07\ a_W + 11.86\ c = +1.94$$
$$c = +0.183$$
$$a_E = -1.342$$
$$a_W = -1.233$$
$$\delta t = -0.02$$

At $8.^h5$ $\Delta T = -20.^s12$

$Q = .79$

$\varepsilon_1 = \pm .^s039$

$\varepsilon = \pm .035$

The remarkably large value for Q arises from the fact that the azimuth errors, a_W and a_E are but feebly determined, see column headed pA and the normal equations.

Sometimes it is assumed that the azimuth error is the same for both halves of a set, and the distinction between a_W and a_E is dropped and a single a derived from the whole set, the normal equations being modified accordingly. This procedure is entirely justifiable if the azimuth error during the two half sets is actually the same. If the two azimuths really differ, some error will be introduced into the computed results by this procedure, and the error so introduced will be larger the greater is said difference. Experience shows that the instability of the instrument in azimuth is in general sufficient to make it desirable to distinguish between the two azimuth errors if accurate results are desired, except when there are but *few* stars observed in the set, say, seven or less.

THE SELECTION OF STARS.

The stars shown in the observing list (p.18) and used in the computation on pages 21, 22 and 26 were chosen by the method now used for longitude work in latitudes less than 50°. In each half set there are five to seven time stars (six stars preferred), a time star being one which has an A factor less than unity. These stars are so selected that the algebraic sum of the A factors in a half set shall not be greater than unity. It is desirable to have the algebraic sum of the A factors of the stars in a half set as small as can be obtained by the use of good judgment in their selection, but it is not desirable to reduce the number of stars per hour to be observed in order to improve the balancing of the A factors, if the balancing is already within the specified limit.

In endeavoring to obtain the maximum number of stars per hour, subject to the condition of the balancing of the A factors, consideration must be given the question of level readings and reversals of the instrument. Ample time should be provided for the performance of these operations. In longitude work allowance must be made for the exchange of time signals, which, if the stations are not very far apart, usually takes place between the two sets—that is, between the second and third half sets. The exchange may be made, however, at any time during the observing period if there is trouble in getting a clear wire between the two observatories or if clouds break up prearranged sets of stars. An observer soon learns from practice how much time must be allowed for the different operations.

It is desirable, but not necessary, to observe the same stars at both stations when determining a difference of longitude. This is of less importance, however, than securing rapid observations with the A factors in each half set well balanced. When the two stations are not distant, many of the stars observed at one station will necessarily be observed at the other.

In longitude work the observations each night consist normally of four half sets of six stars each, with a reversal of the instrument between each two consecutive half sets. The reversal of the instrument after each of the half sets is a precaution which experience has justified, for should only three half sets be observed (through interference of clouds or for other reasons) two sets can still be obtained by combining the first and second and the second and third half sets, thus obtaining two corrections to the chronometer and its rate.

Where it is desired to use the azimuth star method of solution shown on pages 34 and 40, a different selection of stars is to be made. A half set consists of five stars following each other in rapid succession, so chosen that the algebraic sum of the A factors of the four time stars (each near the zenith) will be nearly zero, and that the azimuth star of each half set will have its A factor greater than unity, and yet not be so near the pole as to render the star's transit across the field of observation so slow as to produce long waits between observations. In a time set, chosen as above, observation upon the azimuth star in each half set serves principally to determine the azimuth error of the instrument, but has little effect upon the computed time, since this is almost independent of the azimuth error (the sum of the A factors of the time stars being nearly zero for each half set). Where only approximate time is required, the number of time stars in a half set may be reduced to two, one north and one south of the zenith.

In high latitudes (more than about 50°), it is not feasible to secure time sets with well-balanced A factors, since the stars between the zenith and the pole have comparatively small A factors, which become relatively still smaller after weights are assigned. This condition prevents any but a comparatively weak determination of the azimuth error of the instrument. In such latitudes it is therefore desirable to select sets of stars which will be solved by rigid least-square methods. Under normal conditions there should be six stars in each half set, and while the algebraic sum of the A factors in each half set should be kept as small as can be conveniently done, no very slow-moving stars should be introduced for this purpose. One azimuth star with a declination between 55° and 75° should be selected and observed below the pole.

The preliminary or field computations may be made like that shown on page 26. The final least square computations are made at the office.

As has already been stated (p. 25), the preference is now given to the American Ephemeris over other star lists, as it contains the apparent places of more stars than other available catalogues. It is well to obtain all stars, when possible, from a single catalogue, but this is not essential. It may be considered as almost essential, certainly so from an economic standpoint, to use only stars for which apparent places are published. The time and labor consumed in computing the apparent right ascension of stars for which only mean places are available add to the cost of both the field and office work. Furthermore, it will be found that sufficient stars can be selected for all time work in the northern hemisphere from such catalogues as the American Ephemeris and Nautical Almanac or the Berliner Astronomisches Jahrbuch, and the selection of mean place stars is unnecessary.

DETERMINATION OF EQUATORIAL INTERVALS.

The equatorial intervals of the lines of the diaphragm are needed to reduce incomplete transits. (See p. 32.)

To determine these, select complete transits of stars of large declination.

Let $t_1, t_2, t_3 \ldots\ldots t_n$ be the observed times of transit over the successive lines, t_m, their mean, and $i_1, i_2, i_3 \ldots\ldots i_n$ their equatorial intervals from the mean line and δ the declination of the star:

$$t_m = \frac{1}{n}(t_1 + t_2 + t_3 \ldots\ldots + t_n)$$
$$i_1 = (t_1 - t_m) \cos \delta$$
$$i_2 = (t_2 - t_m) \cos \delta$$
$$\text{etc.}$$
$$i_n = (t_n - t_m) \cos \delta$$

also $$0 = i_1 + i_2 + i_3 \ldots\ldots + i_n.$$

The intervals of the lines $\begin{Bmatrix}\text{east}\\\text{west}\end{Bmatrix}$ of the mean line will then be $\begin{Bmatrix}-\\+\end{Bmatrix}$ at upper culmination.

For stars within 10° of the pole (as for δ Urs. Min., 51 Cephei, Polaris, and λ Urs. Min.) use the formulæ:

$$i_1 = (t_1 - t_m) \cos \delta \sqrt[3]{\cos \tau_1}$$
$$\text{etc.}$$
$$i_n = (t_n - t_m) \cos \delta \sqrt[3]{\cos \tau_n}$$

where $\tau_1, \tau_2, \tau_3 \ldots\ldots t_n$ are the hour angles of the circumpolar star for the successive lines.

When it is necessary to use the more exact formula for circumpolars as given above, the table on page 32 will be found convenient.

If the chronometer rate exceeds 15^s per day it will be desirable to take it into account in computing the equatorial intervals.

A convenient form for the computation of equatorial intervals follows. The observations used were made by Assistant Fremont Morse at Sitka, Alaska, in 1894, with Meridian Telescope No. 7, and by the eye and ear method.

κ Draconis. $\delta = 70° 22' 27''$. Log. cos $\delta = 9.52618$. Clamp West.

Line		May 14	May 15	May 16	May 18	Mean	Log. mean	Log. i	i (equatorial interval)
		s	s	s	s	s			s
1	$t_1 - t_m$	−87.60	−88.00	−87.10	−87.60	−87.575	1.94238	1.46856	−29.414
2	$t_2 - t_m$	−44.60	−44.00	−44.60	−44.60	−44.450	1.64787	1.17405	−14.930
3	$t_3 - t_m$	− 0.10	0.00	+ 0.40	+ 0.40	+ 0.175	9.24304	8.76922	+ 0.059
4	$t_4 - t_m$	+43.90	+44.00	+43.90	+43.40	+43.800	1.64147	1.16765	+14.711
5	$t_5 - t_m$	+88.40	+88.00	+87.40	+88.40	+88.050	1.94473	1.47091	+29.574

The quantities $(t_1 - t_m)$, $(t_2 - t_m)$, etc., for each date were taken directly from the record of observations.

The equatorial intervals were thus computed from observations upon three different stars and the means taken.

It is not necessary to make special observations to determine the equatorial intervals. Complete transits observed during the regular progress of time observations may be utilized for that purpose. If observations upon stars of large declination are not available, observations upon stars of small declination may be used, and will be found to give almost as accurate values for the equatorial intervals.

When pressed for time in the field an incomplete transit of a star may be reduced by assuming that actual intervals between lines on that star are the same as on some preceding date on which a complete transit of that star was observed at that station. The formulæ on page 32 may then be used by dropping the factor sec δ and substituting actual intervals for equatorial intervals.

PIVOT INEQUALITY.

The pivot inequality should be determined with the instrument mounted upon a very stable pier in a room in which the rate of change of temperature is small during the observations. The observations consist of a series of readings of the striding level as indicated in the example of record and computation given below. The notation is the same as on pages 22–23; that is, β_w and β_e indicate the apparent inclination of the telescope axis in each of its two positions as given directly by the readings of the striding level. Then the pivot inequality

$$p=\frac{\beta_e-\beta_w}{4}$$

and is to be expressed in seconds of time.

Observations for inequality of pivots of transit, No. 19.

[Station, Atlanta, Ga., Mar. 12, 1896. G. R. P., observer.]

Zenith distance	Time	Temperature	Band west, Object glass south, Level, W. end	Band west, Object glass south, Level, E. end	$\frac{\Sigma w-\Sigma e}{4}=\beta_w$	Band east, Object glass north, Level, W. end	Band east, Object glass north, Level, E. end	$\frac{\Sigma w-\Sigma e}{4}=\beta_e$	$\frac{\beta_e-\beta_w}{4}=p$
°	h m	°F	div	div	div	div	div	div	div
38	9 43 a. m.	33	33. 5	22. 0		33. 4	21. 7		
			20. 8	34. 8	— . 625	21. 0	34. 0	—. 325	+. 075
43			20. 4	33. 9		21. 0	34. 0		
			32. 4	21. 9	— . 750	33. 1	21. 8	—. 425	+. 081
48			20. 2	33. 9		20. 3	33. 4		
			32. 2	21. 9	— . 850	32. 1	21. 8	—. 700	+. 038
43			31. 8	21. 9		32. 7	21. 1		
			19. 7	33. 9	—1. 075	20. 1	33. 3	—. 400	+. 169
38	10 03 a. m.	35	19. 7	33. 8		20. 1	33. 1		
			31. 9	21. 3	— . 875	32. 0	21. 1	—. 525	+. 088
					Mean, band west, object glass south, and band east, object glass north				+. 090
Zenith distance	Time	Temperature	Band west, Object glass north, Level, W. end	Band west, Object glass north, Level, E. end	$\frac{\Sigma w-\Sigma e}{4}=\beta_w$	Band east, Object glass south, Level, W. end	Band east, Object glass south, Level, E. end	$\frac{\Sigma w-\Sigma e}{4}=\beta_e$	$\frac{\beta_e-\beta_w}{4}=p$
°	h m	°F	div	div	div	div	div	div	div
38	10 07 a. m.	35	19. 7	33. 1		19. 4	33. 6		
			31. 9	20. 9	—. 600	31. 9	20. 9	—. 800	—. 050
43			31. 9	20. 9		31. 7	20. 9		
			19. 1	33. 3	—. 800	19. 1	33. 2	—. 825	—. 006
48			19. 3	33. 0		19. 1	33. 3		
			31. 5	20. 9	—. 775	31. 7	20. 9	—. 850	—. 019
43			31. 3	20. 9		31. 1	21. 0		
			19. 0	33. 2	—. 950	18. 9	33. 2	—1. 050	—. 025
38	10 27 a. m.	36	19. 0	33. 1		18. 8	33. 7		
			31. 7	20. 5	—. 725	31. 2	20. 9	—1. 150	—. 106
					Mean, band west, object glass north, and band east, object glass south				—. 041
					Mean, band west, object glass south, and band east, object glass north				+. 090
					Mean				+. 024
					1 division of striding level=1″.850=0s.123				
					p=+.024 div.=0s.123×.024=+0.003 second of time				

In determining the pivot inequality the level readings are made as in observing time, reversing the telescope between the readings. Observations should be made in two groups, reversing the relation between the positions of the band and object glass as shown in the example. This is done to partially eliminate the effect of the pivots not being truly circular in cross section. In the example shown there is a systematic though unimportant difference in p for the two positions. A complete investigation of the pivots would involve level readings at all angles from the zenith, from 0° to 90°, but the ordinary form of level will not permit readings closer than 30° or 40°, and stars are not often observed more than 50° from the zenith. In the example given the observations were from 38° to 48° zenith distance, less weight being given to the latter angle at which few star observations are made.

A less satisfactory value for the pivot inequality may be obtained from the level readings made in connection with the time observations.

Since the correction for pivot inequality has opposite signs for the two halves of a time set, its effect on the determined clock correction is very small for a set which has the same number of stars in each half. The question of when the pivot inequality correction is to be applied and when not, should be decided after a consideration of the absolute value of the correction but the difference in the sums of the B factors for the two half sets should also be considered. Most of the instruments used at present in this Survey have had their pivots refinished and their pivot inequality made practically zero. With these instruments it is not usually necessary to consider this correction when making the computations for time.

DETERMINATION OF LEVEL VALUE.

The most accurate way of determining the value of one division of a level is by means of a level-trier, which consists of a bar the support of which at one end is a micrometer screw. The level tube to be tested is placed on this bar. The method of observing and computing is shown in the following example. In the level-trier used one division of the micrometer head equals one second of arc; that is, a movement of one division changes the angular position of the bar by one second. The first part of these observations was simply for the purpose of testing the uniformity of the tube, changing the angle by 5″ intervals. In determining the level value about the same length of bubble is employed that is used in the field observations.

Determination of value of one division of stride level of meridian telescope No. 9. Chamber vial 175 mm. by 15 mm., marked 7526, 2″.02 K. and E., mounted by springs. Length of bubble used, 35 div.=70 mm. E. G. F., observer. Mean temperature, 12°.3 C.

Chamber left						Chamber right					
Level-trier reading	Bubble reading		Movement		Value of one division of level	Level-trier reading	Bubble reading		Movement		Value of one division of level
	Left end	Right end	Level-trier	Bubble. Mean of two ends			Left end	Right end	Level-trier	Bubble. Mean of two ends	
″	*div*	*div*	″	*div*	″	″	*div*	*div*	″	*div*	″
25	−0.1	35.2				75	60.4	25.8			
30	2.4	37.7	5	2.5		80	57.7	23.1	5	2.7	
35	4.9	40.2	5	2.5		85	55.3	20.7	5	2.4	
40	7.4	42.7	5	2.5		90	52.9	18.3	5	2.4	
45	10.1	45.4	5	2.7		95	50.2	15.6	5	2.7	
50	12.7	48.0	5	2.6		100	47.5	12.9	5	2.7	
55	15.3	50.6	5	2.6		105	44.9	10.3	5	2.6	
60	17.9	53.2	5	2.6		110	42.2	7.6	5	2.7	
65	20.3	55.6	5	2.4		115	39.6	5.0	5	2.6	
70	22.9	58.2	5	2.6		120	37.0	2.4	5	2.6	
75	25.5	60.8	5	2.6		125	34.5	−0.1	5	2.5	
25	−0.2	35.0	50	25.7	1.945	75	60.9	26.3	50	26.3	1.901
75	25.5	60.7				125	34.6	0.0			
35	4.7	39.9	30	15.8	1.899	85	55.9	21.2	30	16.1	1.863
65	20.5	55.7				115	39.8	5.1			
40	7.4	42.6	20	10.5	1.905	90	53.2	18.5	20	10.8	1.852
60	17.9	53.1				110	42.4	7.7			
45	10.1	45.3	10	5.3	1.887	95	50.4	15.8	10	5.5	1.818
55	15.4	50.6				105	44.9	10.3			
		Mean, chamber left			1.909			Mean, chamber right			1.859
						Final mean 1 div.=2 mm.=1″.884 at 12°.3 C					

If the level vial is so held in its metallic mounting that there is any possibility that it may be put under stress by a change of temperature, it is advisable to determine the value of a division *with the tube in its mounting* at two or more widely different temperatures. Level vials are now usually mounted with springs, so as to avoid such stresses.

If an observer is forced to determine the value of a level division in the field, remote from a level-trier—after some accident, for example—he must devise some method of utilizing whatever apparatus is at his disposal for that purpose.

If a telescope having an eyepiece micrometer fitted for measuring altitudes or zenith distances is available, the unknown angular value of a level division may be found by comparison with the known angular value of a division of the micrometer. Place the level in an extemporized mounting fixed to the telescope so that the level vial is parallel to the plane in which the telescope rotates (about its horizontal axis). Point with the micrometer upon some distant well-defined fixed object and read the micrometer and level. Change the micrometer reading by an integral number of divisions, point to the same object again by a movement of the telescope as a whole, and note the new reading of the level. Every repetition of this process gives a determination of the level value in terms of the micrometer value.

If another level of sufficient sensibility and of which the value is well known is available, it may be used as a standard with which to compare the unknown level. Put the unknown level in an extemporized mounting, fastened to that of the known level in such a way that the two level vials are parallel or nearly so. Adjust so that both bubbles are near the middle at once. Compare corresponding movements of the two bubbles for small changes of inclination common to the two levels.

DISCUSSION OF ERRORS.

The various errors which affect the final result of any astronomic observation may be grouped into three separate classes with respect to their sources, and consequently the precautions which must be taken against them fall under the same general heads. They are: (1) *External errors*, or errors arising from conditions outside the observer; (2) *instrumental errors*, due to the instrument, and arising from imperfect construction[1] or imperfect condition of the instrument, from instability of the relative positions of the different parts, etc.; (3) *observer's errors*, due directly to the observer, arising from his unavoidable errors of judgment as to what he sees and hears and from the fact that nerves and brain do not act instantaneously. By the phrase "*Errors of observation*" is meant the combined errors arising from all these sources.

The principal *external errors* in transit observations for time arise from errors in the assumed right ascensions of the stars and from lateral refraction of the light from the stars.

If the right ascensions of all stars observed are taken from the American Ephemeris and Nautical Almanac or the Berliner Astronomisches Jahrbuch, the probable error of a right ascension will be upon an average about $\pm 0.^s03$, except for stars of large declination, for which this estimate must be increased. The right ascensions are subject also to small constant errors with which the geodesist is hardly concerned, because of their smallness and because they are almost completely eliminated from his final results. When the same stars are used at both stations in determining a difference of longitude the errors of the right ascensions are completely eliminated from the determined difference of longitude.

If one considers how small are the lateral refractions which affect measurements of horizontal angles and azimuth observations, in which lines of sight are close to the ground, it seems certain that the effects of lateral refraction upon transit time observations in which all lines of sight are elevated high above the horizon must be almost or quite inappreciable. This is probably the case whenever proper precautions are taken to avoid local refraction within a few feet of the instrument. If, however, the temperature within the observatory is much above that outside, or if active chimneys or other powerful sources of heat are near the observatory, warm columns of air rising from or passing over the observatory may produce a sensible lateral refraction. The lateral refraction is included, with many other errors from which it can not be separated, in the culmination error, (ε_1), estimated on pages 38–39.

In addition to the lateral refraction referred to in the preceding paragraph and tacitly assumed to be constant during the interval of a few seconds in which a star is being observed upon, there are usually momentary lateral refractions which serve merely to make the apparent rate of progress of the star variable and to make the observer's errors greater than they otherwise would be.

Among the *instrumental errors* in transit observations for time may be mentioned those arising from the chronograph and the reading of the chronograph sheet, from poor focusing, from nonverticality of the micrometer wire or of the lines of the diaphragm, from changes in azimuth and collimation, from errors in the measured collimation, from errors in the measured inclination, from irregularity of pivots, and from changes in the rate of the chronometer.

All of these except the first two are included in the culmination error, (ε_1), as estimated on pages 38 and 39.

As already noted the chronographs of the form now used operate so well that no appreciable error is introduced by the assumption that the speed of the chronograph is constant between successive breaks of the chronometer. The chronograph sheet is read to hundredths of seconds for the exchange of arbitrary signals between stations in telegraphic longitude work. In observations made with an observing key, marking the times of transit across the lines of a diaphragm, the chronograph record of the observations is read for each line to the nearest $0.^s05$.

[1] By imperfect construction is here meant the failure to satisfy fully the rigid geometric conditions imposed by theory, but necessarily attained but imperfectly by the instrument maker, as, for example, the condition that the cross section of a pivot should be a perfect circle and remain so. *Imperfect* construction is therefore not meant to imply *poor* construction, that is, construction much below the attainable degree of excellence.

By so doing, a probable error of about $\pm 0.^s01$ on each single line is introduced into the readings; but this is too small in comparison with the other errors concerned in transit work to warrant a closer reading. In observations made with a transit equipped with a transit micrometer, where 20 observations on each star are recorded, the chronograph record of these observations is read to the nearest $0.^s1$. The probable error of a single record (position of micrometer wire) from this source is about $\pm 0.^s02$, but the number of such records obtained on a star makes the probable error of the mean of these observations less than $\pm 0.^s01$, showing that a closer reading of the chronograph sheet is not justifiable.

Poor focusing of either the objective or the eyepiece leads to increased accidental errors because of poor definition. But poor focusing of the objective is especially objectionable, because it puts the diaphragm (or plane of the micrometer wire) and the star image in different planes, and so produces parallax. The parallax errors may be avoided to a large extent by keeping the eyepiece centered carefully over the part of the diaphragm which is being observed upon, if proper longitudinal motion of the eyepiece is provided for that purpose.

If the lines of the diaphragm do not make an angle of exactly 90° with the horizontal axis of the telescope a star observed above or below the middle of the diaphragm will be observed too late or too early. A similar error will be caused in the case of the transit micrometer if the movable wire does not, in each of its positions, make an angle of 90° with the horizontal axis. Errors from this source may be made very small by careful adjustment and by observing within the narrow limits given by two horizontal lines or wires.

The *mean errors of azimuth and of collimation*, being determined by the time observations themselves, are canceled out from the final result with a thoroughness which depends upon the success attained in selecting stars. The process of elimination depends upon the assumption that the error of azimuth remains constant during each half set and that the collimation error remains constant during the whole set. The *changes* in these errors during the intervals named, arising from changes of temperature, shocks to the instrument, or other causes, produce errors in the final result. These errors will evidently be smaller the more rapidly the observations are made, the more carefully the instrument is handled, and the more symmetrical and constant are the temperature conditions. In general, these errors are small but not inappreciable. In this connection the stability of the pier on which the instrument rests is of especial importance, and also the degree to which it is protected from shocks such as, for instance, the observer's walking in its immediate vicinity, if there is no floor to the observatory or tent.

It is mainly in the light of the preceding paragraph that the number of stars to be observed in a time set must be determined. If the number of stars in a time set and the length of time over which it extends be increased, the errors due to accumulated changes in the azimuth and collimation are increased. On the other hand, if the number of stars is decreased below the present standard (12) the number of observations rapidly approaches equality with the number of unknowns (4), and the accuracy with which the unknowns are determined decreases very rapidly. From these considerations it would seem that 12 stars per set is about the most advantageous number when the highest degree of accuracy is desired.[1] Under normal conditions this number involves the necessity of depending upon the constancy of the instrument in azimuth for about 30 minutes and in collimation for about 1 hour. If greater accuracy is desired than can be obtained from a set of 12 stars, it is necessary to continue observing half sets of 6 stars each, with a reversal of the instrument in its wyes between each two half sets, but the number of stars in a half set should not be materially increased.

To a considerable extent the preceding two paragraphs also apply to the *inclination error*. The *changes* in inclination during each half set produce errors in addition to those arising from uncertainty as to the mean inclination, hence again the desirability of rapid manipulation. The mean inclination is determined from the indications of the striding level, which are more or less in error. Different observers seem to differ radically as to the probable magnitude of

[1] When only a minor degree of accuracy is desired, the number of stars may, of course, be much less than 12.

errors from this source, but the best observers are prone to use the striding level with great care. However small this error may be under the best conditions and most skillful manipulations, there can be no doubt that careless handling of the striding level, or a little heedlessness about bringing a warm reading lamp too near it,[1] may easily make this error one of the largest affecting the result. An error of 0.0002 inch in the determination of the difference of elevation of the two pivots of a transit like that shown in illustration No. 1 produces an error of more than $0^s.1$ in the deduced time of transit of a star near the zenith.

The method of treating the level readings given on page 22 is based upon two assumptions: First, that the indications of the striding level are not sufficiently accurate to determine the small *changes* of inclination during the progress of a half set, and, second, that if (as is generally the case) there is any systematic difference between the inclination as defined by level readings with objective northward and with objective southward the mean of these two inclinations is the required most probable value corresponding to intermediate positions of the telescope in which it points to stars near the zenith (time stars). There may be individual cases in which the first of these assumptions should be reversed and each star transit reduced by using the level reading which is nearest to it in time, upon the supposition that the actual changes of inclination are so large that the level indications furnish a real measure of them. In general, however, the method of treating the level readings shown on pages 21–23 is probably the best.

The errors in the computed time arising from *inequality and irregularity of pivots* are probably negligible for first-class instruments in good condition. Any small error in the adopted mean value of the inequality will appear in the computation with nearly its full value in the derived error of collimation, but will be almost completely eliminated from the computed chronometer correction. It is only the *difference* of the irregularities of the two pivots which affect the observed times, and it should be noted that corresponding points on the two pivots are always under about the same pressure at the same time, and that therefore irregularities due to wear tend to be the same for the two pivots.

Changes in the rate of the chronometer during the progress of a set of observations evidently produce errors in the computed chronometer correction at the mean epoch of the set. Under ordinary circumstances such errors must be exceedingly small. If, however, an observer is forced to use a poor timepiece, or if clouds interfere so as to extend the time required to make a set of observations over several hours, this error may become appreciable.

The *observer's errors* are by far the most serious of any class of errors in transit observations for time. The observer is subject to both accidental and constant[2] errors in his observations of the times of transit and in his readings of the striding level. The level reading errors (such as errors in estimating tenths) are inappreciable in their effect upon the computed time, but the errors in observations of time of transit enter into the computed time with full value. The observer's accidental errors are estimated under the heading "Relative Weights to Transits Depending on the Star's Declination" (pp. 38 and 39). His constant error in estimating the

[1] The longitudinal section of the upper inner surface of a level vial is made as nearly a perfect circle as possible. If an observer will consider how great this radius of curvature is in a sensitive striding level he will understand why very small deformations of the level vial by unequal changes of temperature have a marked effect upon the position of the bubble. The radius of curvature for a level of which each division is 2mm long and equivalent to $1\frac{1}{4}$ seconds of arc is more than 300 m (about 1000 feet).

[2] In discussing errors, and especially when discussing them with reference to their ultimate effects, it is quite important to keep clearly in mind the distinctions between accidental errors, constant errors, and systematic errors. A *constant error* is one which has the same effect upon all the observations of the series or portion of a series under consideration. *Accidental errors* are not constant from observation to observation; they are as apt to be minus as plus, and they presumably follow the law of error which is the basis of the theory of least squares. A *systematic error* is one of which the algebraic sign, and, to a certain extent, the magnitude, bears a fixed relation to some condition or set of conditions. Thus, for example, the phase error in observations of horizontal directions is systematic with respect to the azimuth of the sun and of the line of sight. The expression "constant error" is often used loosely in contradistinction to "accidental error," in such a way as to include both strictly constant errors and systematic errors. The effect of accidental errors upon the final result may be diminished by continued repetition of the observations and by the least square method of computation. The effects of constant errors and of systematic errors must be eliminated by other processes; for example, by changing the method or program of observations, by special investigations or special observations designed to evaluate a constant error or to determine the exact law of a systematic error. The above discussion applies with full force, in so far as the observer is directly concerned, to errors arising from imperfect perception or judgment rather than to blunders or mistakes, such as reading a level five divisions wrong or estimating a time one second wrong. If a mistake is so large that it is caught by the checks which are used for that purpose it is usually without effect upon the computed result, since it is either corrected or the observation concerned is rejected. A mistake which is not caught is, in its effect upon the computed result, an accidental error and, if proper checks have been used to detect mistakes, will lie within the limits of magnitude of the accidental errors. A similar distinction between instrumental errors and instrumental blunders may be drawn; for example, a blunder rather than error is caused by the movement of an objective which is loose in its cell.

time of transit when observing with a key, or by the eye and ear method, is known as personal equation and may amount to half a second or even a whole second in an extreme case. In observations with a transit micrometer this error if it exists at all is very small and may be neglected. The personal equation, and the methods of measuring it and of eliminating it from the final results, will be treated more fully in connection with longitude determinations. In the same place will be found a discussion of the data which indicate that the personal equation in observations made with a transit micrometer is so small that it may be neglected in longitude work.

To sum up, it may be stated that the *accidental* error in the determination of a chronometer correction from observations with a portable transit instrument upon twelve stars may be reduced within limits indicated by a probable error of from $\pm^{s}.01$ to $\pm^{s}.10$. However, in observations made without the transit micrometer the chronometer correction may be subject to a large *constant error*, the observer's absolute personal equation, which may be many times as great as the probable (accidental) error. If the observations have been made with the transit micrometer, there is practically no personal equation, and the results may be considered free from constant errors due to that source.

OTHER METHODS OF DETERMINING TIME.

In the field it is sometimes necessary to use other instruments as transits for the determination of time. A theodolite, when so used, is apt to give results of a higher degree of accuracy than would be expected from an instrument of its size, unless one has in mind that the principal errors in transit time observations are those due directly to the observer. On the other hand, zenith telescopes of the form in which the telescope does not swing in a plane passing through the vertical axis of the instrument have been found to give disappointing results when used in the meridian for time, perhaps because of the asymmetry of the instrument and of the fact that there can be no reversal of the horizontal axis in its bearings, but only of the instrument as a whole. The time may, however, be thus determined with sufficient accuracy for use in connection with determinations of latitude with the zenith telescope.[1]

The determination of time by the use of the transit in any position out of the meridan has been advocated, but has not seemed advisable. The additional difficulty of making the computation, over that for a transit nearly in the meridian, and other incidental inconveniences, much more than offset the fact that the adjustment for putting the transit in the meridian is then unnecessary.

The use of the transit in the vertical plane passing through Polaris at the time of observation has been advocated, and has been used to a considerable extent in Europe and in Canada. It is not used by this Survey. The advantage of this method over the meridian method is that the stability of the instrument is depended upon for only about 5 minutes instead of 30 minutes or more. This method is open, though to a less extent, to the objections stated in the preceding paragraph against the method of observing in any position out of the meridian.

If a mark nearly in the meridian has been established and its azimuth determined the chronometer correction may be determined at noon within a half second by observing the transit of the sun as follows: Point on the meridian mark just before apparent noon; observe the transit of the preceding limb of the sun across the lines of the diaphragm; reverse the horizontal axis of the telescope and observe the transit of the following limb across the lines of the diaphragm. If the transit micrometer is used, the west limb of the sun is followed across the center of the field by the micrometer wire, and then the telescope is reversed and the east limb is followed by the wire. The record of observations on each limb is recorded automatically on the chronograph. The striding level should be read just before the transit of the preceding limb and just after the transit of the following limb. The mean of all the observed times is the chronometer time of transit of the sun's center across the plane of the instrument. This

[1] For methods of determining time with a zenith telescope by using it as an equal-altitude instrument, see Coast Survey Report for 1869, Appendix No. 12, pp. 226-232.

time corrected for azimuth error, as determined by the pointing on the meridian mark, and for inclination, is the chronometer time of the sun's transit across the meridian. During the observations the instrument should be sheltered from the direct rays of the sun. This may be done by hanging in front of it a cloth with a hole cut in it opposite the objective. This method of determining time may sometimes be found desirable in connection with chronometric determinations of longitude in Alaska when continuous cloudy weather prevents star observations.

When setting up a transit at a new station it is sometimes difficult to get a close approximation to the local time with which to make the first setting of the transit in the meridian. The following method has been used to furnish a rough value of the local time, and makes it possible to put the instrument so closely in the meridian on the initial trial that there is almost no time lost from the regular observations. At a little before local noon commence observing the sun, following it by moving the telescope both in azimuth and altitude. While the sun is still rising appreciably, clamp the telescope in altitude, and mark the time of the transit of the sun's limbs across the horizontal wire of the telescope; then keeping the telescope fixed in altitude swing it slightly in azimuth to meet the descending sun and mark the transit of the sun's limbs across the same wire as before. The mean of the times will be approximately the chronometer time of the sun's passage across the local meridian, and the chronometer correction on apparent solar time can be determined, and finally its correction on local sidereal time. With this correction, using an azimuth star first in the final placing of the instrument in azimuth, it will be found that two approximations will usually be all that are required to set the instrument close enough for actual observations. With the meridian telescope form of instrument this method may be easily and accurately followed.

Sextant observations for time by measuring the altitude of the sun give sufficiently accurate results for many purposes.[1] For example, the chronometer correction may thus be determined with sufficient accuracy for use in zenith telescope determinations of latitude or in observations for azimuth made upon a circumpolar star within an hour of elongation. If a specially constructed vertical circle [2] is used, illustration No. 8, the time may be determined from observed altitudes of a star or the sun with sufficient accuracy for all purposes in observations for latitude and azimuth. The sun or star should be observed near the prime vertical if possible. This is the method used at present by nearly all the parties of this Survey engaged in latitude and azimuth observations. With time obtained in this way azimuth observations may be made on Polaris at any hour angle. This method is also used by the field parties engaged in making magnetic observations.[3] As this method is so frequently used a sample record of observations and of the computations is given below with such explanations as are necessary.

DESCRIPTION OF THE VERTICAL CIRCLE AND ITS ADJUSTMENTS.

The vertical circles in use in the Coast and Geodetic Survey are, in general form, like that shown in illustration No. 8.

The instrument is practically a theodolite with the graduated circle in a vertical position and the axis horizontal, with the telescope fastened rigidly to the alidade. The circle and alidade are fastened to a horizontal support which rests upon the top of a vertical axis, the latter fitting into a stand. There is a counterpoise to the circle and alidade on the opposite side of the vertical axis. The stand has three leveling screws, and there may be a graduated circle near its base for measuring horizontal angles approximately.

[1] For convenient instructions, formulæ, and tables for sextant observations for time and other approximate astronomic methods, see Bowditch's American Practical Navigator, published by the U. S. Navy Department.

[2] Such an instrument is used in observing vertical angles or zenith distances in primary triangulation. The circles of these instruments are from 8 to 10 inches in diameter and are graduated very accurately.

[3] See p. 45, Directions for Magnetic Measurements, Coast and Geodetic Survey.

No. 8.

Vertical circle.

Before starting observations the usual adjustments of the eyepiece and object glass should be made and the crosswires should be brought approximately into the center of the field. There is no adjustment for collimation in either the vertical or horizontal plane. A coarse stride level is used to make the horizontal axis of the circle truly horizontal and, consequently, the circle vertical, and a sensitive level is placed parallel with and fastened to the circle to define a horizontal line through the instrument. If, after leveling by the two levels, the instrument is rotated on its vertical axis through 180° and the bubbles remain on the graduated scales of the level vials then the adjustments for level are satisfactory.

TIME FROM OBSERVATIONS ON A STAR WITH A VERTICAL CIRCLE.

When making the observations the star's image is brought into the field of the telescope and the telescope clamped with the horizontal wire slightly ahead of the star. As the star crosses the horizontal wire the observer notes the time of the chronometer by the eye-and-ear method, or, at the instant of crossing, he calls "Mark" to the recorder, who notes the chronometer time. Readings are made of the bubble of the fixed level and of the verniers of the vertical circle. The telescope is then rotated on its horizontal axis and revolved 180° about the vertical axis of the instrument. A second observation is made on the star and the level and vertical circle are read again. These observations constitute one complete determination of the time. It is advisable to take at least four such sets of observations for the determination of the chronometer correction if the results are used for primary azimuth work where Polaris or some other close circumpolar star is observed at any hour angle.

If, upon revolving the instrument through 180° in azimuth for the second reading on the star for any one set, it is found that one end of the bubble extends beyond the graduations of the level vial, it may be brought back by the foot screws of the instrument. It should *never* be brought back to the graduations by moving the tangent screw which controls the relation between the bubble and the graduations of the circle. In other words, the relation between the fixed level and the vertical circle of the instrument should remain undisturbed during a set. If the level is badly out of adjustment, it should be adjusted between sets. Whenever practicable one half of the sets of observations should be made on a star in the east and the other half on a west star, both stars being nearly in the prime vertical and at about the same elevation, in order to eliminate instrumental errors and errors due to refraction.

The above two paragraphs apply also to observations on the sun, except, of course, the last sentence of the second paragraph. The instrumental and refraction errors may be minimized by observing the sun in the morning and again in the afternoon at about the same angular distance from the meridian.

RECORD OF OBSERVATIONS ON STARS.

The following record shows four sets of observations with the vertical circle, all on an eastern star. These observations were made in connection with primary azimuth observations at Sears triangulation station in Texas. The azimuth observations and computations are shown on pages 147 to 149 of this publication. It will be noticed that the zenith distances of the star corrected for level are computed in the record.

*Double zenith distances.**

Form 252.

[Station: Sears triangulation station. Observer: W. Bowie. State: Texas. County: Jones. Instrument: Vertical circle No. 46. Date: Dec. 22, 1908.]

Object observed	Time	Level		Circle, right or left	Circle reading	Verniers					Zenith distance	Remarks
		O	E			A	B	C	D*	Mean		
	h m s	d	d		° ′	″	″	″	″	″	° ′ ″	
α Tauri	1 03 49 0	14 1	12 0	R	49 57	40	60	20	......	40.0	49 59 46.6	Sidereal chronometer No. 1769 was used. Temperature, 5° C. Barometer, 716 mm
	1 06 02 5	14 4	11 8	L	50 01	50	50	60	......	53.3	− 3.0	
											43.6	
	1 04 55.8	28.5	23.8									
		−4.7										
α Tauri	1 07 05.0	14.3	11.8	L	49 49	00	40	$\overline{40}$	......	06.7	49 24 01.7	
	1 08 23.5	11 8	14.3	R	48 59	00	20	$\overline{30}$	......	56.7	0.0	
											01.7	
	1 07 46.8	26.1	26.1									
		0.0										
α Tauri	1 10 06.5	16.8	09.5	R	48 36	30	60	10	......	33.3	48 41 55.0	Value of one division of level bubble=2″.58
	1 12 00.5	13.4	12.9	L	48 47	20	40	$\overline{50}$	......	16.7	− 5.0	
											50.0	
	1 11 03.5	30.2	22.4									
		−7.8										
α Tauri	1 13 14.5	13.2	12.8	L	48 31	20	60	40	......	40.0	48 03 03.4	
	1 15 13.0	14.1	12.2	R	47 34	20	40	20	......	26.7	− 1.5	
											01.9	
	1 14 13.8	27.3	25.0									
		−2.3										

* Vertical circle No. 46 differs from the usual type of this instrument in use by the Survey in the number of verniers and in the numbering of the graduations of the circle. There are four verniers as a rule, and the circle graduations are generally numbered continuously, so that the difference of the two circle readings, Circle R and Circle L, gives the double zenith distance. No. 46 has only three verniers and the vertical circle graduations are numbered from 0° to 180° both ways from the zenith.

In the column of remarks is given such information as is necessary for the proper interpretation of the record by the computer. In this column should also be given notes on any unusual occurrence, such as the jarring of the instrument or the adjustment of the instrument during the period of observations.

The above form is bound in books of octavo size, which are furnished to field parties upon request.

The level correction, which is shown in the column headed "Level" and is applied to the observed zenith distance in the next to the last column, is computed by the formula:

$$C=\frac{1}{4}\{(\mathrm{E}+\mathrm{E}_1)-(\mathrm{O}+\mathrm{O}_1)\}d.$$

When the level graduations are numbered continuously, the formula is:

$$C=\frac{1}{4}\{(\mathrm{E}_1-\mathrm{E})-(\mathrm{O}-\mathrm{O}_1)\}d;$$

in which O and E are the readings of the level when the larger numbers are at the object end of the level vial, and d is the value in seconds of arc of one division of the vial.

The formula used in computing time from observations with a vertical circle on a star or on the sun is

$$\sin\frac{1}{2}t=\sqrt{\frac{\sin\frac{1}{2}[\zeta+(\phi-\delta)]\sin\frac{1}{2}[\zeta-(\phi-\delta)]}{\cos\phi\cos\delta}};$$

in which t is the hour angle, δ the declination, ζ the zenith distance of the object observed, and ϕ is the latitude of the station.

In the following form (No. 381a) the usual method of computation is shown. This form is designed especially for the computation of time from the observed altitudes of a star.

Computation of time, observations on a star with vertical circle.

Form 381a.

[State, Texas. Station, Sears triangulation station. Chronometer, 1769 Sidereal. Date, Dec. 22, 1908. Barometer, 716 mm. Temperature, 5° C.]

		Star: α Tauri		Star: α Tauri	
		h m s	° ′ ″	h m s	° ′ ″
Chron. reading,	Zenith dist.	1 04 55.8	49 59 44	1 07 46.8	49 24 02
	Refraction		+1 06		+1 05
	Corrected Z. D.=ζ		50 00 50		49 25 07
log cos ϕ,	ϕ	9.9257458	32 33 31	9.9257458	32 33 31
log cos δ,	δ	9.9821234	16 19 37	9.9821234	16 19 37
log cos ϕ+log cos δ=log D,	$\phi-\delta$	9.9078692	16 13 54	9.9078692	16 13 54
log sin $\frac{1}{2}$ [$\zeta+(\phi-\delta)$],	$\frac{1}{2}$ [$\zeta+(\phi-\delta)$]	9.7375385	33 07 22	9.7340593	32 49 30
log sin $\frac{1}{2}$ [$\zeta-(\phi-\delta)$],	$\frac{1}{2}$ [$\zeta-(\phi-\delta)$]	9.4632265	16 53 28	9.4557230	16 35 36
Sum two log sines=log N,		9.2007650		9.1897823	
log N−log D=log sin$^2 \frac{1}{2} t$,		9.2928958		9.2819131	
log sin $\frac{1}{2} t$,	$\frac{1}{2} t$ (arc)	9.6464479	26 17 54	9.6409566	25 56 35
		h m s		h m s	
t (time),	t (arc)	3 30 23.2	52 35 48	3 27 32.7	51 53 10
Right ascension of star,		4 30 41.9		4 30 41.9	
Sidereal time,		1 00 18.7		1 03 09.2	
Chronometer reading,		1 04 55.8		1 07 46.8	
Chronometer correction,		−04 37.1		−04 37.6	

The correction is plus if the chronometer is slow and minus if fast.
Carry all angles to seconds only, all times to tenths of seconds, and all logarithms to seven decimal places.
In space below, compute rate of chronometer, etc.

Mean Epoch	Star	Chronometer correction
h m		m s
1 10	α Tauri	−4 37.7
4 58	β Geminor.	−4 36.7

Clock rate=$0^s.263$ per hour losing.

In the above computation the correction for refraction was obtained from the tables on pages 58–59 of this publication.

The apparent declination and right ascension of the star were obtained from the American Ephemeris and Nautical Almanac for 1908 (the year of observation).

TIME FROM OBSERVATIONS ON THE SUN WITH THE VERTICAL CIRCLE.

When the sun is the object observed upon a slightly different program of observations is required. The telescope is pointed on the sun's upper limb (the horizontal wire of the telescope made tangent to the disk of the sun) with the circle right and immediately afterward with the circle left. At each pointing the time of contact, the level reading, and the reading of the vertical circle are noted. The letters R and L (right and left) are used to designate the position of the circle with reference to the vertical axis of the instrument. Two quarter sets similar to the above are then made in quick succession on the sun's lower limb, and finally another quarter set on the upper limb. These are recorded on the form shown below, on which are also computed the zenith distances of the sun's limbs corrected for level.

Double zenith distances.

Form 252.

[Station, Tilden. Observer, W. Bowie. State, Minnesota. County, Polk. Instrument, Vertical circle No. 63. Date, Sept. 6, 1906.]

Object observed		Time	Level O	Level E	Circle right or left	Circle reading	Verniers A	Verniers B	Verniers C	Verniers D	Verniers Mean	Zenith distance	Remarks
		h m s	d	d		° ′	″	″	″	″	″	° ′ ″	
Sun's upper limb	$\bar{\odot}$	8 47 39.5	32.3	11.0	R	49 02	24	54	45	30	38.2		Value of one division of
	$\bar{\odot}$	8 48 47.0	07.6	29.2	L	147 30	36	06	15	06	15.8	49 13 48.8	the level vial = 4″.00
			24.7	18.2								−6.5	
			−6.5									49 13 42.3	
Sun's lower limb	$\underline{\odot}$	8 50 12.5	32.5	11.2	R								
	$\underline{\odot}$	8 51 17.5	07.8	29.3	L	246 26	24	21	00	36	20.2	49 28 02.2	Chronometer, Sidereal
			24.7	18.1								−6.6	No. 102
			−6.6									49 27 55.6	Temperature, 27° C
Sun's lower limb	$\underline{\odot}$	8 52 57.0	31.2	09.8	R								Barometer not read
	$\underline{\odot}$	8 53 45.2	08.0	29.8	L	344 46	15	06	30	00	12.8	49 09 56.3	
			23.2	20.0								−3.2	
			−3.2									49 09 53.1	
Sun's upper limb	$\bar{\odot}$	8 55 08.2	31.5	10.0	R								
	$\bar{\odot}$	8 55 52.0	09.3	31.0	L	81 32	54	81	48	45	57.0	48 23 22.1	
			22.2	21.0								−1.2	
			−1.2									48 23 20.9	

The observations on the upper limb are computed separately from those on the lower limb in order that one may make more exact corrections for refraction.

Computation of time, observations on sun with vertical circle.

Form 381.

[Station, Tilden. Date, Sept. 6, 1906. Chronometer, Sidereal 102. Temperature, 27° C. Barometer (not read).]

		Sun's upper limb		Sun's lower limb	
		h m s	° ′ ″	h m s	° ′ ″
Chron. reading,	Zenith dist.	8 48 13.2	49 13 42	8 50 45.0	49 27 56
Chron. reading,	Zenith dist.	8 55 30.1	48 23 21	8 53 21.1	49 09 53
Mean,	Mean	8 51 51.6	48 48 32	8 52 03.0	49 18 54
	Parallax		— 07		— 07
	Refraction		+ 1 03		+ 1 04
	Semidiameter		+15 54		−15 54
	Corrected Z. D. = ζ		49 05 22		49 03 57
log cos ϕ,	ϕ	9.8279861	47 42 16	9.8279861	47 42 16
log cos δ,	δ	9.9970883	6 37 38	9.9970883	6 37 38
log cos ϕ + log cos δ = log D,	$\phi-\delta$	9.8250744	41 04 38	9.8250744	41 04 38
log sin $\frac{1}{2}[\zeta+(\phi-\delta)]$,	$\frac{1}{2}[\zeta+(\phi-\delta)]$	9.8501157	45 05 00	8.8500254	45 04 17
log sin $\frac{1}{2}[\zeta-(\phi-\delta)]$,	$\frac{1}{2}[\zeta-(\phi-\delta)]$	8.8442464	4 00 22	8.8429819	3 59 40
Sum two log sines = log N,		8.6943621		8.6930073	
log N − log D = log sin$^2 \frac{1}{2} t$,		8.8692897		8.8679329	
log sin $\frac{1}{2} t$,	$\frac{1}{2} t$ (arc)	9.4346438	15 47 10	9.4339664	15 45 39
		h m s		h m s	
t (time),	t (arc)	2 06 17.3	31 34 20	2 06 05.2	31 31 18
Local apparent time,		21 53 42.7		21 53 54.8	
Equation of time,		−1 31.4		−1 31.4	
Local mean time,		21 52 11.3		21 52 23.4	
Local sidereal time,		8 51 33.8		8 51 45.9	
Chronometer reading,		8 51 51.6		8 52 03.0	
Chronometer correction		−17.8		−17.1	

	h m	h m
Longitude from Greenwich,	=6 25.3	=6 25.3
Estimated local mean time of observation,	=9 52	=9 53
Greenwich mean time of observation,	=4 17	=4 18
Interpolation interval, from Greenwich mean noon,	=4.3 hours	=4.3 hours.

In this computation the correction for refraction was obtained from the tables on pages 58–59 of this publication. The argument used was the apparent altitude.

The first table gives the mean refraction, or the refraction under an assumed standard condition of 760 mm. (=29. 9 in.) pressure and 10° C. (=50° F.) temperature.

The second table gives the factor C_B, by which the mean refraction as obtained from the first table must be multiplied, on account of a barometer reading different from 760 mm.

In the third table is obtained the factor C_T by which the mean refraction must be multiplied on account of a temperature different from the standard (10° C.).

The resulting refraction is then $r = r_M \times C_B \times C_T$ in which r_M is the refraction under standard conditions obtained from the first table and C_B and C_T are the factors obtained from the second and third tables, respectively.[1]

The reduction for semidiameter, and the values for the sun's declination and for the equation of time were obtained from the American Ephemeris and Nautical Almanac for 1906 (the year of observations).

The parallax was obtained from the table on page 60, which was also taken from Hayford's Geodetic Astronomy.

The semidiameter was obtained from page 405 of the Ephemeris.

The declination and the equation of time were obtained from pages 146 and 147 of the Ephemeris. The interpolation of these quantities for the time of observation is made by the use of the interpolation interval obtained at the bottom of the computation.

The mean of the observations on either limb, reduced for parallax, refraction, and semidiameter gives the true zenith distance of the sun's center. The computation is by the same formula as is given for the reduction of the observations on a star. (See p. 54.)

As the above observations were made using a sidereal chronometer, and as the correction on sidereal time was required, it was necessary to reduce the computed mean time of the observation to its corresponding local sidereal time before a comparison was made with the time as read from the chronometer face. The following computation shows the various steps of this reduction for the observations on the sun's upper limb:

	h	m	s
Local mean time of observation (Sept. 5, 1906)[2]	21	52	11. 3
Reduction to sidereal interval (Table III, Ephemeris)		3	35. 6
Right ascension of mean sun, Greenwich mean noon September 5, 1906	10	54	43. 6
Increase in right ascension of mean sun, at Tilden mean noon September 5, 1906 (Table III, Ephemeris, 6^h $25^m.3$ west)		1	03. 3
Sum, local sidereal time of observation at Tilden	8	51	33. 8

For several reasons the observations on a star are more satisfactory than those on the sun. When used in connection with other astronomic observations, such as the determination of azimuth, a chronometer correction from observations on a star may be obtained close to the epoch of the observations, since any one of many available stars may be used. The computation is more easily made as there is no reduction for semidiameter or for parallax, and the declination and right ascension of a star are practically constant during an entire set of observations and therefore easily and quickly obtained from a star list. No equation of time is introduced.

The observer should have a star chart[3] for use in identifying the stars observed upon.

---

[1] These tables were copied from A Text Book of Geodetic Astronomy by John F. Hayford, formerly inspector of geodetic work and Chief of the Computing Division, U. S. Coast and Geodetic Survey. John Wiley & Sons, 1898.

[2] It must be remembered that the day of the Ephermis is astronomic, and begins at noon of the civil day of the same date. Sept. 5, 21^h 52^m $11^s.3$, astronomic mean time is the forenoon of Sept. 6, civil time.

[3] Star Charts are published by the Hydrographic Office of the U. S. Navy and may be obtained from the Navy Department, Washington, D. C. Star Charts are also contained in A Field Book of the Stars, by W. T. Olcott (G. P. Putnam's Sons, publishers).

Mean refraction (r_M)

[Barometer, 760 millimeters (=29.9 inches). Temperature, 10° C. (=50° F).]

Altitude	Mean refraction	Change per minute	Altitude	Mean refraction	Change per minute	Altitude	Mean refraction	Change per minute	Altitude	Mean refraction	Change per minute	Altitude	Mean refraction	Change per minute
° ′	′ ″	″	° ′	′ ″	″	° ′	′ ″	″	° ′	′ ″	″	° ′	′ ″	″
0 00	34 08.6	11.66	7 00	7 24.2	0.95	19 00	2 47.6	0.16	33 00	1 29.4	0.06	52 30	0 44.7	0.03
10	32 15.9	10.88	10	7 14.9	0.91	20	2 44.6	0.15	20	1 28.2	0.06	53 00	0 43.9	0.03
20	30 31.1	10.10	20	7 06.0	0.88	40	2 41.6	0.15	40	1 27.1	0.06	30	0 43.1	0.03
30	28 53.9	9.64	30	6 57.4	0.84	20 00	2 38.7	0.14	34 00	1 26.1	0.05	54 00	0 42.3	0.03
40	27 18.2	9.20	40	6 49.1	0 81	20	2 35.9	0.14	20	1 25.0	0.05	30	0 41.6	0.03
50	25 49.8	8.50	50	6 41.2	0.78	40	2 33.2	0.13	40	1 24.0	0.05	55 00	0 40.8	0.03
1 00	24 28.3	7.82	8 00	6 33.5	0.76	21 00	2 30.6	0.13	35 00	1 23.0	0.05	30	0 40.0	0.03
10	23 13.5	7.17	10	6 26.0	0.73	20	2 28.1	0.13	20	1 22.0	0.05	56 00	0 39.3	0.025
20	22 04.9	6.58	20	6 18.9	0.70	40	2 25.6	0.12	40	1 21.0	0.05	57 00	0 37.8	0.024
30	21 01.8	6.06	30	6 12.0	0.68	22 00	2 23.2	0.12	36 00	1 20.0	0.05	58 00	0 36.4	0.023
40	20 03.7	5.60	40	6 05.3	0.66	20	2 20.9	0.12	30	1 18.5	0.05	59 00	0 35.0	0.023
50	19 09.8	5.20	50	5 58.9	0 63	40	2 18.6	0.11	37 00	1 17.1	0.04	60 00	0 33.6	0.022
2 00	18 19.7	4.84	9 00	5 52.7	0.61	23 00	2 16.4	0.11	30	1 15.7	0 04	61 00	0 32.3	0.022
10	17 33.1	4.50	20	5 40.8	0.58	20	2 14.2	0.11	38 00	1 14.4	0.04	62 00	0 31.0	0.022
20	16 49.7	4.18	40	5 29.7	0.54	40	2 12.1	0.10	30	1 13.1	0.04	63 00	0 29.7	0.022
30	16 09.5	3.88	10 00	5 19.2	0.51	24 00	2 10.1	0.10	39 00	1 11.8	0.04	64 00	0 28.4	0.021
40	15 32.1	3.62	20	5 09.4	0 48	20	2 08.1	0.10	30	1 10.5	0.04	65 00	0 27.2	0.021
50	14 57.1	3.39	40	5 00.1	0.46	40	2 06.1	0.10	40 00	1 09.3	0.04	66 00	0 25.9	0.021
3 00	14 24.3	3.18	11 00	4 51.2	0.43	25 00	2 04.2	0.09	30	1 08.1	0.04	67 00	0 24.7	0.020
10	13 53.6	2.98	20	4 42.8	0.40	20	2 02.4	0.09	41 00	1 06.9	0.04	68 00	0 23.6	0.020
20	13 24.8	2.79	40	4 35.0	0.38	40	2 00.6	0.09	30	1 05.7	0.04	69 00	0 22.4	0.020
30	12 57.8	2.61	12 00	4 27.5	0.37	26 00	1 58.8	0.09	42 00	1 04.6	0.04	70 00	0 21.2	0.019
40	12 32.5	2.46	20	4 20.3	0.35	20	1 57.1	0.09	30	1 03.5	0.04	71 00	0 20.1	0.019
50	12 08.7	2.33	40	4 13.5	0.33	40	1 55.4	0.08	43 00	1 02.4	0.04	72 00	0 18.9	0.019
4 00	11 46.0	2.20	13 00	4 07.1	0.32	27 00	1 53.8	0.08	30	1 01.3	0.04	73 00	0 17.8	0.018
10	11 24.6	2.09	20	4 00.9	0.30	20	1 52.2	0.08	44 00	1 00.2	0.03	74 00	0 16.7	0.018
20	11 04.2	1.98	40	3 55.1	0.28	40	1 50.6	0.08	30	0 59.2	0.03	75 00	0 15.6	0.018
30	10 44.9	1.88	14 00	3 49.5	0.27	28 00	1 49.1	0 08	45 00	0 58.2	0.03	76 00	0 14.5	0.018
40	10 26.5	1.79	20	3 44.2	0.26	20	1 47.6	0.07	30	0 57.2	0.03	77 00	0 13.5	0.018
50	10 09.1	1.70	40	3 39.1	0.25	40	1 46.1	0.07	46 00	0 56.2	0.03	78 00	0 12.4	0.018
5 00	9 52.6	1.61	15 00	3 34.1	0.24	29 00	1 44.6	0.07	30	0 55.2	0.03	79 00	0 11.3	0.018
10	9 36.9	1.54	20	3 29.4	0.23	20	1 43.2	0.07	47 00	0 54.2	0.03	80 00	0 10.3	0.018
20	9 21 9	1.46	40	3 24.8	0.23	40	1 41.8	0.07	30	0 53.3	0.03	81 00	0 09.2	0.018
30	9 07.6	1.40	16 00	3 20.4	0.22	30 00	1 40.5	0.07	48 00	0 52.5	0.03	82 00	0 08.2	0.018
40	8 54.0	1.33	20	3 16.1	0.21	20	1 39.1	0.07	30	0 51.6	0.03	83 00	0 07.2	0.018
50	8 41.0	1.27	40	3 12.0	0.20	40	1 37.8	0.06	49 00	0 50.7	0.03	84 00	0 06.1	0.018
6 00	8 28.6	1.22	17 00	3 08.2	0.19	31 00	1 36.6	0.06	30	0 49.8	0.03	85 00	0 05.1	0.018
10	8 16.7	1.16	20	3 04.5	0.19	20	1 35.3	0.06	50 00	0 48.9	0.03	86 00	0 04.1	0.017
20	8 05.3	1.12	40	3 00.9	0.18	40	1 34.1	0.06	30	0 48.0	0.03	87 00	0 03.1	0.017
30	7 54.3	1.07	18 00	2 57.4	0.17	32 00	1 32.0	0.06	51 00	0 47.2	0.03	88 00	0 02.0	0.017
40	7 43.9	1.02	20	2 54.0	0.17	20	1 31.8	0.06	30	0 46.3	0.03	89 00	0 01.0	0.017
50	7 33.9	0.98	40	2 50.7	0.16	40	1 30.6	0.06	52 00	0 45.5	0.03	90 00	0 00.0	0.017

Correction to mean refraction as given on page 58, depending upon the reading of the barometer.

[$r=(r_M)(C_B)(C_T)$.]

Barometer		C_B	Barometer		C_B	Barometer		C_B	Barometer		C_B	Barometer		C_B
Inches	*mm*		*Inches*	*mm*		*Inches*	*mm*		*Inches*	*mm*		*Inches*	*mm*	
20.0	508	0.670	22.4	569	0.749	24.8	630	0.829	27.2	691	0.909	29.6	752	0.989
20.1	511	0.673	22.5	572	0.752	24.9	632	0.832	27.3	693	0.912	29.7	754	0.992
20.2	513	0.676	22.6	574	0.755	25.0	635	0.835	27.4	696	0.916	29.8	757	0.996
20.3	516	0.679	22.7	576	0.759	25.1	637	0.838	27.5	699	0.920	29.9	759	0.999
20.4	518	0.682	22.8	579	0.762	25.2	640	0.842	27.6	701	0.923	30.0	762	1.003
20.5	521	0.685	22.9	582	0.766	25.3	643	0.846	27.7	704	0.926	30.1	765	1.007
20.6	523	0.688	23.0	584	0.770	25.4	645	0.849	27.8	706	0.929	30.2	767	1.010
20.7	526	0.692	23.1	587	0.773	25.5	648	0.853	27.9	709	0.933	30.3	770	1.013
20.8	528	0.696	23.2	589	0.776	25.6	650	0.856	28.0	711	0.936	30.4	772	1.016
20.9	531	0.699	23.3	592	0.779	25.7	653	0.859	28.1	714	0.939	30.5	775	1.020
21.0	533	0.703	23.4	594	0.783	25.8	655	0.862	28.2	716	0.942	30.6	777	1.023
21.1	536	0.706	23.5	597	0.786	25.9	658	0.866	28.3	719	0.946	30.7	780	1.026
21.2	538	0.709	23.6	599	0.789	26.0	660	0.869	28.4	721	0.949	30.8	782	1.029
21.3	541	0.712	23.7	602	0.792	26.1	663	0.872	28.5	724	0.953	30.9	785	1.033
21.4	544	0.716	23.8	605	0.796	26.2	665	0.875	28.6	726	0.956	31.0	787	1.036
21.5	546	0.719	23.9	607	0.799	26.3	668	0.879	28.7	729	0.959			
21.6	549	0.722	24.0	610	0.803	26.4	671	0.882	28.8	732	0.963			
21.7	551	0.725	24.1	612	0.806	26.5	673	0.885	28.9	734	0.966			
21.8	554	0.729	24.2	615	0.809	26.6	676	0.889	29.0	737	0.970			
21.9	556	0.732	24.3	617	0.813	26.7	678	0.892	29.1	739	0.973			
22.0	559	0.735	24.4	620	0.816	26.8	681	0.896	29.2	742	0.976			
22.1	561	0.739	24.5	622	0.820	26.9	683	0.899	29.3	744	0.979			
22.2	564	0.742	24.6	625	0.823	27.0	686	0.902	29.4	747	0.983			
22.3	566	0.746	24.7	627	0.826	27.1	688	0.905	29.5	749	0.986			

Correction to mean refraction as given on page 58, depending upon the reading of the detached thermometer.

[$r=(r_M)(C_B)(C_T)$.]

Temperature		C_T	Temperature		C_T	Temperature		C_T	Temperature		C_T	Temperature		C_T
Fahrenheit	Centigrade		Fahrenheit	Centigrade		Fahrenheit	Centigrade		Fahrenheit	Centigrade		Fahrenheit	Centigrade	
°	°		°	°		°	°		°	°		°	°	
−25	−31.7	1.172	8	−13.3	1.089	41	5.0	1.018	74	23.3	0.955	107	41.7	0.900
−24	−31.1	1.169	9	−12.8	1.087	42	5.6	1.016	75	23.9	0.953	108	42.2	0.899
−23	−30.6	1.166	10	−12.2	1.085	43	6.1	1.014	76	24.4	0.952	109	42.8	0.897
−22	−30.0	1.164	11	−11.7	1.082	44	6.7	1.012	77	25.0	0.950	110	43.3	0.895
−21	−29.4	1.161	12	−11.1	1.080	45	7.2	1.010	78	25.6	0.948	111	43.9	0.894
−20	−28.9	1.158	13	−10.6	1.078	46	7.8	1.008	79	26.1	0.946	112	44.4	0.892
−19	−28.3	1.156	14	−10.0	1.076	47	8.3	1.006	80	26.7	0.945	113	45.0	0.891
−18	−27.8	1.153	15	− 9.4	1.073	48	8.9	1.004	81	27.2	0.943	114	45.6	0.890
−17	−27.2	1.151	16	− 8.9	1.071	49	9.4	1.002	82	27.8	0.941	115	46.1	0.888
−16	−26.7	1.148	17	− 8.3	1.069	50	10.0	1.000	83	28.3	0.939	116	46.7	0.886
−15	−26.1	1.145	18	− 7.8	1.067	51	10.6	0.998	84	28.9	0.938	117	47.2	0.885
−14	−25.6	1.143	19	− 7.2	1.064	52	11.1	0.996	85	29.4	0.936	118	47.8	0.884
−13	−25.0	1.140	20	− 6.7	1.062	53	11.7	0.994	86	30.0	0.934	119	48.3	0.882
−12	−24.4	1.138	21	− 6.1	1.060	54	12.2	0.992	87	30.6	0.933	120	48.9	0.881
−11	−23.9	1.135	22	− 5.6	1.058	55	12.8	0.990	88	31.1	0.931	121	49.4	0.880
−10	−23.3	1.133	23	− 5.0	1.056	56	13.3	0.988	89	31.7	0.929	122	50.0	0.878
− 9	−22.8	1.130	24	− 4.4	1.054	57	13.9	0.986	90	32.2	0.928	123	50.6	0.877
− 8	−22.2	1.128	25	− 3.9	1.051	58	14.4	0.985	91	32.8	0.926	124	51.1	0.876
− 7	−21.7	1.125	26	− 3.3	1.049	59	15.0	0.983	92	33.3	0.924	125	51.7	0.874
− 6	−21.1	1.123	27	− 2.8	1.047	60	15.6	0.981	93	33.9	0.923	126	52.2	0.873
− 5	−20.6	1.120	28	− 2.2	1.045	61	16.1	0.979	94	34.4	0.921	127	52.8	0.871
− 4	−20.0	1.118	29	− 1.7	1.043	62	16.7	0.977	95	35.0	0.919	128	53.3	0.870
− 3	−19.4	1.115	30	− 1.1	1.041	63	17.2	0.975	96	35.6	0.917	129	53.9	0.868
− 2	−18.9	1.113	31	− 0.6	1.039	64	17.8	0.973	97	36.1	0.916	130	54.4	0.867
− 1	−18.3	1.111	32	0.0	1.036	65	18.3	0.972	98	36.7	0.914			
0	−17.8	1.108	33	+ 0.6	1.034	66	18.9	0.970	99	37.2	0.912			
+ 1	−17.2	1.106	34	1.1	1.032	67	19.4	0.968	100	37.8	0.911			
2	−16.7	1.103	35	1.7	1.030	68	20.0	0.966	101	38.3	0.909			
3	−16.1	1.101	36	2.2	1.028	69	20.6	0.964	102	38.9	0.908			
4	−15.6	1.099	37	2.8	1.026	70	21.1	0.962	103	39.4	0.906			
5	−15.0	1.096	38	3.3	1.024	71	21.7	0.961	104	40.0	0.905			
6	−14.4	1.094	39	3.9	1.022	72	22.2	0.959	105	40.6	0.903			
7	−13.9	1.092	40	4.4	1.020	73	22.8	0.957	106	41.1	0.902			

The parallax of the sun (p) for the first day of each month.

Altitude	Jan. 1	Feb. 1 Dec. 1	Mar. 1 Nov. 1	Apr. 1 Oct. 1	May 1 Sept. 1	June 1 Aug. 1	July 1	Zenith distance
°	″	″	″	″	″	″	″	°
0	9.0	9.0	8.9	8.9	8.8	8.7	8.7	90
3	9.0	9.0	8.9	8.8	8.8	8.7	8.7	87
6	9.0	8.9	8.9	8.8	8.7	8.7	8.7	84
9	8.9	8.9	8.8	8.8	8.7	8.6	8.6	81
12	8.8	8.8	8.7	8.7	8.6	8.5	8.5	78
15	8.7	8.7	8.6	8.6	8.5	8.4	8.4	75
18	8.6	8.6	8.5	8.4	8.4	8.3	8.3	72
21	8.4	8.4	8.3	8.3	8.2	8.2	8.1	69
24	8.2	8.2	8.2	8.1	8.0	8.0	8.0	66
27	8.0	8.0	8.0	7.9	7.8	7.8	7.8	63
30	7.8	7.8	7.7	7.7	7.6	7.6	7.6	60
33	7.6	7.5	7.5	7.4	7.4	7.3	7.3	57
36	7.3	7.3	7.2	7.2	7.1	7.1	7.0	54
39	7.0	7.0	6.9	6.9	6.8	6.8	6.8	51
42	6.7	6.7	6.6	6.6	6.5	6.5	6.5	48
44	6.5	6.5	6.4	6.4	6.3	6.3	6.3	46
46	6.3	6.2	6.2	6.2	6.1	6.1	6.0	44
48	6.0	6.0	6.0	5.9	5.9	5.8	5.8	42
50	5.8	5.8	5.7	5.7	5.6	5.6	5.6	40
52	5.6	5.5	5.5	5.4	5.4	5.4	5.4	38
54	5.3	5.3	5.2	5.2	5.2	5.1	5.1	36
56	5.0	5.0	5.0	5.0	4.9	4.9	4.9	34
58	4.8	4.8	4.7	4.7	4.7	4.6	4.6	32
60	4.5	4.5	4.5	4.4	4.4	4.4	4.4	30
62	4.2	4.2	4.2	4.2	4.1	4.1	4.1	28
64	4.0	3.9	3.9	3.9	3.8	3.8	3.8	26
66	3.7	3.7	3.6	3.6	3.6	3.6	3.5	24
68	3.4	3.4	3.4	3.3	3.3	3.3	3.3	22
70	3.1	3.1	3.1	3.0	3.0	3.0	3.0	20
72	2.8	2.8	2.8	2.7	2.7	2.7	2.7	18
74	2.5	2.5	2.5	2.4	2.4	2.4	2.4	16
76	2.2	2.2	2.2	2.1	2.1	2.1	2.1	14
78	1.9	1.9	1.9	1.8	1.8	1.8	1.8	12
80	1.6	1.6	1.6	1.5	1.5	1.5	1.5	10
82	1.2	1.2	1.2	1.2	1.2	1.2	1.2	8
84	0.9	0.9	0.9	0.9	0.9	0.9	0.9	6
86	0.6	0.6	0.6	0.6	0.6	0.6	0.6	4
88	0.3	0.3	0.3	0.3	0.3	0.3	0.3	2
90	0.0	0.0	0.0	0.0	0.0	0.0	0.0	0

A, B, C, FACTORS.

These factors are referred to in the computations of time from observations with the transit on pages 23 and 25. Their arithmetical values are as follows:

Azimuth factor $= A = \sin \zeta \sec \delta$

Level factor $= B = \cos \zeta \sec \delta$

Collimation factor $= C = \sec \delta$

where δ = declination and ζ = zenith distance $= \phi - \delta$ or $\phi - (180° - \delta)$ for stars observed at upper or lower culmination respectively.

The signs of the factors are as follows:

A is plus except for stars between the zenith and the pole.

B is plus except for stars observed at lower culmination.

C is plus for stars at upper culmination and minus for stars at lower culmination, when observations are made with the instrument in the position, band (clamp or illumination) *west.*

C is minus for stars at upper culmination and plus for stars at lower culmination when observations are made with the instrument in the position, band (clamp or illumination) *east.*

These factors are given to two decimal places in the tables on pages 62 to 77, and will be found sufficiently accurate whenever the errors of adjustment, a, b, and c, are not allowed to exceed one second of time. In 1874 this Survey published more extended tables, giving these factors to three decimal places. Where, from any cause, observations are made with an instrumental error abnormally large it is desirable to take the corresponding star factors from the more extended table or to compute them.

No. 9.

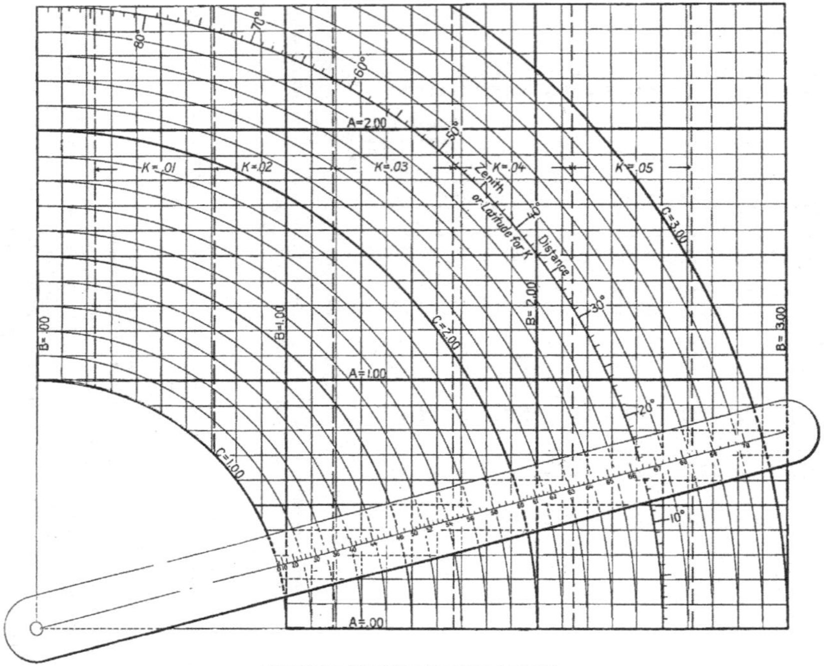

NOMOGRAM FOR OBTAINING STAR FACTORS.

STAR FACTORS OBTAINED GRAPHICALLY.

For a number of years there has been in use in the Survey a nomogram for obtaining graphically the star factors *A*, *B*, and *C*, and also κ, the correction for diurnal aberration. This nomogram was devised by Mr. C. R. Duvall, a computer in the Survey. It is not only more expeditious than the tables, but the elimination of the double interpolation which the use of the tables necessitates adds to the accuracy of the derived factor in many cases.

The nomogram is shown in illustration No. 9, reduced in size. It consists of two systems of equidistant parallel lines perpendicular to each other, a system of arcs of equidistant concentric circles, and a transparent arm, carrying a graduated straight line which revolves about the common center of the circles. The decimeter has been the unit of length in the nomograms used. The three systems of lines are drawn at a common distance apart of 1 centimeter. The estimated tenth of this centimeter space gives the second decimal place in the required factors.

The graduated line on the under surface of the transparent arm passes through the center of the axis about which the arm revolves. A secant graduation is made upon this line, measured from the center of the axis of revolution. That is, the graduation corresponding to any angle is at a distance from the center equal to the secant of the angle in question. This center of the axis of revolution is the common center of the concentric circles and also the origin of the two systems of parallel lines.

The graduations on the arm are for the declinations. In the nomograms used the graduations have not been carried beyond three decimeters from the center, which limits the use of the instrument to declinations from 0° to slightly over 70°.

The zenith distances are graduated on one of the concentric circles at a convenient distance from the center. In the instrument shown in the illustration the distance is 25 centimeters. Since stars are never observed at zenith distances approaching 90°, the upper part of the quadrant is not used.

To determine the factors *A*, *B*, and *C* of a given star, revolve the transparent arm until the graduated line of the arm coincides with the star's zenith distance on the graduated arc. Holding the arm in this position, place a needle point at that point of the graduated line which corresponds to the star's declination. The position of this point in the three systems of equidistant lines gives the three factors, *A* being the ordinate, *B* the abscissa, and *C* the radius vector.

The nomogram shown in the illustration is of thin bristol board pasted smoothly on thick cardboard. The transparent arm is of celluloid one-sixteenth of an inch thick. The axis of the arm is a solid metal cylinder with a head which fits against the back of the cardboard. The axis is made long so that the arm can be placed on it and revolved without being made fast.

The correction for aberration may be taken from the same nomogram, as follows: Set the revolving arm at that angle on the graduated circle which is equal to the latitude of the given station. From the graduated line of the arm read off the declination at each intersection with a broken-line ordinate. These declinations are the limits between which κ has the values $0^s.00$, $0^s.01$, $0^s.02$, etc., for the latitude of the station in question. By means of these limits the κ of any star can be immediately written down from its declination. The broken-line ordinates are drawn at distances from the origin equal to $\frac{.005}{.021}, \frac{.015}{.021}, \frac{.025}{.021}$, etc. . . . decimeters.

Table of factors for reduction of transit observations.

TOP ARGUMENT = STAR'S DECLINATION (δ).

SIDE ARGUMENT = STAR'S ZENITH DISTANCE (ζ).

[For factor *A* use left-hand argument. For factor *B* use right-hand argument. For factor *C* use bottom line on *opposite* page.]

ζ	0°	10°	15°	20°	22°	24°	26°	28°	30°	32°	34°	36°	38°	40°	41°	42°	ζ
°																	°
1	.02	.02	.02	.02	.02	.02	.02	.02	.02	.02	.02	.02	.02	.02	.02	.02	89
2	.04	.04	.04	.04	.04	.04	.04	.04	.04	.04	.04	.04	.04	.05	.05	.05	88
3	.05	.05	.05	.06	.06	.06	.06	.06	.06	.06	.06	.06	.07	.07	.07	.07	87
4	.07	.07	.07	.07	.08	.08	.08	.08	.08	.08	.08	.09	.09	.09	.09	.09	86
5	.09	.09	.09	.09	.09	.10	.10	.10	.10	.10	.10	.11	.11	.11	.11	.12	85
6	.11	.11	.11	.11	.11	.11	.12	.12	.12	.12	.13	.13	.13	.14	.14	.14	84
7	.12	.12	.13	.13	.13	.13	.14	.14	.14	.14	.15	.15	.15	.16	.16	.16	83
8	.14	.14	.14	.15	.15	.15	.16	.16	.16	.16	.17	.17	.18	.18	.18	.19	82
9	.16	.16	.16	.17	.17	.17	.17	.18	.18	.18	.19	.19	.20	.20	.21	.21	81
10	.17	.18	.18	.19	.19	.19	.19	.20	.20	.21	.21	.21	.22	.23	.23	.23	80
11	.19	.19	.20	.20	.21	.21	.21	.22	.22	.23	.23	.24	.24	.25	.25	.26	79
12	.21	.21	.22	.22	.22	.23	.23	.24	.24	.25	.25	.26	.26	.27	.27	.28	78
13	.22	.23	.23	.24	.24	.25	.25	.26	.26	.27	.27	.28	.29	.29	.30	.30	77
14	.24	.25	.25	.26	.26	.27	.27	.27	.28	.29	.29	.30	.31	.32	.32	.33	76
15	.26	.26	.27	.28	.28	.28	.29	.29	.30	.31	.31	.32	.33	.34	.34	.35	75
16	.28	.28	.29	.29	.30	.30	.31	.31	.32	.33	.33	.34	.35	.36	.37	.37	74
17	.29	.30	.30	.31	.31	.32	.33	.33	.34	.34	.35	.36	.37	.38	.39	.39	73
18	.31	.31	.32	.33	.33	.33	.34	.35	.36	.36	.37	.38	.39	.40	.41	.42	72
19	.33	.33	.34	.35	.35	.36	.36	.37	.38	.38	.39	.40	.41	.42	.43	.44	71
20	.34	.35	.35	.36	.37	.37	.38	.39	.40	.40	.41	.42	.43	.45	.45	.46	70
21	.36	.36	.37	.38	.39	.39	.40	.41	.41	.42	.43	.44	.45	.47	.47	.48	69
22	.37	.38	.39	.40	.40	.41	.42	.42	.43	.44	.45	.46	.48	.49	.50	.50	68
23	.39	.40	.41	.42	.42	.43	.44	.44	.45	.46	.47	.48	.50	.51	.52	.53	67
24	.41	.41	.42	.43	.44	.45	.45	.46	.47	.48	.49	.50	.52	.53	.54	.55	66
25	.42	.43	.44	.45	.46	.46	.47	.48	.49	.50	.51	.52	.54	.55	.56	.57	65
26	.44	.45	.45	.47	.47	.48	.49	.50	.51	.52	.53	.54	.56	.57	.58	.59	64
27	.45	.46	.47	.48	.49	.50	.51	.51	.52	.54	.55	.56	.58	.59	.60	.61	63
28	.47	.48	.49	.50	.51	.51	.52	.53	.54	.55	.57	.58	.60	.61	.62	.63	62
29	.48	.49	.50	.52	.52	.53	.54	.55	.56	.57	.58	.60	.61	.63	.64	.65	61
30	.50	.51	.52	.53	.54	.55	.56	.57	.58	.59	.60	.62	.63	.65	.66	.67	60
31	.52	.52	.53	.55	.56	.56	.57	.58	.59	.61	.62	.64	.65	.67	.68	.69	59
32	.53	.54	.55	.56	.57	.58	.59	.60	.61	.63	.64	.65	.67	.69	.70	.71	58
33	.55	.55	.56	.58	.59	.60	.61	.62	.63	.64	.66	.67	.69	.71	.72	.73	57
34	.56	.57	.58	.59	.60	.61	.62	.63	.65	.66	.67	.69	.71	.73	.74	.75	56
35	.57	.58	.59	.61	.62	.63	.64	.65	.66	.68	.69	.71	.73	.75	.76	.77	55
36	.59	.60	.61	.63	.63	.64	.65	.67	.68	.69	.71	.73	.75	.77	.78	.79	54
37	.60	.61	.62	.64	.65	.66	.67	.68	.70	.71	.73	.74	.76	.79	.80	.81	53
38	.62	.63	.64	.66	.66	.67	.69	.70	.71	.73	.74	.76	.78	.80	.82	.83	52
39	.63	.64	.65	.67	.68	.69	.70	.71	.73	.74	.76	.78	.80	.82	.83	.85	51
40	.64	.65	.67	.68	.69	.70	.72	.73	.74	.76	.77	.79	.82	.84	.85	.86	50
41	.66	.67	.68	.70	.71	.72	.73	.74	.76	.77	.79	.81	.83	.86	.87	.88	49
42	.67	.68	.69	.71	.72	.73	.74	.76	.77	.79	.81	.83	.85	.87	.89	.90	48
43	.68	.69	.71	.73	.74	.75	.76	.77	.79	.80	.82	.84	.86	.89	.90	.92	47
44	.69	.71	.72	.74	.75	.76	.77	.79	.80	.82	.84	.86	.88	.91	.92	.93	46
45	.71	.72	.73	.75	.76	.77	.79	.80	.82	.83	.85	.87	.90	.92	.94	.95	45
	0°	10°	15°	20°	22°	24°	26°	28°	30°	32°	34°	36°	38°	40°	41°	42°	

Table of factors for reduction of transit observations.

TOP ARGUMENT=STAR'S DECLINATION (δ).

SIDE ARGUMENT=STAR'S ZENITH DISTANCE (ζ).

[For factor *A* use left-hand argument. For factor *B* use right-hand argument. For factor *C* use bottom line on *this* page.]

ζ	0°	10°	15°	20°	22°	24°	26°	28°	30°	32°	34°	36°	38°	40°	41°	42°	ζ
°																	°
46	.72	.73	.74	.77	.78	.79	.80	.82	.83	.85	.87	.89	.91	.94	.95	.97	44
47	.73	.74	.76	.78	.79	.80	.81	.83	.84	.86	.88	.90	.93	.95	.97	.98	43
48	.74	.76	.77	.79	.80	.81	.83	.84	.86	.88	.90	.92	.94	.97	.98	1.00	42
49	.75	.77	.78	.80	.81	.83	.84	.86	.87	.89	.91	.93	.96	.99	1.00	1.02	41
50	.77	.78	.79	.82	.83	.84	.85	.87	.89	.90	.92	.95	.97	1.00	1.01	1.03	40
51	.78	.79	.80	.83	.84	.85	.87	.88	.90	.92	.94	.96	.99	1.01	1.03	1.05	39
52	.79	.80	.82	.84	.85	.86	.88	.89	.91	.93	.95	.97	1.00	1.03	1.04	1.06	38
53	.80	.81	.83	.85	.86	.87	.89	.91	.92	.94	.96	.99	1.01	1.04	1.06	1.07	37
54	.81	.82	.84	.86	.87	.89	.90	.92	.93	.95	.98	1.00	1.03	1.06	1.07	1.09	36
55	.82	.83	.85	.87	.88	.90	.91	.93	.95	.97	.99	1.01	1.04	1.07	1.08	1.10	35
56	.83	.84	.86	.88	.89	.91	.92	.94	.96	.98	1.00	1.02	1.05	1.08	1.10	1.12	34
57	.84	.85	.87	.89	.90	.92	.93	.95	.97	.99	1.01	1.04	1.06	1.09	1.11	1.13	33
58	.85	.86	.88	.90	.91	.93	.94	.96	.98	1.00	1.02	1.05	1.08	1.11	1.12	1.14	32
59	.86	.87	.89	.91	.92	.94	.95	.97	.99	1.01	1.03	1.06	1.09	1.12	1.14	1.15	31
60	.87	.88	.90	.92	.93	.95	.96	.98	1.00	1.02	1.04	1.07	1.10	1.13	1.15	1.17	30
61	.87	.89	.91	.93	.94	.96	.97	.99	1.01	1.03	1.05	1.08	1.11	1.14	1.16	1.18	29
62	.88	.90	.91	.94	.95	.97	.98	1.00	1.02	1.04	1.06	1.09	1.12	1.15	1.17	1.19	28
63	.89	.91	.92	.95	.96	.98	.99	1.01	1.03	1.05	1.07	1.10	1.13	1.16	1.18	1.20	27
64	.90	.91	.93	.96	.97	.98	1.00	1.02	1.04	1.06	1.08	1.11	1.14	1.17	1.19	1.21	26
65	.91	.92	.94	.96	.98	.99	1.01	1.03	1.05	1.07	1.09	1.12	1.15	1.18	1.20	1.22	25
66	.91	.93	.95	.97	.99	1.00	1.02	1.04	1.06	1.08	1.10	1.13	1.16	1.19	1.21	1.23	24
67	.92	.94	.95	.98	.99	1.01	1.02	1.04	1.06	1.09	1.11	1.14	1.17	1.20	1.22	1.24	23
68	.93	.94	.96	.99	1.00	1.02	1.03	1.05	1.07	1.09	1.12	1.15	1.18	1.21	1.23	1.25	22
69	.93	.95	.97	.99	1.01	1.02	1.04	1.06	1.08	1.10	1.13	1.15	1.18	1.22	1.24	1.26	21
70	.94	.95	.97	1.00	1.01	1.03	1.05	1.06	1.09	1.11	1.13	1.16	1.19	1.23	1.25	1.26	20
71	.95	.96	.98	1.01	1.02	1.04	1.05	1.07	1.09	1.12	1.14	1.17	1.20	1.23	1.25	1.27	19
72	.95	.97	.98	1.01	1.03	1.04	1.06	1.08	1.10	1.12	1.15	1.17	1.21	1.24	1.26	1.28	18
73	.96	.97	.99	1.02	1.03	1.05	1.06	1.08	1.10	1.13	1.15	1.18	1.21	1.25	1.27	1.29	17
74	.96	.98	1.00	1.02	1.04	1.05	1.07	1.09	1.11	1.13	1.16	1.19	1.22	1.25	1.27	1.29	16
75	.97	.98	1.00	1.03	1.04	1.06	1.08	1.09	1.12	1.14	1.16	1.19	1.23	1.26	1.28	1.30	15
76	.97	.99	1.00	1.03	1.05	1.06	1.08	1.10	1.12	1.14	1.17	1.20	1.23	1.27	1.29	1.31	14
77	.97	.99	1.01	1.04	1.05	1.07	1.08	1.10	1.13	1.15	1.17	1.20	1.24	1.27	1.29	1.31	13
78	.98	.99	1.01	1.04	1.05	1.07	1.09	1.11	1.13	1.15	1.18	1.21	1.24	1.28	1.30	1.32	12
79	.98	1.00	1.02	1.04	1.06	1.08	1.09	1.11	1.13	1.16	1.18	1.21	1.25	1.28	1.30	1.32	11
80	.98	1.00	1.02	1.05	1.06	1.08	1.10	1.12	1.14	1.16	1.19	1.22	1.25	1.29	1.30	1.33	10
81	.99	1.00	1.02	1.05	1.07	1.08	1.10	1.12	1.14	1.17	1.19	1.22	1.25	1.29	1.31	1.33	9
82	.99	1.01	1.03	1.05	1.07	1.08	1.10	1.12	1.14	1.17	1.19	1.22	1.26	1.29	1.31	1.33	8
83	.99	1.01	1.03	1.06	1.07	1.09	1.10	1.12	1.15	1.17	1.20	1.23	1.26	1.30	1.32	1.34	7
84	.99	1.01	1.03	1.06	1.07	1.09	1.11	1.13	1.15	1.17	1.20	1.23	1.26	1.30	1.32	1.34	6
85	1.00	1.01	1.03	1.06	1.07	1.09	1.11	1.13	1.15	1.17	1.20	1.23	1.26	1.30	1.32	1.34	5
86	1.00	1.01	1.03	1.06	1.08	1.09	1.11	1.13	1.15	1.18	1.20	1.23	1.27	1.30	1.32	1.34	4
87	1.00	1.01	1.03	1.06	1.08	1.09	1.11	1.13	1.15	1.18	1.20	1.23	1.27	1.30	1.32	1.34	3
88	1.00	1.01	1.03	1.06	1.08	1.09	1.11	1.13	1.15	1.18	1.20	1.23	1.27	1.30	1.32	1.34	2
89	1.00	1.02	1.04	1.06	1.08	1.09	1.11	1.13	1.15	1.18	1.21	1.24	1.27	1.31	1.32	1.35	1
90	1.00	1.02	1.04	1.06	1.08	1.09	1.11	1.13	1.15	1.18	1.21	1.24	1.27	1.31	1.32	1.35	0
	0°	10°	15°	20°	22°	24°	26°	28°	30°	32°	34°	36°	38°	40°	41°	42°	

Table of factors for reduction of transit observations.

TOP ARGUMENT=STAR'S DECLINATION (δ).

SIDE ARGUMENT=STAR'S ZENITH DISTANCE (ζ).

[For factor *A* use left-hand argument. For factor *B* use right-hand argument. For factor *C* use bottom line on *opposite* page.]

ζ	42°	43°	44°	45°	46°	47°	48°	49°	50°	51°	52°	53°	54°	55°	56°	57°	ζ
°																	°
1	.02	.02	.02	.02	.02	.03	.03	.03	.03	.03	.03	.03	.03	.03	.03	.03	89
2	.05	.05	.05	.05	.05	.05	.05	.05	.05	.06	.06	.06	.06	.06	.06	.06	88
3	.07	.07	.07	.07	.07	.08	.08	.08	.08	.08	.08	.09	.09	.09	.09	.10	87
4	.09	.10	.10	.10	.10	.10	.10	.11	.11	.11	.11	.12	.12	.12	.12	.13	86
5	.12	.12	.12	.12	.13	.13	.13	.13	.13	.14	.14	.14	.15	.15	.16	.16	85
6	.14	.14	.15	.15	.15	.15.	.16	.16	.16	.17	.17	.17	.18	.18	.19	.19	84
7	.16	.17	.17	.17	.18	.18	.18	.19	.19	.19	.20	.20	.21	.21	.22	.22	83
8	.19	.19	.19	.20	.20	.20	.21	.21	.22	.22	.23	.23	.24	.24	.25	.26	82
9	.21	.21	.22	.22	.22	.23	.23	.24	.24	.25	.25	.26	.27	.27	.28	.29	81
10	.23	.24	.24	.25	.25	.25	.26	.26	.27	.28	.28	.29	.30	.30	.31	.32	80
11	.26	.26	.27	.27	.28	.28	.28	.29	.30	.30	.31	.32	.32	.33	.34	.35	79
12	.28	.28	.29	.29	.30	.30	.31	.32	.32	.33	.34	.35	.35	.36	.37	.38	78
13	.30	.31	.31	.32	.32	.33	.34	.34	.35	.36	.36	.37	.38	.39	.40	.41	77
14	.33	.33	.34	.34	.35	.35	.36	.37	.38	.38	.39	.40	.41	.42	.43	.44	76
15	.35	.35	.36	.37	.37	.38	.39	.39	.40	.41	.42	.43	.44	.45	.46	.48	75
16	.37	.38	.38	.39	.40	.40	.41	.42	.43	.44	.45	.46	.47	.48	.49	.51	74
17	.39	.40	.41	.41	.42	.43	.44	.45	.45	.46	.47	.49	.50	.51	.52	.54	73
18	.42	.42	.43	.44	.44	.45	.46	.47	.48	.49	.50	.51	.53	.54	.55	.57	72
19	.44	.45	.45	.46	.47	.48	.49	.50	.51	.52	.53	.54	.55	.57	.58	.60	71
20	.46	.47	.48	.48	.49	.50	.51	.52	.53	.54	.56	.57	.58	.60	.61	.63	70
21	.48	.49	.50	.51	.52	.52	.54	.55	.56	.57	.58	.59	.61	.62	.64	.66	69
22	.50	.51	.52	.53	.54	.55	.56	.57	.58	.60	.61	.62	.64	.65	.67	.69	68
23	.53	.53	.54	.55	.56	.57	.58	.60	.61	.62	.63	.65	.66	.68	.70	.72	67
24	.55	.56	.57	.58	.59	.60	.61	.62	.63	.65	.66	.68	.69	.71	.73	.75	66
25	.57	.58	.59	.60	.61	.62	.63	.64	.66	.67	.69	.70	.72	.74	.76	.78	65
26	.59	.60	.61	.62	.63	.64	.65	.67	.68	.70	.71	.73	.75	.76	.78	.80	64
27	.61	.62	.63	.64	.65	.67	.68	.69	.71	.72	.74	.75	.77	.79	.81	.83	63
28	.63	.64	.65	.66	.68	.69	.70	.72	.73	.75	.76	.78	.80	.82	.84	.86	62
29	.65	.66	.67	.69	.70	.71	.72	.74	.75	.77	.79	.81	.82	.84	.87	.89	61
30	.67	.68	.69	.71	.72	.73	.75	.76	.78	.79	.81	.83	.85	.87	.89	.92	60
31	.69	.70	.72	.73	.74	.75	.77	.78	.80	.82	.84	.86	.88	.90	.92	.95	59
32	.71	.72	.74	75	.76	.78	.79	.81	.82	.84	.86	.88	.90	.92	.95	.97	58
33	.73	.74	.76	.77	.78	.80	.81	.83	.85	.87	.88	.91	.93	.95	.97	1.00	57
34	.75	.76	.78	.79	.80	.82	.84	.85	.87	.89	.91	.93	.95	.97	1.00	1.03	56
35	.77	.78	.80	.81	.83	.84	.86	.87	.89	.91	.93	.95	.98	1.00	1.03	1.05	55
36	.79	.80	.82	.83	.85	.86	.88	.90	.91	.93	.95	.98	1.00	1.03	1.06	1.08	54
37	.81	.82	.84	.85	.87	.88	.90	.92	.94	.96	.98	1.00	1.02	1.05	1.08	1.10	53
38	.83	.84	.86	.87	.89	.90	.92	.94	.96	.98	1.00	1.02	1.05	1.07	1.10	1.13	52
39	.85	.86	.87	.89	.91	.92	.94	.96	.98	1.00	1.02	1.05	1.07	1.10	1.12	1.15	51
40	.86	.88	.89	.91	.93	.94	.96	.98	1.00	1.02	1.04	1.07	1.09	1.12	1.15	1.18	50
41	.88	.90	.91	.93	.94	.96	.98	1.00	1.02	1.04	1.07	1.09	1.12	1.14	1.17	1.20	49
42	.90	.91	.93	.95	.96	.98	1.00	1.02	1.04	1.06	1.09	1.11	1.14	1.17	1.20	1.23	48
43	.92	.93	.95	.96	.98	1.00	1.02	1.04	1.06	1.08	1.11	1.13	1.16	1.19	1.22	1.25	47
44	.93	.95	.97	.98	1.00	1.02	1.04	1.06	1.08	1.10	1.13	1.15	1.18	1.21	1.24	1.28	46
45	.95	.97	.98	1.00	1.02	1.04	1.06	1.08	1.10	1.12	1.15	1.17	1.20	1.23	1.26	1.30	45
	42°	43°	44°	45°	46°	47°	48°	49°	50°	51°	52°	53°	54°	55°	56°	57°	

Table of factors for reduction of transit observations.

TOP ARGUMENT=STAR'S DECLINATION (δ).

SIDE ARGUMENT=STAR'S ZENITH DISTANCE (ζ).

[For factor *A* use left-hand argument. For factor *B* use right-hand argument. For factor *C* use bottom line on *this* page.]

ζ	42°	43°	44°	45°	46°	47°	48°	49°	50°	51°	52°	53°	54°	55°	56°	57°	ζ
°																	°
46	.97	.98	1.00	1.02	1.04	1.05	1.07	1.10	1.12	1.14	1.17	1.19	1.22	1.25	1.29	1.32	44
47	.98	1.00	1.02	1.03	1.05	1.07	1.09	1.11	1.14	1.16	1.19	1.21	1.24	1.27	1.31	1.34	43
48	1.00	1.02	1.03	1.05	1.07	1.09	1.11	1.13	1.16	1.18	1.21	1.23	1.26	1.30	1.33	1.36	42
49	1.02	1.03	1.05	1.07	1.09	1.11	1.13	1.15	1.17	1.20	1.23	1.25	1.28	1.32	1.35	1.39	41
50	1.03	1.05	1.06	1.08	1.10	1.12	1.14	1.17	1.19	1.22	1.24	1.27	1.30	1.34	1.37	1.41	40
51	1.05	1.06	1.08	1.10	1.12	1.14	1.16	1.18	1.21	1.23	1.26	1.29	1.32	1.35	1.39	1.43	39
52	1.06	1.08	1.10	1.11	1.13	1.15	1.18	1.20	1.23	1.25	1.28	1.31	1.34	1.37	1.41	1.45	38
53	1.07	1.09	1.11	1.13	1.15	1.17	1.19	1.22	1.24	1.27	1.30	1.33	1.36	1.39	1.43	1.47	37
54	1.09	1.11	1.12	1.14	1.16	1.19	1.21	1.23	1.26	1.29	1.31	1.34	1.38	1.41	1.45	1.49	36
55	1.10	1 12	1 14	1 16	1.18	1.20	1.22	1.25	1.27	1.30	1.33	1.36	1.39	1.43	1.46	1.50	35
56	1.12	1.13	1.15	1.17	1.19	1.22	1.24	1.26	1.29	1.32	1.35	1.38	1.41	1.45	1.48	1.52	34
57	1.13	1.15	1.17	1.19	1.21	1.23	1.25	1.28	1.31	1.33	1.36	1.39	1.43	1.46	1.50	1.54	33
58	1.14	1.16	1.18	1.20	1.22	1.24	1.27	1.29	1.32	1.35	1.38	1.41	1.44	1.48	1.52	1.56	32
59	1.15	1.17	1.19	1.21	1.23	1.26	1.28	1.31	1.33	1.36	1.39	1.42	1.46	1.49	1.53	1.57	31
60	1.17	1.18	1.20	1.22	1.25	1.27	1.29	1.32	1.35	1.38	1.41	1.44	1.47	1.51	1.55	1.59	30
61	1.18	1.20	1.22	1.24	1.26	1.28	1.31	1.33	1.36	1.39	1.42	1.45	1.49	1.53	1.56	1.61	29
62	1.19	1.21	1.23	1.25	1.27	1.29	1.32	1.35	1.37	1.40	1.43	1.47	1.50	1.54	1.58	1.62	28
63	1.20	1.22	1.24	1.26	1.28	1.31	1.33	1.36	1.39	1.42	1.45	1.48	1.52	1.55	1.59	1.64	27
64	1.21	1.23	1.25	1.27	1.29	1.32	1.34	1.37	1.40	1.43	1.46	1.49	1.53	1.57	1.61	1.65	26
65	1.22	1.24	1.26	1.28	1.30	1.33	1.35	1.38	1.41	1.44	1.47	1.51	1.54	1.58	1.62	1.66	25
66	1.23	1.25	1.27	1.29	1.32	1.34	1.37	1.39	1.42	1.45	1.48	1.52	1.55	1.59	1.63	1.68	24
67	1.24	1.26	1.28	1.30	1.33	1.35	1.38	1.40	1.43	1.46	1.50	1.53	1.57	1.60	1.65	1.69	23
68	1.25	1.27	1.29	1.31	1.33	1.36	1.39	1.41	1.44	1.47	1.51	1.54	1.58	1.62	1.66	1.70	22
69	1.26	1.28	1.30	1.32	1.34	1.37	1.40	1.42	1.45	1.48	1.52	1.55	1.59	1.63	1.67	1.71	21
70	1.26	1.28	1.31	1.33	1.35	1.38	1.40	1.43	1.46	1.49	1.53	1.56	1.60	1.64	1.68	1.73	20
71	1.27	1.29	1.31	1.34	1.36	1.39	1.41	1.44	1.47	1.50	1.54	1.57	1.61	1.65	1.69	1.74	19
72	1.28	1.30	1.32	1.34	1.37	1.39	1.42	1.45	1.48	1.51	1.54	1.58	1.62	1.66	1.70	1.75	18
73	1.29	1.31	1.33	1.35	1.38	1.40	1.43	1.46	1.49	1.52	1.55	1.59	1.63	1.67	1.71	1.76	17
74	1.29	1.31	1.34	1.36	1.38	1.41	1.44	1.46	1.49	1.53	1.56	1.60	1.63	1.68	1.72	1.76	16
75	1.30	1.32	1.34	1.37	1.39	1.42	1.44	1.47	1.50	1.53	1.57	1.60	1.64	1.68	1.73	1.77	15
76	1.31	1.33	1.35	1.37	1.40	1.42	1.45	1.48	1.51	1.54	1.58	1.61	1.65	1.69	1.73	1.78	14
77	1.31	1.33	1.35	1.38	1.40	1.43	1.46	1.48	1.52	1.55	1.58	1.62	1.66	1.70	1.74	1.79	13
78	1.32	1.34	1.36	1.38	1.41	1.43	1.46	1.49	1.52	1.55	1.59	1.62	1.66	1.70	1.75	1.80	12
79	1.32	1.34	1.36	1.39	1.41	1.44	1.47	1.50	1.53	1.56	1.59	1.63	1.67	1.71	1.76	1.80	11
80	1.33	1.35	1.37	1.39	1.42	1.44	1.47	1.50	1.53	1.56	1.60	1.64	1.67	1.72	1.76	1.81	10
81	1.33	1.35	1.37	1.40	1.42	1.45	1.48	1.51	1.54	1.57	1.60	1.64	1.68	1.72	1.77	1.81	9
82	1.33	1.35	1.38	1.40	1.43	1.45	1.48	1.51	1.54	1.57	1.61	1.64	1.68	1.73	1.77	1.82	8
83	1.34	1.36	1.38	1.40	1.43	1.46	1.48	1.51	1.54	1.58	1.61	1.65	1.69	1.73	1.77	1.82	7
84	1.34	1.36	1.38	1.41	1.43	1.46	1.49	1.52	1.55	1.58	1.62	1.65	1.69	1.73	1.78	1.83	6
85	1.34	1.36	1.38	1.41	1.43	1.46	1.49	1.52	1.55	1.58	1.62	1.65	1.69	1.74	1.78	1.83	5
86	1.34	1.36	1.39	1.41	1.44	1.46	1.49	1.52	1.55	1.59	1.62	1.66	1.70	1.74	1.78	1.83	4
87	1.34	1.37	1.39	1.41	1.44	1.46	1.49	1.52	1.55	1.59	1.62	1.66	1.70	1.74	1.79	1.83	3
88	1.34	1.37	1.39	1.41	1.44	1.46	1.49	1.52	1.55	1.59	1.62	1.66	1.70	1.74	1.79	1.83	2
89	1.35	1.37	1.39	1.41	1.44	1.47	1.49	1.52	1.56	1.59	1.62	1.66	1.70	1.74	1.79	1.84	1
90	1.35	1.37	1.39	1.41	1.44	1.47	1.49	1.52	1.56	1.59	1.62	1.66	1.70	1.74	1.79	1.84	0
	42°	43°	44°	45°	46°	47°	48°	49°	50°	51°	52°	53°	54°	55°	56°	57°	

Table of factors for reduction of transit observations.

TOP ARGUMENT=STAR'S DECLINATION (δ).

SIDE ARGUMENT=STAR'S ZENITH DISTANCE (ζ).

[For factor *A* use left-hand argument. For factor *B* use right-hand argument. For factor *C* use bottom line on *opposite* page.]

ζ	57°	58°	58½°	59°	59½°	60°	60½°	61°	61½°	62°	62½°	63°	63½°	64°	64½°	65°	ζ
°																	°
1	.03	.03	.03	.03	.03	.03	.04	.04	.04	.04	.04	.04	.04	.04	.04	.04	89
2	.06	.07	.07	.07	.07	.07	.07	.07	.07	.07	.08	.08	.08	.08	.08	.08	88
3	.10	.10	.10	.10	.10	.10	.11	.11	.11	.11	.11	.12	.12	.12	.12	.12	87
4	.13	.13	.13	.14	.14	.14	.14	.14	.15	.15	.15	.15	.16	.16	.16	.17	86
5	.16	.16	.17	.17	.17	.17	.18	.18	.18	.19	.19	.19	.19	.20	.20	.21	85
6	.19	.20	.20	.20	.21	.21	.21	.22	.22	.22	.23	.23	.23	.24	.24	.25	84
7	.22	.23	.23	.24	.24	.24	.25	.25	.26	.26	.26	.27	.27	.28	.28	.29	83
8	.26	.26	.27	.27	.27	.28	.28	.29	.29	.30	.30	.31	.31	.32	.32	.33	82
9	.29	.29	.30	.30	.31	.31	.32	.32	.33	.33	.34	.35	.35	.36	.36	.37	81
10	.32	.33	.33	.34	.34	.35	.35	.36	.36	.37	.38	.38	.39	.40	.40	.41	80
11	.35	.36	.36	.37	.38	.38	.39	.39	.40	.41	.41	.42	.43	.44	.44	.45	79
12	.38	.39	.40	.40	.41	.42	.42	.43	.44	.44	.45	.46	.47	.47	.48	.49	78
13	.41	.42	.43	.44	.44	.45	.46	.46	.47	.48	.49	.50	.50	.51	.52	.53	77
14	.44	.46	.46	.47	.48	.48	.49	.50	.51	.52	.52	.53	.54	.55	.56	.57	76
15	.48	.49	.50	.50	.51	.52	.53	.53	.54	.55	.56	.57	.58	.59	.60	.61	75
16	.51	.52	.53	.54	.54	.55	.56	.57	.58	.59	.60	.61	.62	.63	.64	.65	74
17	.54	.55	.56	.57	.58	.58	.59	.60	.61	.62	.63	.64	.66	.67	.68	.69	73
18	.57	.58	.59	.60	.61	.62	.63	.64	.65	.66	.67	.68	.69	.70	.72	.73	72
19	.60	.61	.62	.63	.64	.65	.66	.67	.68	.69	.70	.72	.73	.74	.76	.77	71
20	.63	.64	.65	.66	.67	.68	.69	.70	.72	.73	.74	.75	.77	.78	.79	.81	70
21	.66	.68	.69	.70	.71	.72	.73	.74	.75	.76	.78	.79	.80	.82	.83	.85	69
22	.69	.71	.72	.73	.74	.75	.76	.77	.78	.80	.81	.82	.84	.85	.87	.89	68
23	.72	.74	.75	.76	.77	.78	.79	.81	.82	.83	.85	.86	.88	.89	.91	.92	67
24	.75	.77	.78	.79	.80	.81	.83	.84	.85	.87	.88	.90	.91	.93	.94	.96	66
25	.78	.80	.81	.82	.83	.85	.86	.87	.89	.90	.92	.93	.95	.96	.98	1.00	65
26	.80	.83	.84	.85	.86	.88	.89	.90	.92	.93	.95	.97	.98	1.00	1.02	1.04	64
27	.83	.86	.87	.88	.89	.91	.92	.94	.95	.97	.98	1.00	1.02	1.04	1.05	1.07	63
28	.86	.89	.90	.91	.93	.94	.95	.97	.98	1.00	1.02	1.03	1.05	1.07	1.09	1.11	62
29	.89	.91	.93	.94	.96	.97	.98	1.00	1.02	1.03	1.05	1.07	1.09	1.11	1.13	1.15	61
30	.92	.94	.96	.97	.99	1.00	1.01	1.03	1.05	1.07	1.08	1.10	1.12	1.14	1.16	1.18	60
31	.95	.97	.99	1.00	1.01	1.03	1.05	1.06	1.08	1.10	1.11	1.13	1.15	1.17	1.20	1.22	59
32	.97	1.00	1.01	1.03	1.04	1.06	1.08	1.09	1.11	1.13	1.15	1.17	1.19	1.21	1.23	1.25	58
33	1.00	1.03	1.04	1.06	1.07	1.09	1.11	1.12	1.14	1.16	1.18	1.20	1.22	1.24	1.26	1.29	57
34	1.03	1.05	1.07	1.09	1.10	1.12	1.14	1.15	1.17	1.19	1.21	1.23	1.25	1.27	1.30	1.32	56
35	1.05	1.08	1.10	1.11	1.13	1.15	1.16	1.18	1.20	1.22	1.24	1.26	1.29	1.31	1.33	1.36	55
36	1.08	1.11	1.12	1.14	1.16	1.18	1.19	1.21	1.23	1.25	1.27	1.30	1.32	1.34	1.37	1.39	54
37	1.10	1.14	1.15	1.17	1.19	1.20	1.22	1.24	1.26	1.28	1.30	1.33	1.35	1.37	1.40	1.42	53
38	1.13	1.16	1.18	1.20	1.21	1.23	1.25	1.27	1.29	1.31	1.33	1.36	1.38	1.40	1.43	1.46	52
39	1.15	1.19	1.20	1.22	1.24	1.26	1.28	1.30	1.32	1.34	1.36	1.39	1.41	1.43	1.46	1.49	51
40	1.18	1.21	1.23	1.25	1.27	1.29	1.31	1.33	1.35	1.37	1.39	1.42	1.44	1.47	1.49	1.52	50
41	1.20	1.24	1.26	1.27	1.29	1.31	1.33	1.35	1.37	1.40	1.42	1.45	1.47	1.50	1.52	1.55	49
42	1.23	1.26	1.28	1.30	1.32	1.34	1.36	1.38	1.40	1.42	1.45	1.47	1.50	1.53	1.55	1.58	48
43	1.25	1.29	1.30	1.32	1.34	1.36	1.39	1.41	1.43	1.45	1.48	1.50	1.53	1.56	1.58	1.61	47
44	1.28	1.31	1.33	1.35	1.37	1.39	1.41	1.43	1.46	1.48	1.50	1.53	1.56	1.58	1.61	1.64	46
45	1.30	1.33	1.35	1.37	1.39	1.41	1.44	1.46	1.48	1.51	1.53	1.56	1.58	1.61	1.64	1.67	45
	57°	58°	58½°	59°	59½°	60°	60½°	61°	61½°	62°	62½°	63°	63½°	64°	64½°	65°	

Table of factors for reduction of transit observations.

TOP ARGUMENT=STAR'S DECLINATION (δ).

SIDE ARGUMENT=STAR'S ZENITH DISTANCE (ζ).

[For factor *A* use left-hand argument. For factor *B* use right-hand argument. For factor *C* use bottom line on *this* page.]

ζ	57°	58°	58½°	59°	59½°	60°	60½°	61°	61½°	62°	62½°	63°	63½°	64°	64½°	65°	ζ
°																	°
46	1.32	1.36	1.38	1.40	1.42	1.44	1.46	1.48	1.51	1.53	1.56	1.58	1.61	1.64	1.67	1.70	44
47	1.34	1.38	1.40	1.42	1.44	1.46	1.49	1.51	1.53	1.56	1.58	1.61	1.64	1.67	1.70	1.73	43
48	1.36	1.40	1.42	1.44	1.46	1.48	1.51	1.53	1.55	1.58	1.60	1.63	1.66	1.69	1.72	1.76	42
49	1.39	1.42	1.44	1.47	1.49	1.51	1.53	1.56	1.58	1.61	1.63	1.66	1.69	1.72	1.75	1.79	41
50	1.41	1.44	1.47	1.49	1.51	1.53	1.56	1.58	1.60	1.63	1.66	1.69	1.72	1.75	1.78	1.81	**40**
51	1.43	1.47	1.49	1.51	1.53	1.55	1.58	1.60	1.63	1.66	1.68	1.71	1.74	1.77	1.80	1.84	39
52	1.45	1.49	1.51	1.53	1.55	1.58	1.60	1.63	1.65	1.68	1.71	1.74	1.77	1.80	1.83	1.86	38
53	1.47	1.51	1.53	1.55	1.57	1.60	1.62	1.65	1.67	1.70	1.73	1.76	1.79	1.82	1.85	1.89	37
54	1.49	1.53	1.55	1.57	1.59	1.62	1.64	1.67	1.69	1.72	1.75	1.78	1.81	1.85	1.88	1.91	36
55	1.50	1.55	1.57	1.59	1.61	1.64	1.66	1.69	1.72	1.74	1.77	1.80	1.84	1.87	1.90	1.94	35
56	1.52	1.56	1.59	1.61	1.63	1.66	1.68	1.71	1.74	1.77	1.80	1.83	1.86	1.89	1.93	1.96	34
57	1.54	1.58	1.61	1.63	1.65	1.68	1.70	1.73	1.76	1.79	1.82	1.85	1.88	1.91	1.95	1.98	33
58	1.56	1.60	1.62	1.65	1.67	1.70	1.72	1.75	1.78	1.81	1.84	1.87	1.90	1.93	1.97	2.01	32
59	1.57	1.62	1.64	1.66	1.69	1.71	1.74	1.77	1.80	1.83	1.86	1.89	1.92	1.96	1.99	2.03	31
60	1.59	1.63	1.66	1.68	1.71	1.73	1.76	1.79	1.81	1.84	1.88	1.91	1.94	1.98	2.01	2.05	**30**
61	1.61	1.65	1.67	1.70	1.72	1.75	1.78	1.80	1.83	1.86	1.89	1.93	1.96	2.00	2.03	2.07	29
62	1.62	1.67	1.69	1.71	1.74	1.77	1.79	1.82	1.85	1.88	1.91	1.94	1.98	2.01	2.05	2.09	28
63	1.64	1.68	1.70	1.73	1.76	1.78	1.81	1.84	1.87	1.90	1.93	1.96	2.00	2.03	2.07	2.11	27
64	1.65	1.70	1.72	1.75	1.77	1.80	1.83	1.85	1.88	1.91	1.95	1.98	2.02	2.05	2.09	2.13	26
65	1.66	1.71	1.73	1.76	1.79	1.81	1.84	1.87	1.90	1.93	1.96	2.00	2.03	2.07	2.11	2.14	25
66	1.68	1.72	1.75	1.77	1.80	1.83	1.85	1.88	1.91	1.95	1.98	2.01	2.05	2.08	2.12	2.16	24
67	1.69	1.74	1.76	1.79	1.81	1.84	1.87	1.90	1.93	1.96	1.99	2.03	2.06	2.10	2.14	2.18	23
68	1.70	1.75	1.77	1.80	1.83	1.85	1.88	1.91	1.94	1.97	2.01	2.04	2.08	2.11	2.15	2.19	22
69	1.71	1.76	1.79	1.81	1.84	1.87	1.90	1.93	1.96	1.99	2.02	2.06	2.09	2.13	2.17	2.21	21
70	1.73	1.77	1.80	1.82	1.85	1.88	1.91	1.94	1.97	2.00	2.03	2.07	2.11	2.14	2.18	2.22	**20**
71	1.74	1.78	1.81	1.84	1.86	1.89	1.92	1.95	1.98	2.01	2.05	2.08	2.12	2.16	2.20	2.24	19
72	1.75	1.79	1.82	1.85	1.87	1.90	1.93	1.96	1.99	2.03	2.06	2.09	2.13	2.17	2.21	2.25	18
73	1.76	1.80	1.83	1.86	1.88	1.91	1.94	1.97	2.00	2.04	2.07	2.11	2.14	2.18	2.22	2.26	17
74	1.76	1.81	1.84	1.87	1.89	1.92	1.95	1.98	2.01	2.05	2.08	2.12	2.15	2.19	2.23	2.27	16
75	1.77	1.82	1.85	1.88	1.90	1.93	1.96	1.99	2.02	2.06	2.09	2.13	2.16	2.20	2.24	2.29	15
76	1.78	1.83	1.86	1.88	1.91	1.94	1.97	2.00	2.03	2.07	2.10	2.14	2.17	2.21	2.25	2.30	14
77	1.79	1.84	1.86	1.89	1.92	1.95	1.98	2.01	2.04	2.07	2.11	2.15	2.18	2.22	2.26	2.31	13
78	1.80	1.85	1.87	1.90	1.98	1.96	1.99	2.02	2.05	2.08	2.12	2.15	2.19	2.23	2.27	2.31	12
79	1.80	1.85	1.88	1.91	1.93	1.96	1.99	2.02	2.06	2.09	2.13	2.16	2.20	2.24	2.28	2.32	11
80	1.81	1.86	1.88	1.91	1.94	1.97	2.00	2.03	2.06	2.10	2.13	2.17	2.21	2.25	2.29	2.33	**10**
81	1.81	1.86	1.89	1.92	1.95	1.98	2.01	2.04	2.07	2.10	2.14	2.18	2.21	2.25	2.29	2.34	9
82	1.82	1.87	1.90	1.92	1.95	1.98	2.01	2.04	2.08	2.11	2.15	2.18	2.22	2.26	2.30	2.34	8
83	1.82	1.87	1.90	1.93	1.96	1.99	2.02	2.05	2.08	2.12	2.15	2.19	2.22	2.26	2.31	2.35	7
84	1.83	1.88	1.90	1.93	1.96	1.99	2.02	2.05	2.08	2.12	2.15	2.19	2.23	2.27	2.31	2.35	6
85	1.83	1.88	1.91	1.93	1.96	1.99	2.02	2.05	2.09	2.12	2.16	2.19	2.23	2.27	2.31	2.36	5
86	1.83	1.88	1.91	1.94	1.97	2.00	2.03	2.06	2.09	2.13	2.16	2.20	2.24	2.28	2.32	2.36	4
87	1.83	1.88	1.91	1.94	1.97	2.00	2.03	2.06	2.09	2.13	2.16	2.20	2.24	2.28	2.32	2.36	3
88	1.83	1.89	1.91	1.94	1.97	2.00	2.03	2.06	2.09	2.13	2.16	2.20	2.24	2.28	2.32	2.36	2
89	1.84	1.89	1.91	1.94	1.97	2.00	2.03	2.06	2.10	2.13	2.17	2.20	2.24	2.28	2.32	2.37	1
90	1.84	1.89	1.91	1.94	1.97	2.00	2.03	2.06	2.10	2.13	2.17	2.20	2.24	2.28	2.32	2.37	**0**
	57°	58°	58½°	59°	59½°	60°	60½°	61°	61½°	62°	62½°	63°	63½°	64°	64½°	65°	

Table of factors for reduction of transit observations.

TOP ARGUMENT=STAR'S DECLINATION (δ).

SIDE ARGUMENT=STAR'S ZENITH DISTANCE (ζ).

[For factor *A* use left-hand argument. For factor *B* use right-hand argument. For factor *C* use bottom line on *opposite* page.]

ζ	65°	65½°	66°	66½°	67°	67½°	68°	68½°	69°	69° 10′	69° 20′	69° 30′	69° 40′	69° 50′	70°	70° 10′	ζ
°																	°
1	.04	.04	.04	.04	.04	.05	.05	.05	.05	.05	.05	.05	.05	.05	.05	.05	89
2	.08	.08	.09	.09	.09	.09	.09	.10	.10	.10	.10	.10	.10	.10	.10	.10	88
3	.12	.13	.13	.13	.13	.14	.14	.14	.15	.15	.15	.15	.15	.15	.15	.15	87
4	.17	.17	.17	.18	.18	.18	.19	.19	.20	.20	.20	.20	.20	.20	.20	.20	86
5	.21	.21	.21	.22	.22	.23	.23	.24	.24	.24	.25	.25	.25	.25	.25	.26	85
6	.25	.25	.26	.26	.27	.27	.28	.28	.29	.29	.30	.30	.30	.30	.31	.31	84
7	.29	.29	.30	.31	.31	.32	.33	.33	.34	.34	.34	.35	.35	.35	.36	.36	83
8	.33	.34	.34	.35	.36	.36	.37	.38	.39	.39	.39	.40	.40	.40	.41	.41	82
9	.37	.38	.39	.39	.40	.41	.42	.43	.44	.44	.44	.45	.45	.45	.46	.46	81
10	.41	.42	.43	.43	.44	.45	.46	.47	.48	.49	.49	.50	.50	.50	.51	.51	80
11	.45	.46	.47	.48	.49	.50	.51	.52	.53	.54	.54	.54	.55	.55	.56	.56	79
12	.49	.50	.51	.52	.53	.54	.56	.57	.58	.58	.59	.59	.60	.60	.61	.61	78
13	.53	.54	.55	.56	.58	.59	.60	.61	.63	.63	.64	.64	.65	.65	.66	.66	77
14	.57	.58	.59	.61	.62	.63	.65	.66	.67	.68	.68	.69	.70	.70	.71	.71	76
15	.61	.62	.64	.65	.66	.68	.69	.71	.72	.73	.73	.74	.74	.75	.76	.76	75
16	.65	.66	.68	.69	.71	.72	.74	.75	.77	.78	.78	.79	.79	.80	.81	.81	74
17	.69	.70	.72	.73	.75	.76	.78	.80	.81	.82	.83	.83	.84	.85	.85	.86	73
18	.73	.74	.76	.77	.79	.81	.83	.84	.86	.87	.88	.88	.89	.90	.90	.91	72
19	.77	.78	.80	.82	.83	.85	.87	.89	.91	.92	.92	.93	.94	.94	.95	.96	71
20	.81	.82	.84	.86	.88	.89	.91	.93	.95	.96	.97	.98	.98	.99	1.00	1.01	70
21	.85	.86	.88	.90	.92	.94	.96	.98	1.00	1.01	1.02	1.02	1.03	1.04	1.05	1.06	69
22	.89	.90	.92	.94	.96	.98	1.00	1.02	1.05	1.05	1.06	1.07	1.08	1.09	1.09	1.10	68
23	.92	.94	.96	.98	1.00	1.02	1.04	1.07	1.09	1.10	1.11	1.12	1.12	1.13	1.14	1.15	67
24	.96	.98	1.00	1.02	1.04	1.06	1.09	1.11	1.14	1.14	1.15	1.16	1.17	1.18	1.19	1.20	66
25	1.00	1.02	1.04	1.06	1.08	1.10	1.13	1.15	1.18	1.19	1.20	1.21	1.22	1.23	1.24	1.25	65
26	1.04	1.06	1.08	1.10	1.12	1.15	1.17	1.20	1.22	1.23	1.24	1.25	1.26	1.27	1.28	1.29	64
27	1.07	1.09	1.12	1.14	1.16	1.19	1.21	1.24	1.27	1.28	1.29	1.30	1.31	1.32	1.33	1.34	63
28	1.11	1.13	1.15	1.18	1.20	1.23	1.25	1.28	1.31	1.32	1.33	1.34	1.35	1.36	1.37	1.38	62
29	1.15	1.17	1.19	1.22	1.24	1.27	1.29	1.32	1.35	1.36	1.37	1.38	1.40	1.41	1.42	1.43	61
30	1.18	1.21	1.23	1.25	1.28	1.31	1.33	1.36	1.39	1.41	1.42	1.43	1.44	1.45	1.46	1.47	60
31	1.22	1.24	1.27	1.29	1.32	1.35	1.38	1.40	1.44	1.45	1.46	1.47	1.48	1.49	1.51	1.52	59
32	1.25	1.28	1.30	1.33	1.36	1.39	1.42	1.45	1.48	1.49	1.50	1.51	1.52	1.54	1.55	1.56	58
33	1.29	1.31	1.34	1.37	1.39	1.42	1.45	1.49	1.52	1.53	1.54	1.55	1.57	1.58	1.59	1.60	57
34	1.32	1.35	1.37	1.40	1.43	1.46	1.49	1.53	1.56	1.57	1.58	1.60	1.61	1.62	1.63	1.65	56
35	1.36	1.38	1.41	1.44	1.47	1.50	1.53	1.56	1.60	1.61	1.62	1.64	1.65	1.66	1.68	1.69	55
36	1.39	1.42	1.45	1.47	1.51	1.54	1.57	1.60	1.64	1.65	1.66	1.68	1.69	1.70	1.72	1.73	54
37	1.42	1.45	1.48	1.51	1.54	1.57	1.61	1.64	1.68	1.69	1.70	1.72	1.73	1.74	1.76	1.77	53
38	1.46	1.48	1.51	1.54	1.58	1.61	1.64	1.68	1.72	1.73	1.74	1.76	1.77	1.79	1.80	1.82	52
39	1.49	1.52	1.55	1.58	1.61	1.65	1.68	1.72	1.75	1.77	1.78	1.80	1.81	1.82	1.84	1.86	51
40	1.52	1.55	1.58	1.61	1.65	1.68	1.72	1.75	1.79	1.81	1.82	1.84	1.85	1.86	1.88	1.89	50
41	1.55	1.58	1.61	1.64	1.68	1.71	1.75	1.79	1.83	1.84	1.86	1.87	1.89	1.90	1.92	1.93	49
42	1.58	1.61	1.64	1.68	1.71	1.75	1.79	1.83	1.87	1.88	1.90	1.91	1.93	1.94	1.96	1.97	48
43	1.61	1.64	1.68	1.71	1.75	1.78	1.82	1.86	1.90	1.92	1.93	1.95	1.96	1.98	1.99	2.01	47
44	1.64	1.67	1.71	1.74	1.78	1.82	1.85	1.90	1.94	1.95	1.97	1.98	2.00	2.02	2.03	2.05	46
45	1.67	1.70	1.74	1.77	1.81	1.85	1.89	1.93	1.97	1.99	2.00	2.02	2.04	2.05	2.07	2.08	45
	65°	65½°	66°	66½°	67°	67½°	68°	68½°	69°	69° 10′	69° 20′	69° 30′	69° 40′	69° 50′	70°	70° 10′	

Table of factors for reduction of transit observations.

TOP ARGUMENT=STAR'S DECLINATION (δ).

SIDE ARGUMENT=STAR'S ZENITH DISTANCE (ζ)

[For factor *A* use left-hand argument. For factor *B* use right-hand argument. For factor *C* use bottom line on *this* page.]

ζ	65°	65½°	66°	66½°	67°	67½°	68°	68½°	69°	69° 10′	69° 20′	69° 30′	69° 40′	69° 50′	70°	70° 10′	ζ
°																	°
46	1.70	1.74	1.77	1.80	1.84	1.88	1.92	1.96	2.01	2.02	2.04	2.05	2.07	2.09	2.10	2.12	44
47	1.73	1.76	1.80	1.83	1.87	1.91	1.95	2.00	2.04	2.06	2.07	2.09	2.10	2.12	2.14	2.16	43
48	1.76	1.79	1.83	1.86	1.90	1.94	1.98	2.03	2.07	2.09	2.11	2.12	2.14	2.16	2.17	2.19	42
49	1.79	1.82	1.86	1.89	1.93	1.97	2.01	2.06	2.11	2.12	2.14	2.16	2.17	2.19	2.21	2.22	41
50	1.81	1.85	1.88	1.92	1.96	2.00	2.04	2.09	2.14	2.15	2.17	2.19	2.20	2.22	2.24	2.26	40
51	1.84	1.87	1.91	1.95	1.99	2.03	2.07	2.12	2.17	2.18	2.20	2.22	2.24	2.25	2.27	2.29	39
52	1.86	1.90	1.94	1.98	2.02	2.06	2.10	2.15	2.20	2.22	2.23	2.25	2.27	2.29	2.30	2.32	38
53	1.89	1.93	1.96	2.00	2.04	2.09	2.13	2.18	2.23	2.25	2.26	2.28	2.30	2.32	2.33	2.35	37
54	1.91	1.95	1.99	2.03	2.07	2.11	2.16	2.21	2.26	2.28	2.29	2.31	2.33	2.35	2.37	2.38	36
55	1.94	1.98	2.01	2.05	2.10	2.14	2.19	2.23	2.29	2.30	2.32	2.34	2.36	2.38	2.40	2.41	35
56	1.96	2.00	2.04	2.08	2.12	2.17	2.21	2.26	2.31	2.33	2.35	2.37	2.39	2.40	2.42	2.44	34
57	1.98	2.02	2.06	2.10	2.15	2.19	2.24	2.29	2.34	2.36	2.38	2.39	2.41	2.43	2.45	2.47	33
58	2.01	2.05	2.08	2.13	2.17	2.22	2.26	2.31	2.37	2.38	2.40	2.42	2.44	2.46	2.48	2.50	32
59	2.03	2.07	2.11	2.15	2.19	2.24	2.29	2.34	2.39	2.41	2.43	2.45	2.47	2.49	2.51	2.53	31
60	2.05	2.09	2.13	2.17	2.22	2.26	2.31	2.36	2.42	2.44	2.45	2.47	2.49	2.51	2.53	2.55	30
61	2.07	2.11	2.15	2.19	2.24	2.29	2.33	2.39	2.44	2.46	2.48	2.50	2.52	2.54	2.56	2.58	29
62	2.09	2.13	2.17	2.21	2.26	2.31	2.36	2.41	2.46	2.48	2.50	2.52	2.54	2.56	2.58	2.60	28
63	2.11	2.15	2.19	2.23	2.28	2.33	2.38	2.43	2.49	2.50	2.52	2.54	2.56	2.58	2.60	2.63	27
64	2.13	2.17	2.21	2.25	2.30	2.35	2.40	2.45	2.51	2.53	2.55	2.57	2.59	2.61	2.63	2.65	26
65	2.14	2.19	2.23	2.27	2.32	2.37	2.42	2.47	2.53	2.55	2.57	2.59	2.61	2.63	2.65	2.67	25
66	2.16	2.20	2.25	2.29	2.34	2.39	2.44	2.49	2.55	2.57	2.59	2.61	2.63	2.65	2.67	2.69	24
67	2.18	2.22	2.26	2.31	2.36	2.41	2.46	2.51	2.57	2.59	2.61	2.63	2.65	2.67	2.69	2.71	23
68	2.19	2.24	2.28	2.32	2.37	2.42	2.47	2.53	2.59	2.61	2.63	2.65	2.67	2.69	2.71	2.73	22
69	2.21	2.25	2.30	2.34	2.39	2.44	2.49	2.55	2.61	2.62	2.64	2.67	2.69	2.71	2.73	2.75	21
70	2.22	2.27	2.31	2.36	2.40	2.46	2.51	2.56	2.62	2.64	2.66	2.68	2.70	2.73	2.75	2.77	20
71	2.24	2.28	2.32	2.37	2.42	2.47	2.52	2.58	2.64	2.66	2.68	2.70	2.72	2.74	2.77	2.79	19
72	2.25	2.29	2.34	2.38	2.43	2.49	2.54	2.59	2.65	2.67	2.70	2.72	2.74	2.76	2.78	2.80	18
73	2.26	2.31	2.35	2.40	2.45	2.50	2.55	2.61	2.67	2.69	2.71	2.73	2.75	2.77	2.80	2.82	17
74	2.27	2.32	2.36	2.41	2.46	2.51	2.57	2.62	2.68	2.70	2.72	2.74	2.77	2.79	2.81	2.83	16
75	2.29	2.33	2.37	2.42	2.47	2.52	2.58	2.64	2.70	2.72	2.74	2.76	2.78	2.80	2.82	2.85	15
76	2.30	2.34	2.39	2.43	2.48	2.54	2.59	2.65	2.71	2.73	2.75	2.77	2.79	2.81	2.84	2.86	14
77	2.31	2.35	2.40	2.44	2.49	2.55	2.60	2.66	2.72	2.74	2.76	2.78	2.80	2.83	2.85	2.87	13
78	2.31	2.36	2.40	2.45	2.50	2.56	2.61	2.67	2.73	2.75	2.77	2.79	2.81	2.84	2.86	2.88	12
79	2.32	2.37	2.41	2.46	2.51	2.57	2.62	2.68	2.74	2.76	2.78	2.80	2.82	2.85	2.87	2.89	11
80	2.33	2.38	2.42	2.47	2.52	2.57	2.63	2.69	2.75	2.77	2.79	2.81	2.83	2.86	2.88	2.90	10
81	2.34	2.38	2.43	2.48	2.53	2.58	2.64	2.69	2.76	2.78	2.80	2.82	2.84	2.86	2.89	2.91	9
82	2.34	2.39	2.43	2.48	2.53	2.59	2.64	2.70	2.76	2.78	2.81	2.83	2.85	2.87	2.90	2.92	8
83	2.35	2.39	2.44	2.49	2.54	2.59	2.65	2.71	2.77	2.79	2.81	2.83	2.86	2.88	2.90	2.92	7
84	2.35	2.40	2.45	2.49	2.55	2.60	2.66	2.71	2.78	2.80	2.82	2.84	2.86	2.88	2.91	2.93	6
85	2.36	2.40	2.45	2.50	2.55	2.60	2.66	2.72	2.78	2.80	2.82	2.84	2.87	2.89	2.91	2.94	5
86	2.36	2.41	2.45	2.50	2.55	2.61	2.66	2.72	2.78	2.80	2.83	2.85	2.87	2.89	2.92	2.94	4
87	2.36	2.41	2.46	2.50	2.56	2.61	2.67	2.72	2.79	2.81	2.83	2.85	2.87	2.90	2.92	2.94	3
88	2.36	2.41	2.46	2.51	2.56	2.61	2.67	2.73	2.79	2.81	2.83	2.85	2.88	2.90	2.92	2.95	2
89	2.37	2.41	2.46	2.51	2.56	2.61	2.67	2.73	2.79	2.81	2.83	2.86	2.88	2.90	2.92	2.95	1
90	2.37	2.41	2.46	2.51	2.56	2.61	2.67	2.73	2.79	2.81	2.83	2.86	2.88	2.90	2.92	2.95	0
	65°	65½°	66°	66½°	67°	67½°	68°	68½°	69°	69° 10′	69° 20′	69° 30′	69 °40′	69° 50′	70°	70° 10′	

Table of factors for reduction of transit observations.

TOP ARGUMENT=STAR'S DECLINATION (δ).

SIDE ARGUMENT=STAR'S ZENITH DISTANCE (ζ).

[For factor *A* use left-hand argument. For factor *B* use right-hand argument. For factor *C* use bottom line on *opposite* page.]

ζ	70° 10′	70° 20′	70° 30′	70° 40′	70° 50′	71°	71° 10′	71° 20′	71° 30′	71° 40′	71° 50′	72°	72° 10′	72° 20′	72° 30′	72° 40′	ζ
°																	°
1	.05	.05	.05	.05	.05	.05	.05	.05	.05	.06	.06	.06	.06	.06	.06	.06	89
2	.10	.10	.10	.10	.11	.11	.11	.11	.11	.11	.11	.11	.11	.12	.12	.12	88
3	.15	.16	.16	.16	.16	.16	.16	.16	.16	.17	.17	.17	.17	.17	.17	.18	87
4	.20	.21	.21	.21	.21	.21	.22	.22	.22	.22	.22	.23	.23	.23	.23	.23	86
5	.26	.26	.26	.26	.26	.27	.27	.27	.27	.28	.28	.28	.28	.29	.29	.29	85
6	.31	.31	.31	.32	.32	.32	.32	.33	.33	.33	.34	.34	.34	.34	.35	.35	84
7	.36	.36	.37	.37	.37	.37	.38	.38	.38	.39	.39	.39	.40	.40	.41	.41	83
8	.41	.41	.42	.42	.42	.43	.43	.44	.44	.44	.45	.45	.45	.46	.46	.47	82
9	.46	.46	.47	.47	.48	.48	.48	.49	.49	.50	.50	.51	.51	.52	.52	.52	81
10	.51	.52	.52	.52	.53	.53	.54	.54	.55	.55	.56	.56	.57	.57	.58	.58	80
11	.56	.57	.57	.58	.58	.59	.59	.60	.60	.61	.61	.62	.62	.63	.63	.64	79
12	.61	.62	.62	.63	.63	.64	.64	.65	.66	.66	.67	.67	.68	.68	.69	.70	78
13	.66	.67	.67	.68	.68	.69	.70	.70	.71	.72	.72	.73	.74	.74	.75	.76	77
14	.71	.72	.72	.73	.74	.74	.75	.76	.76	.77	.78	.78	.79	.80	.80	.81	76
15	.76	.77	.78	.78	.79	.79	.80	.81	.81	.82	.83	.84	.84	.85	.86	.87	75
16	.81	.82	.83	.83	.84	.85	.85	.86	.87	.88	.88	.89	.90	.91	.92	.92	74
17	.86	.87	.88	.88	.89	.90	.90	.91	.92	.93	.94	.95	.96	.96	.97	.98	73
18	.91	.92	.93	.93	.94	.95	.96	.96	.97	.98	.99	1.00	1.01	1.02	1.03	1.04	72
19	.96	.97	.98	.98	.99	1.00	1.01	1.02	1.03	1.04	1.04	1.05	1.06	1.07	1.08	1.09	71
20	1.01	1.02	1.02	1.03	1.04	1.05	1.06	1.07	1.08	1.09	1.10	1.11	1.12	1.13	1.14	1.15	70
21	1.06	1.06	1.07	1.08	1.09	1.10	1.11	1.12	1.13	1.14	1.15	1.16	1.17	1.18	1.19	1.20	69
22	1.10	1.11	1.12	1.13	1.14	1.15	1.16	1.17	1.18	1.19	1.20	1.21	1.22	1.24	1.25	1.26	68
23	1.15	1.16	1.17	1.18	1.19	1.20	1.21	1.22	1.23	1.24	1.25	1.26	1.28	1.29	1.30	1.31	67
24	1.20	1.21	1.22	1.23	1.24	1.25	1.26	1.27	1.28	1.29	1.30	1.32	1.33	1.34	1.35	1.36	66
25	1.25	1.26	1.27	1.28	1.29	1.30	1.31	1.32	1.33	1.34	1.36	1.37	1.38	1.39	1.41	1.42	65
26	1.29	1.30	1.31	1.32	1.34	1.35	1.36	1.37	1.38	1.39	1.41	1.42	1.43	1.44	1.46	1.47	64
27	1.34	1.35	1.36	1.37	1.38	1.39	1.41	1.42	1.43	1.44	1.46	1.47	1.48	1.50	1.51	1.52	63
28	1.38	1.40	1.41	1.42	1.43	1.44	1.45	1.47	1.48	1.49	1.51	1.52	1.53	1.55	1.56	1.58	62
29	1.43	1.44	1.45	1.46	1.48	1.49	1.50	1.52	1.53	1.54	1.56	1.57	1.58	1.60	1.61	1.63	61
30	1.47	1.49	1.50	1.51	1.52	1.54	1.55	1.56	1.58	1.59	1.60	1.62	1.63	1.65	1.66	1.68	60
31	1.52	1.53	1.54	1.56	1.57	1.58	1.60	1.61	1.62	1.64	1.65	1.67	1.68	1.70	1.71	1.73	59
32	1.56	1.57	1.59	1.60	1.61	1.63	1.64	1.66	1.67	1.68	1.70	1.71	1.73	1.75	1.76	1.78	58
33	1.60	1.62	1.63	1.64	1.66	1.67	1.69	1.70	1.72	1.73	1.75	1.76	1.78	1.80	1.81	1.83	57
34	1.65	1.66	1.68	1.69	1.70	1.72	1.73	1.75	1.76	1.78	1.79	1.81	1.83	1.84	1.86	1.88	56
35	1.69	1.70	1.72	1.73	1.75	1.76	1.78	1.79	1.81	1.82	1.84	1.86	1.87	1.89	1.91	1.92	55
36	1.73	1.75	1.76	1.78	1.79	1.80	1.82	1.84	1.85	1.87	1.88	1.90	1.92	1.94	1.95	1.97	54
37	1.77	1.79	1.80	1.82	1.83	1.85	1.86	1.88	1.90	1.91	1.93	1.95	1.96	1.98	2.00	2.02	53
38	1.82	1.83	1.84	1.86	1.88	1.89	1.91	1.92	1.94	1.96	1.98	1.99	2.01	2.03	2.05	2.07	52
39	1.86	1.87	1.89	1.90	1.92	1.93	1.95	1.97	1.98	2.00	2.02	2.04	2.06	2.07	2.09	2.11	51
40	1.89	1.91	1.93	1.94	1.96	1.97	1.99	2.01	2.03	2.04	2.06	2.08	2.10	2.12	2.14	2.16	50
41	1.93	1.95	1.96	1.98	2.00	2.01	2.03	2.05	2.07	2.09	2.10	2.12	2.14	2.16	2.18	2.20	49
42	1.97	1.99	2.00	2.02	2.04	2.05	2.07	2.09	2.11	2.13	2.15	2.16	2.18	2.20	2.22	2.25	48
43	2.01	2.03	2.04	2.06	2.08	2.09	2.11	2.13	2.15	2.17	2.19	2.21	2.23	2.25	2.27	2.29	47
44	2.05	2.06	2.08	2.10	2.12	2.13	2.15	2.17	2.19	2.21	2.23	2.25	2.27	2.29	2.31	2.33	46
45	2.08	2.10	2.12	2.14	2.15	2.17	2.19	2.21	2.23	2.25	2.27	2.29	2.31	2.33	2.35	2.37	45
	70° 10′	70° 20′	70° 30′	70° 40′	70° 50′	71°	71° 10′	71° 20′	71° 30′	71° 40′	71° 50′	72°	72° 10′	72° 20′	72° 30′	72° 40′	

Table of factors for reduction of transit observations.

TOP ARGUMENT=STAR'S DECLINATION (δ).

SIDE ARGUMENT=STAR'S ZENITH DISTANCE (ζ).

[For factor *A* use left-hand argument. For factor *B* use right-hand argument. For factor *C* use bottom line on *this* page.]

ζ	70° 10′	70° 20′	70° 30′	70° 40′	70° 50′	71°	71° 10′	71° 20′	71° 30′	71° 40′	71° 50′	72°	72° 10′	72° 20′	72° 30′	72° 40′	ζ
°																	°
46	2.12	2.14	2.15	2.17	2.19	2.21	2.23	2.25	2.27	2.29	2.31	2.33	2.35	2.37	2.39	2.41	44
47	2.16	2.17	2.19	2.21	2.23	2.25	2.27	2.28	2.30	2.32	2.35	2.37	2.39	2.41	2.43	2.45	43
48	2.19	2.21	2.22	2.24	2.26	2.28	2.30	2.32	2.34	2.36	2.38	2.40	2.43	2.45	2.47	2.49	42
49	2.22	2.24	2.26	2.28	2.30	2.32	2.34	2.36	2.38	2.40	2.42	2.44	2.46	2.49	2.51	2.53	41
50	2.26	2.28	2.29	2.31	2.33	2.35	2.37	2.39	2.41	2.44	2.46	2.48	2.50	2.52	2.55	2.57	40
51	2.29	2.31	2.33	2.35	2.37	2.39	2.41	2.43	2.45	2.47	2.49	2.51	2.54	2.56	2.58	2.61	39
52	2.32	2.34	2.36	2.38	2.40	2.42	2.44	2.46	2.48	2.50	2.53	2.55	2.57	2.60	2.62	2.64	38
53	2.35	2.37	2.39	2.41	2.43	2.45	2.47	2.50	2.52	2.54	2.56	2.58	2.61	2.63	2.66	2.68	37
54	2.38	2.40	2.42	2.44	2.46	2.48	2.51	2.53	2.55	2.57	2.60	2.62	2.64	2.67	2.69	2.72	36
55	2.41	2.43	2.45	2.47	2.50	2.52	2.54	2.56	2.58	2.60	2.63	2.65	2.68	2.70	2.72	2.75	35
56	2.44	2.46	2.48	2.50	2.52	2.55	2.57	2.59	2.61	2.64	2.66	2.68	2.71	2.73	2.76	2.78	34
57	2.47	2.49	2.51	2.53	2.55	2.58	2.60	2.62	2.64	2.67	2.69	2.71	2.74	2.76	2.79	2.82	33
58	2.50	2.52	2.54	2.56	2.58	2.61	2.63	2.65	2.67	2.70	2.72	2.74	2.77	2.79	2.82	2.85	32
59	2.53	2.55	2.57	2.59	2.61	2.63	2.66	2.68	2.70	2.72	2.75	2.77	2.80	2.82	2.85	2.88	31
60	2.55	2.57	2.59	2.62	2.64	2.66	2.68	2.71	2.73	2.75	2.78	2.80	2.83	2.85	2.88	2.91	30
61	2.58	2.60	2.62	2.64	2.66	2.69	2.71	2.73	2.76	2.78	2.80	2.83	2.86	2.88	2.91	2.94	29
62	2.60	2.62	2.64	2.67	2.69	2.71	2.74	2.76	2.78	2.81	2.83	2.86	2.88	2.91	2.94	2.96	28
63	2.63	2.65	2.67	2.69	2.71	2.74	2.76	2.78	2.81	2.83	2.86	2.88	2.91	2.94	2.96	2.99	27
64	2.65	2.67	2.69	2.72	2.74	2.76	2.78	2.81	2.83	2.86	2.88	2.91	2.94	2.96	2.99	3.02	26
65	2.67	2.69	2.71	2.74	2.76	2.78	2.81	2.83	2.86	2.88	2.91	2.93	2.96	2.99	3.01	3.04	25
66	2.69	2.71	2.74	2.76	2.78	2.81	2.83	2.85	2.88	2.90	2.93	2.96	2.98	3.01	3.04	3.07	24
67	2.71	2.74	2.76	2.78	2.80	2.83	2.85	2.88	2.90	2.93	2.95	2.98	3.01	3.03	3.06	3.09	23
68	2.73	2.76	2.78	2.80	2.82	2.85	2.87	2.90	2.92	2.95	2.97	3.00	3.03	3.06	3.08	3.11	22
69	2.75	2.77	2.80	2.82	2.84	2.87	2.89	2.92	2.94	2.97	2.99	3.02	3.05	3.08	3.10	3.13	21
70	2.77	2.79	2.81	2.84	2.86	2.89	2.91	2.94	2.96	2.99	3.01	3.04	3.07	3.10	3.12	3.15	20
71	2.79	2.81	2.83	2.86	2.88	2.90	2.93	2.95	2.98	3.01	3.03	3.06	3.09	3.12	3.14	3.17	19
72	2.80	2.83	2.85	2.87	2.90	2.92	2.95	2.97	3.00	3.02	3.05	3.08	3.10	3.13	3.16	3.19	18
73	2.82	2.84	2.86	2.89	2.91	2.94	2.96	2.99	3.01	3.04	3.07	3.09	3.12	3.15	3.18	3.21	17
74	2.83	2.86	2.88	2.90	2.93	2.95	2.98	3.00	3.03	3.06	3.08	3.11	3.14	3.17	3.20	3.23	16
75	2.85	2.87	2.89	2.92	2.94	2.97	2.99	3.02	3.04	3.07	3.10	3.13	3.15	3.18	3.21	3.24	15
76	2.86	2.88	2.91	2.93	2.96	2.98	3.01	3.03	3.06	3.08	3.11	3.14	3.17	3.20	3.23	3.26	14
77	2.87	2.90	2.92	2.94	2.97	2.99	3.02	3.04	3.07	3.10	3.12	3.15	3.18	3.21	3.24	3.27	13
78	2.88	2.91	2.93	2.95	2.98	3.00	3.03	3.06	3.08	3.11	3.14	3.16	3.19	3.22	3.25	3.28	12
79	2.89	2.92	2.94	2.96	2.99	3.02	3.04	3.07	3.09	3.12	3.15	3.18	3.20	3.23	3.26	3.29	11
80	2.90	2.93	2.95	2.97	3.00	3.02	3.05	3.08	3.10	3.13	3.16	3.19	3.22	3.24	3.27	3.31	10
81	2.91	2.94	2.96	2.98	3.01	3.03	3.06	3.09	3.11	3.14	3.17	3.20	3.23	3.25	3.28	3.32	9
82	2.92	2.94	2.97	2.99	3.02	3.04	3.07	3.09	3.12	3.15	3.18	3.20	3.23	3.26	3.29	3.32	8
83	2.92	2.95	2.97	3.00	3.02	3.05	3.08	3.10	3.13	3.16	3.18	3.21	3.24	3.27	3.30	3.33	7
84	2.93	2.96	2.98	3.00	3.03	3.06	3.08	3.11	3.13	3.16	3.19	3.22	3.25	3.28	3.31	3.34	6
85	2.94	2.96	2.98	3.01	3.03	3.06	3.09	3.11	3.14	3.17	3.20	3.22	3.25	3.28	3.31	3.34	5
86	2.94	2.96	2.99	3.01	3.04	3.06	3.09	3.12	3.14	3.17	3.20	3.23	3.26	3.29	3.32	3.35	4
87	2.94	2.97	2.99	3.02	3.04	3.07	3.09	3.12	3.15	3.18	3.20	3.23	3.26	3.29	3.32	3.35	3
88	2.95	2.97	2.99	3.02	3.04	3.07	3.10	3.12	3.15	3.18	3.20	3.23	3.26	3.29	3.32	3.35	2
89	2.95	2.97	3.00	3.02	3.04	3.07	3.10	3.12	3.15	3.18	3.21	3.24	3.27	3.30	3.33	3.36	1
90	2.95	2.97	3.00	3.02	3.05	3.07	3.10	3.12	3.15	3.18	3.21	3.24	3.27	3.30	3.33	3.36	0
	70° 10′	70° 20′	70° 30′	70° 40′	70° 50′	71°	71° 10′	71° 20′	71° 30′	71° 40′	71° 50′	72°	72° 10′	72° 20′	72° 30′	72° 40′	

Table of factors for reduction of transit observations.

TOP ARGUMENT=STAR'S DECLINATION (δ).

SIDE ARGUMENT=STAR'S ZENITH DISTANCE (ζ).

[For factor *A* use left-hand argument. For factor *B* use right-hand argument. For factor *C* use bottom line on *opposite* page.]

ζ	72° 40′	72° 50′	73°	73° 10′	73° 20′	73° 30′	73° 40′	73° 50′	74°	74° 10′	74° 20′	74° 30′	74° 40′	74° 50′	75°	75° 10′	ζ
°																	°
1	.06	.06	.06	.06	.06	.06	.06	.06	.06	.06	.06	.06	.07	.07	.07	.07	89
2	.12	.12	.12	.12	.12	.12	.12	.12	.13	.13	.13	.13	.13	.13	.13	.14	88
3	.18	.18	.18	.18	.18	.18	.19	.19	.19	.19	.19	.20	.20	.20	.20	.20	87
4	.23	.24	.24	.24	.24	.24	.25	.25	.25	.26	.26	.26	.26	.27	.27	.27	86
5	.29	.30	.30	.30	.30	.31	.31	.31	.32	.32	.32	.33	.33	.33	.34	.34	85
6	.35	.85	.36	.36	.36	.37	.37	.38	.38	.38	.39	.39	.40	.40	.40	.41	84
7	.41	.41	.42	.42	.42	.43	.43	.44	.44	.45	.45	.46	.46	.47	.47	.48	83
8	.47	.47	.48	.48	.48	.49	.50	.50	.50	.51	.52	.52	.53	.53	.54	.54	82
9	.52	.53	.53	.54	.54	.55	.56	.56	.57	.57	.58	.58	.59	.60	.60	.61	81
10	.58	.59	.59	.60	.60	.61	.62	.62	.63	.64	.64	.65	.66	.66	.67	.68	**80**
11	.64	.65	.65	.66	.66	.67	.68	.68	.69	.70	.71	.71	.72	.73	.74	.74	79
12	.70	.70	.71	.72	.72	.73	.74	.75	.75	.76	.77	.78	.79	.79	.80	.81	78
13	.76	.76	.77	.78	.78	.79	.80	.81	.82	.82	.83	.84	.85	.86	.87	.88	77
14	.81	.82	.83	.84	.84	.85	.86	.87	.88	.89	.90	.91	.92	.93	.94	.95	76
15	.87	.88	.89	.89	.90	.91	.92	.93	.94	.95	.96	.97	.98	.99	1.00	1.01	75
16	.92	.93	.94	.95	.96	.97	.98	.99	1.00	1.01	1.02	1.03	1.04	1.05	1.06	1.08	74
17	.98	.99	1.00	1.01	1.02	1.03	1.04	1.05	1.06	1.07	1.08	1.09	1.11	1.12	1.13	1.14	73
18	1.04	1.05	1.06	1.07	1.08	1.09	1.10	1.11	1.12	1.13	1.14	1.16	1.17	1.18	1.19	1.21	72
19	1.09	1.10	1.11	1.12	1.14	1.15	1.16	1.17	1.18	1.19	1.21	1.22	1.23	1.24	1.26	1.27	71
20	1.15	1.16	1.17	1.18	1.19	1.20	1.22	1.23	1.24	1.25	1.27	1.28	1.29	1.31	1.32	1.34	**70**
21	1.20	1.21	1.22	1.24	1.25	1.26	1.27	1.29	1.30	1.31	1.33	1.34	1.36	1.37	1.38	1.40	69
22	1.26	1.27	1.28	1.29	1.31	1.32	1.33	1.34	1.36	1.37	1.39	1.40	1.42	1.43	1.45	1.46	68
23	1.31	1.32	1.34	1.35	1.36	1.38	1.39	1.40	1.42	1.43	1.45	1.46	1.48	1.49	1.51	1.53	67
24	1.36	1.38	1.39	1.40	1.42	1.43	1.45	1.46	1.48	1.49	1.51	1.52	1.54	1.55	1.57	1.59	66
25	1.42	1.43	1.45	1.46	1.47	1.49	1.50	1.52	1.53	1.55	1.56	1.58	1.60	1.62	1.63	1.65	65
26	1.47	1.48	1.50	1.51	1.53	1.54	1.56	1.58	1.59	1.61	1.62	1.64	1.66	1.68	1.69	1.71	64
27	1.52	1.54	1.55	1.57	1.58	1.60	1.61	1.63	1.65	1.66	1.68	1.70	1.72	1.74	1.75	1.77	63
28	1.58	1.59	1.60	1.62	1.64	1.65	1.67	1.69	1.70	1.72	1.74	1.76	1.78	1.79	1.81	1.83	62
29	1.63	1.64	1.66	1.67	1.69	1.71	1.72	1.74	1.76	1.78	1.80	1.81	1.83	1.85	1.87	1.89	61
30	1.68	1.69	1.71	1.73	1.74	1.76	1.78	1.80	1.81	1.83	1.85	1.87	1.89	1.91	1.93	1.95	**60**
31	1.73	1.74	1.76	1.78	1.80	1.81	1.83	1.85	1.87	1.89	1.91	1.93	1.95	1.97	1.99	2.01	59
32	1.78	1.80	1.81	1.83	1.85	1.87	1.88	1.90	1.92	1.94	1.96	1.98	2.00	2.02	2.05	2.07	58
33	1.83	1.85	1.86	1.88	1.90	1.92	1.94	1.96	1.98	2.00	2.02	2.04	2.06	2.08	2.10	2.13	57
34	1.88	1.89	1.91	1.93	1.95	1.97	1.99	2.01	2.03	2.05	2.07	2.09	2.12	2.14	2.16	2.18	56
35	1.92	1.94	1.96	1.98	2.00	2.02	2.04	2.06	2.08	2.10	2.12	2.15	2.17	2.19	2.22	2.24	55
36	1.97	1.99	2.01	2.03	2.05	2.07	2.09	2.11	2.13	2.15	2.18	2.20	2.22	2.25	2.27	2.30	54
37	2.02	2.04	2.06	2.08	2.10	2.12	2.14	2.16	2.18	2.21	2.23	2.25	2.28	2.30	2.33	2.35	53
38	2.07	2.09	2.11	2.13	2.15	2.17	2.19	2.21	2.23	2.26	2.28	2.30	2.33	2.35	2.38	2.40	52
39	2.11	2.13	2.15	2.17	2.19	2.22	2.24	2.26	2.28	2.31	2.33	2.35	2.38	2.40	2.43	2.46	51
40	2.16	2.18	2.20	2.22	2.24	2.26	2.29	2.31	2.33	2.36	2.38	2.40	2.43	2.46	2.48	2.51	**50**
41	2.20	2.22	2.24	2.26	2.29	2.31	2.33	2.36	2.38	2.40	2.43	2.45	2.48	2.51	2.53	2.56	49
42	2.25	2.27	2.29	2.31	2.33	2.36	2.38	2.40	2.43	2.45	2.48	2.50	2.53	2.56	2.58	2.61	48
43	2.29	2.31	2.33	2.36	2.38	2.40	2.42	2.45	2.47	2.50	2.53	2.55	2.58	2.61	2.63	2.66	47
44	2.33	2.35	2.38	2.40	2.42	2.45	2.47	2.50	2.52	2.55	2.57	2.60	2.63	2.66	2.68	2.71	46
45	2.37	2.40	2.42	2.44	2.46	2.49	2.51	2.54	2.56	2.59	2.62	2.65	2.67	2.70	2.73	2.76	45
	72° 40′	72° 50′	73°	73° 10′	73° 20′	73° 30′	73° 40′	73° 50′	74°	74° 10′	74° 20′	74° 30′	74° 40′	74° 50′	75°	75° 10′	

Table of factors for reduction of transit observations.

TOP ARGUMENT = STAR'S DECLINATION (δ).

SIDE ARGUMENT = STAR'S ZENITH DISTANCE (ζ).

[For factor *A* use left-hand argument. For factor *B* use right-hand argument. For factor *C* use bottom line on *this* page.]

ζ	72° 40′	72° 50′	73°	73° 10′	73° 20′	73° 30′	73° 40′	73° 50′	74°	74° 10′	74° 20′	74° 30′	74° 40′	74° 50′	75°	75° 10′	ζ
°																	°
46	2.41	2.44	2.46	2.48	2.51	2.53	2.56	2.58	2.61	2.64	2.66	2.69	2.72	2.75	2.78	2.81	44
47	2.45	2.48	2.50	2.52	2.55	2.57	2.60	2.63	2.65	2.68	2.71	2.74	2.77	2.80	2.83	2.86	43
48	2.49	2.52	2.54	2.57	2.59	2.62	2.64	2.67	2.70	2.72	2.75	2.78	2.81	2.84	2.87	2.90	42
49	2.53	2.56	2.58	2.61	2.63	2.66	2.68	2.71	2.74	2.77	2.80	2.82	2.85	2.88	2.92	2.95	41
50	2.57	2.60	2.62	2.64	2.67	2.70	2.72	2.75	2.78	2.81	2.84	2.87	2.90	2.93	2.96	2.99	40
51	2.61	2.63	2.66	2.68	2.71	2.74	2.76	2.79	2.82	2.85	2.88	2.91	2.94	2.97	3.00	3.04	39
52	2.64	2.67	2.69	2.72	2.75	2.77	2.80	2.83	2.86	2.89	2.92	2.95	2.98	3.01	3.04	3.08	38
53	2.68	2.71	2.73	2.76	2.78	2.81	2.84	2.87	2.90	2.93	2.96	2.99	3.02	3.05	3.09	3.12	37
54	2.72	2.74	2.77	2.79	2.82	2.85	2.88	2.91	2.94	2.97	3.00	3.03	3.06	3.09	3.13	3.16	36
55	2.75	2.78	2.80	2.83	2.86	2.88	2.91	2.94	2.97	3.00	3.03	3.07	3.10	3.13	3.16	3.20	35
56	2.78	2.81	2.84	2.86	2.89	2.92	2.95	2.98	3.01	3.04	3.07	3.10	3.14	3.17	3.20	3.24	34
57	2.82	2.84	2.87	2.90	2.92	2.95	2.98	3.01	3.04	3.07	3.11	3.14	3.17	3.21	3.24	3.28	33
58	2.85	2.87	2.90	2.93	2.96	2.99	3.02	3.05	3.08	3.11	3.14	3.17	3.21	3.24	3.28	3.31	32
59	2.88	2.90	2.93	2.96	2.99	3.02	3.05	3.08	3.11	3.14	3.17	3.21	3.24	3.28	3.31	3.35	31
60	2.91	2.93	2.96	2.99	3.02	3.05	3.08	3.11	3.14	3.17	3.21	3.24	3.28	3.31	3.35	3.38	30
61	2.94	2.96	2.99	3.02	3.05	3.08	3.11	3.14	3.17	3.21	3.24	3.27	3.31	3.34	3.38	3.42	29
62	2.96	2.99	3.02	3.05	3.08	3.11	3.14	3.17	3.20	3.24	3.27	3.30	3.34	3.38	3.41	3.45	28
63	2.99	3.02	3.05	3.08	3.11	3.14	3.17	3.20	3.23	3.27	3.30	3.33	3.37	3.41	3.44	3.48	27
64	3.02	3.04	3.07	3.10	3.13	3.16	3.20	3.23	3.26	3.29	3.33	3.36	3.40	3.44	3.47	3.51	26
65	3.04	3.07	3.10	3.13	3.16	3.19	3.22	3.26	3.29	3.32	3.36	3.39	3.43	3.46	3.50	3.54	25
66	3.07	3.10	3.13	3.16	3.18	3.22	3.25	3.28	3.31	3.35	3.38	3.42	3.46	3.49	3.53	3.57	24
67	3.09	3.12	3.15	3.18	3.21	3.24	3.27	3.31	3.34	3.37	3.41	3.44	3.48	3.52	3.56	3.60	23
68	3.11	3.14	3.17	3.20	3.23	3.26	3.30	3.33	3.36	3.40	3.43	3.47	3.51	3.54	3.58	3.62	22
69	3.13	3.16	3.19	3.22	3.26	3.29	3.32	3.35	3.39	3.42	3.46	3.49	3.53	3.57	3.61	3.65	21
70	3.15	3.18	3.21	3.24	3.28	3.31	3.34	3.38	3.41	3.44	3.48	3.52	3.55	3.59	3.63	3.67	20
71	3.17	3.20	3.23	3.26	3.30	3.33	3.36	3.40	3.43	3.47	3.50	3.54	3.58	3.61	3.65	3.69	19
72	3.19	3.22	3.25	3.28	3.32	3.35	3.38	3.42	3.45	3.49	3.52	3.56	3.60	3.63	3.67	3.72	18
73	3.21	3.24	3.27	3.30	3.33	3.37	3.40	3.44	3.47	3.50	3.54	3.58	3.62	3.65	3.69	3.74	17
74	3.23	3.26	3.29	3.32	3.35	3.38	3.42	3.45	3.49	3.52	3.56	3.60	3.64	3.67	3.71	3.76	16
75	3.24	3.27	3.30	3.34	3.37	3.40	3.44	3.47	3.50	3.54	3.58	3.61	3.65	3.69	3.73	3.77	15
76	3.26	3.29	3.32	3.35	3.38	3.42	3.45	3.48	3.52	3.56	3.59	3.63	3.67	3.71	3.75	3.79	14
77	3.27	3.30	3.33	3.36	3.40	3.43	3.46	3.50	3.54	3.57	3.61	3.65	3.68	3.72	3.76	3.81	13
78	3.28	3.31	3.34	3.38	3.41	3.44	3.48	3.51	3.55	3.58	3.62	3.66	3.70	3.74	3.78	3.82	12
79	3.29	3.33	3.36	3.39	3.42	3.46	3.49	3.53	3.56	3.60	3.64	3.67	3.71	3.75	3.79	3.83	11
80	3.31	3.34	3.37	3.40	3.43	3.47	3.50	3.54	3.57	3.61	3.65	3.68	3.72	3.76	3.81	3.85	10
81	3.32	3.35	3.38	3.41	3.44	3.48	3.51	3.55	3.58	3.62	3.66	3.70	3.74	3.78	3.82	3.86	9
82	3.32	3.36	3.39	3.42	3.45	3.49	3.52	3.56	3.59	3.63	3.67	3.71	3.75	3.79	3.83	3.87	8
83	3.33	3.36	3.40	3.43	3.46	3.49	3.53	3.56	3.60	3.64	3.68	3.72	3.75	3.79	3.84	3.88	7
84	3.34	3.37	3.40	3.43	3.47	3.50	3.54	3.57	3.61	3.64	3.68	3.72	3.76	3.80	3.84	3.88	6
85	3.34	3.38	3.41	3.44	3.47	3.51	3.54	3.58	3.61	3.65	3.69	3.73	3.77	3.81	3.85	3.89	5
86	3.35	3.38	3.41	3.44	3.48	3.51	3.55	3.58	3.62	3.66	3.69	3.78	3.77	3.81	3.85	3.90	4
87	3.35	3.38	3.42	3.45	3.48	3.52	3.55	3.59	3.62	3.66	3.70	3.74	3.78	3.82	3.86	3.90	3
88	3.35	3.39	3.42	3.45	3.48	3.52	3.55	3.59	3.62	3.66	3.70	3.74	3.78	3.82	3.86	3.90	2
89	3.36	3.39	3.42	3.45	3.49	3.52	3.56	3.59	3.63	3.66	3.70	3.74	3.78	3.82	3.86	3.91	1
90	3.36	3.39	3.42	3.45	3.49	3.52	3.56	3.59	3.63	3.66	3.70	3.74	3.78	3.82	3.86	3.91	0
	72° 40′	72° 50′	73°	73° 10′	73° 20′	73° 30′	73° 40′	73° 50′	74°	74° 10′	74° 20′	74° 30′	74° 40′	74° 50′	75°	75° 10′	

Table of factors for reduction of transit observations.

TOP ARGUMENT=STAR'S DECLINATION (δ).

SIDE ARGUMENT=STAR'S ZENITH DISTANCE (ζ).

[For factor *A* use left-hand argument. For factor *B* use right-hand argument. For factor *C* use bottom line on *opposite* page.]

ζ	75° 10′	75° 20′	75° 30′	75° 40′	75° 50′	76°	76° 10′	76° 20′	76° 30′	76° 40′	76° 50′	77°	77° 10′	77° 20′	77° 30′	77° 40′	ζ
°																	°
1	.07	.07	.07	.07	.07	.07	.07	.07	.07	.08	.08	.08	.08	.08	.08	.08	89
2	.14	.14	.14	.14	.14	.14	.15	.15	.15	.15	.15	.16	.16	.16	.16	.16	88
3	.20	.21	.21	.21	.21	.22	.22	.22	.22	.23	.23	.23	.24	.24	.24	.24	87
4	.27	.28	.28	.28	.28	.29	.29	.30	.30	.30	.31	.31	.31	.32	.32	.33	86
5	.34	.34	.35	.35	.36	.36	.36	.37	.37	.38	.38	.39	.39	.40	.40	.41	85
6	.41	.41	.42	.42	.43	.43	.44	.44	.45	.45	.46	.46	.47	.48	.48	.49	84
7	.48	.48	.49	.49	.50	.50	.51	.52	.52	.53	.54	.54	.55	.56	.56	.57	83
8	.54	.55	.56	.56	.57	.58	.58	.59	.60	.60	.61	.62	.63	.64	.64	.65	82
9	.61	.62	.62	.63	.64	.65	.65	.66	.67	.68	.69	.70	.70	.71	.72	.73	81
10	.68	.69	.69	.70	.71	.72	.73	.74	.74	.75	.76	.77	.78	.79	.80	.81	80
11	.74	.75	.76	.77	.78	.79	.80	.81	.82	.83	.84	.85	.86	.87	.88	.89	79
12	.81	.82	.83	.84	.85	.86	.87	.88	.89	.90	.91	.92	.94	.95	.96	.97	78
13	.88	.89	.90	.91	.92	.93	.94	.95	.96	.98	.99	1.00	1.01	1.03	1.04	1.05	77
14	.95	.96	.97	.98	.99	1.00	1.01	1.02	1.04	1.05	1.06	1.08	1.09	1.10	1.12	1.13	76
15	1.01	1.02	1.03	1.04	1.06	1.07	1.08	1.10	1.11	1.12	1.14	1.15	1.16	1.18	1.20	1.21	75
16	1.08	1.09	1.10	1.11	1.13	1.14	1.15	1.17	1.18	1.20	1.21	1.23	1.24	1.26	1.28	1.29	74
17	1.14	1.16	1.17	1.18	1.20	1.21	1.22	1.24	1.25	1.27	1.28	1.30	1.32	1.33	1.35	1.37	73
18	1.21	1.22	1.23	1.25	1.26	1.28	1.29	1.31	1.32	1.34	1.36	1.37	1.39	1.41	1.43	1.45	72
19	1.27	1.29	1.30	1.32	1.33	1.35	1.36	1.38	1.39	1.41	1.43	1.45	1.47	1.48	1.50	1.52	71
20	1.34	1.35	1.37	1.38	1.40	1.41	1.43	1.45	1.47	1.48	1.50	1.52	1.54	1.56	1.58	1.60	70
21	1.40	1.42	1.43	1.45	1.46	1.48	1.50	1.52	1.54	1.55	1.57	1.59	1.61	1.63	1.65	1.68	69
22	1.46	1.48	1.50	1.51	1.53	1.55	1.57	1.58	1.60	1.62	1.64	1.66	1.69	1.71	1.73	1.75	68
23	1.53	1.54	1.56	1.58	1.60	1.62	1.63	1.65	1.67	1.69	1.72	1.74	1.76	1.78	1.81	1.83	67
24	1.59	1.61	1.63	1.64	1.66	1.68	1.70	1.72	1.74	1.76	1.79	1.81	1.83	1.86	1.88	1.90	66
25	1.65	1.67	1.69	1.71	1.73	1.75	1.77	1.79	1.81	1.83	1.86	1.88	1.90	1.93	1.95	1.98	65
26	1.71	1.73	1.75	1.77	1.79	1.81	1.83	1.86	1.88	1.90	1.92	1.95	1.97	2.00	2.02	2.05	64
27	1.77	1.79	1.81	1.83	1.86	1.88	1.90	1.92	1.95	1.97	1.99	2.02	2.04	2.07	2.10	2.12	63
28	1.83	1.85	1.87	1.90	1.92	1.94	1.96	1.99	2.01	2.04	2.06	2.09	2.11	2.14	2.17	2.20	62
29	1.89	1.92	1.94	1.96	1.98	2.00	2.03	2.05	2.08	2.10	2.13	2.15	2.18	2.21	2.24	2.27	61
30	1.95	1.98	2.00	2.02	2.04	2.07	2.09	2.12	2.14	2.17	2.20	2.22	2.25	2.28	2.31	2.34	60
31	2.01	2.03	2.06	2.08	2.10	2.13	2.15	2.18	2.21	2.23	2.26	2.29	2.32	2.35	2.38	2.41	59
32	2.07	2.09	2.12	2.14	2.16	2.19	2.22	2.24	2.27	2.30	2.33	2.36	2.39	2.42	2.45	2.48	58
33	2.13	2.15	2.18	2.20	2.22	2.25	2.28	2.30	2.33	2.36	2.39	2.42	2.45	2.48	2.52	2.55	57
34	2.18	2.21	2.23	2.26	2.28	2.31	2.34	2.37	2.40	2.42	2.46	2.49	2.52	2.55	2.58	2.62	56
35	2.24	2.26	2.29	2.32	2.34	2.37	2.40	2.43	2.46	2.49	2.52	2.55	2.58	2.62	2.65	2.68	55
36	2.30	2.32	2.35	2.37	2.40	2.43	2.46	2.49	2.52	2.55	2.58	2.61	2.65	2.68	2.72	2.75	54
37	2.35	2.38	2.40	2.43	2.46	2.49	2.52	2.55	2.58	2.61	2.64	2.67	2.71	2.74	2.78	2.82	53
38	2.40	2.43	2.46	2.49	2.52	2.55	2.58	2.61	2.64	2.67	2.70	2.74	2.77	2.81	2.85	2.88	52
39	2.46	2.49	2.51	2.54	2.57	2.60	2.63	2.66	2.70	2.73	2.76	2.80	2.83	2.87	2.91	2.95	51
40	2.51	2.54	2.57	2.60	2.63	2.66	2.69	2.72	2.75	2.79	2.82	2.86	2.89	2.93	2.97	3.01	50
41	2.56	2.59	2.62	2.65	2.68	2.71	2.74	2.78	2.81	2.84	2.88	2.92	2.95	2.99	3.03	3.07	49
42	2.61	2.64	2.67	2.70	2.73	2.77	2.80	2.83	2.87	2.90	2.94	2.97	3.01	3.05	3.09	3.13	48
43	2.66	2.69	2.72	2.76	2.79	2.82	2.85	2.89	2.92	2.96	2.99	3.03	3.07	3.11	3.15	3.19	47
44	2.71	2.74	2.77	2.81	2.84	2.87	2.90	2.94	2.98	3.01	3.05	3.09	3.13	3.17	3.21	3.25	46
45	2.76	2.79	2.82	2.86	2.89	2.92	2.96	2.99	3.03	3.07	3.10	3.14	3.18	3.22	3.27	3.31	45
	75° 10′	75° 20′	75° 30′	75° 40′	75° 50′	76°	76° 10′	76° 20′	76° 30′	76° 40′	76° 50′	77°	77° 10′	77° 20′	77° 30′	77° 40′	

Table of factors for reduction of transit observations.

TOP ARGUMENT=STAR'S DECLINATION (δ).

SIDE ARGUMENT=STAR'S ZENITH DISTANCE (ζ).

[For factor *A* use left-hand argument. For factor *B* use right-hand argument. For factor *C* use bottom line on *this* page.]

ζ	75° 10′	75° 20′	75° 30′	75° 40′	75° 50′	76°	76° 10′	76° 20′	76° 30′	76° 40′	76° 50′	77°	77° 10′	77° 20′	77° 30′	77° 40′	ζ
°																	°
46	2.81	2.84	2.87	2.91	2.94	2.97	3.01	3.04	3.08	3.12	3.16	3.20	3.24	3.28	3.32	3.37	44
47	2.86	2.89	2.92	2.95	2.99	3.02	3.06	3.10	3.13	3.17	3.21	3.25	3.29	3.34	3.38	3.42	43
48	2.90	2.94	2.97	3.00	3.04	3.07	3.11	3.15	3.18	3.22	3.26	3.30	3.35	3.39	3.43	3.48	42
49	2.95	2.98	3.01	3.05	3.08	3.12	3.16	3.19	3.23	3.27	3.31	3.36	3.40	3.44	3.49	3.53	41
50	2.99	3.02	3.06	3.09	3.13	3.17	3.20	3.24	3.28	3.32	3.36	3.41	3.45	3.49	3.54	3.59	40
51	3.04	3.07	3.10	3.14	3.18	3.21	3.25	3.29	3.33	3.37	3.41	3.45	3.50	3.54	3.59	3.64	39
52	3.08	3.11	3.15	3.18	3.22	3.26	3.30	3.34	3.38	3.42	3.46	3.50	3.55	3.59	3.64	3.69	38
53	3.12	3.15	3.19	2.23	3.26	3.30	3.34	3.38	3.42	3.46	3.51	3.55	3.60	3.64	3.69	3.74	37
54	3.16	3.20	3.23	3.27	3.31	3.34	3.38	3.42	3.47	3.51	3.55	3.60	3.64	3.69	3.74	3.79	36
55	3.20	3.24	3.27	3.31	3.35	3.39	3.43	3.47	3.51	3.55	3.60	3.64	3.69	3.74	3.78	3.83	35
56	3.24	3.27	3.31	3.35	3.39	3.43	3.47	3.51	3.55	3.60	3.64	3.68	3.73	2.78	3.83	3.88	34
57	3.28	3.31	3.35	3.39	3.43	3.47	3.51	3.55	3.59	3.64	3.68	3.73	3.78	3.83	3.88	3.93	33
58	3.31	3.35	3.39	3.43	3.47	3.51	3.55	3.59	3.63	3.68	3.72	3.77	3.82	3.87	3.92	3.97	32
59	3.35	3.38	3.42	3.46	3.50	3.54	3.58	3.63	3.67	3.72	3.76	3.81	3.86	3.91	3.96	4.01	31
60	3.38	3.42	3.46	3.50	3.54	3.58	3.62	3.66	3.71	3.76	3.80	3.85	3.90	3.95	4.00	4.05	30
61	3.42	3.45	3.49	3.53	3.57	3.62	3.66	3.70	3.75	3.79	3.84	3.89	3.94	3.99	4.04	4.09	29
62	3.45	3.49	3.53	3.57	3.61	3.65	3.69	3.74	3.78	3.83	3.88	3.93	3.98	4.03	4.08	4.13	28
63	3.48	3.52	3.56	3.60	3.64	3.68	3.73	3.77	3.82	3.86	3.91	3.96	4.01	4.06	4.12	4.17	27
64	3.51	3.55	3.59	3.63	3.67	3.72	3.76	3.80	3.85	3.90	3.95	4.00	4.05	4.10	4.15	4.21	26
65	3.54	3.58	3.62	3.66	3.70	3.75	3.79	3.84	3.88	3.93	3.98	4.03	4.08	4.13	4.19	4.24	25
66	3.57	3.61	3.65	3.69	3.73	3.78	3.82	3.87	3.91	3.96	4.01	4.06	4.11	4.17	4.22	4.28	24
67	3.60	3.64	3.68	3.72	3.76	3.81	3.85	3.90	3.94	3.99	4.04	4.09	4.14	4.20	4.25	4.31	23
68	3.62	3.66	3.70	3.74	3.79	3.83	3.88	3.92	3.97	4.02	4.07	4.12	4.17	4.23	4.28	4.34	22
69	3.65	3.69	3.73	3.77	3.82	3.86	3.90	3.95	4.00	4.05	4.10	4.15	4.20	4.26	4.31	4.37	21
70	3.67	3.71	3.75	3.80	3.84	3.89	3.93	3.98	4.03	4.08	4.12	4.18	4.23	4.28	4.34	4.40	20
71	3.69	3.73	3.78	3.82	3.86	3.91	3.96	4.00	4.05	4.10	4.15	4.20	4.26	4.31	4.37	4.43	19
72	3.72	3.76	3.80	3.84	3.89	3.93	3.98	4.02	4.07	4.12	4.18	4.23	4.28	4.34	4.39	4.45	18
73	3.74	3.78	3.82	3.86	3.91	3.95	4.00	4.05	4.10	4.15	4.20	4.25	4.30	4.36	4.42	4.48	17
74	3.76	3.80	3.84	3.88	3.93	3.97	4.02	4.07	4.12	4.17	4.22	4.27	4.33	4.38	4.44	4.50	16
75	3.77	3.82	3.86	3.90	3.95	3.99	4.04	4.09	4.14	4.19	4.24	4.29	4.35	4.40	4.46	4.52	15
76	3.79	3.83	3.88	3.92	3.96	4.01	4.06	4.11	4.16	4.21	4.26	4.31	4.37	4.42	4.48	4.54	14
77	3.81	3.85	3.89	3.94	3.98	4.03	4.08	4.12	4.17	4.22	4.28	4.33	4.39	4.44	4.50	4.56	13
78	3.82	3.86	3.91	3.95	4.00	4.04	4.09	4.14	4.19	4.24	4.29	4.35	4.40	4.46	4.52	4.58	12
79	3.83	3.88	3.92	3.96	4.01	4.06	4.11	4.16	4.21	4.26	4.31	4.36	4.42	4.48	4.54	4.60	11
80	3.85	3.89	3.93	3.98	4.02	4.07	4.12	4.17	4.22	4.27	4.32	4.38	4.43	4.49	4.55	4.61	10
81	3.86	3.90	3.94	3.99	4.04	4.08	4.13	4.18	4.23	4.28	4.34	4.39	4.45	4.50	4.56	4.62	9
82	3.87	3.91	3.96	4.00	4.05	4.09	4.14	4.19	4.24	4.29	4.35	4.40	4.46	4.52	4.58	4.64	8
83	3.88	3.92	3.96	4.01	4.06	4.10	4.15	4.20	4.25	4.30	4.36	4.41	4.47	4.53	4.59	4.65	7
84	3.88	3.93	3.97	4.02	4.06	4.11	4.16	4.21	4.26	4.31	4.37	4.42	4.48	4.54	4.60	4.66	6
85	3.89	3.93	3.98	4.02	4.07	4.12	4.17	4.22	4.27	4.32	4.37	4.43	4.48	4.54	4.60	4.66	5
86	3.90	3.94	3.98	4.03	4.08	4.12	4.17	4.22	4.27	4.33	4.38	4.43	4.49	4.55	4.61	4.67	4
87	3.90	3.94	3.99	4.03	4.08	4.13	4.18	4.23	4.28	4.33	4.38	4.44	4.50	4.55	4.61	4.68	3
88	3.90	3.95	3.99	4.04	4.08	4.13	4.18	4.23	4.28	4.33	4.39	4.44	4.50	4.56	4.62	4.68	2
89	3.91	3.95	3.99	4.04	4.08	4.13	4.18	4.23	4.28	4.34	4.39	4.44	4.50	4.56	4.62	4.68	1
90	3.91	3.95	3.99	4.04	4.09	4.13	4.18	4.23	4.28	4.34	4.39	4.44	4.50	4.56	4.62	4.68	0
	75° 10′	75° 20′	75° 30′	75° 40′	75° 50′	76°	76° 10′	76° 20′	76° 30′	76° 40′	76° 50′	77°	77° 10′	77° 20′	77° 50′	77° 40′	

Table of factors for reduction of transit observations.

TOP ARGUMENT=STAR'S DECLINATION (δ).

SIDE ARGUMENT=STAR'S ZENITH DISTANCE (ζ).

[For factor *A* use left-hand argument. For factor *B* use right-hand argument. For factor *C* use bottom line on *opposite* page.]

ζ	77° 40′	77° 50′	78°	78° 10′	78° 20′	78° 30′	78° 40′	78° 50′	79°	79° 10′	79° 20′	79° 30′	79° 40′	79° 50′	80°	ζ
°																°
1	.08	.08	.08	.08	.09	.09	.09	.09	.09	.09	.09	.10	.10	.10	.10	89
2	.16	.17	.17	.17	.17	.18	.18	.18	.18	.19	.19	.19	.20	.20	.20	88
3	.24	.25	.25	.26	.26	.26	.27	.27	.27	.28	.28	.29	.29	.30	.30	87
4	.33	.33	.34	.34	.34	.35	.36	.36	.37	.37	.38	.38	.39	.40	.40	86
5	.41	.41	.42	.42	.43	.44	.44	.45	.46	.46	.47	.48	.49	.49	.50	85
6	.49	.50	.51	.51	.52	.52	.53	.54	.55	.56	.56	.57	.58	.59	.60	84
7	.57	.58	.59	.59	.60	.61	.62	.63	.64	.65	.66	.67	.68	.69	.70	83
8	.65	.66	.67	.68	.69	.70	.71	.72	.73	.74	.75	.76	.78	.79	.80	82
9	.73	.74	.75	.76	.77	.78	.80	.81	.82	.83	.84	.86	.87	.89	.90	81
10	.81	.82	.84	.85	.86	.87	.88	.90	.91	.92	.94	.95	.97	.98	1.00	80
11	.89	.90	.92	.93	.94	.96	.97	.98	1.00	1.02	1.03	1.05	1.06	1.08	1.10	79
12	.97	.99	1.00	1.01	1.03	1.04	1.06	1.07	1.09	1.11	1.12	1.14	1.16	1.18	1.20	78
13	1.05	1.07	1.08	1.10	1.11	1.13	1.14	1.16	1.18	1.20	1.22	1.23	1.25	1.27	1.30	77
14	1.13	1.15	1.16	1.18	1.20	1.21	1.23	1.25	1.27	1.29	1.31	1.33	1.35	1.37	1.39	76
15	[illegible]	[illegible]	[illegible]	1.26	1.28	1.30	1.32	1.34	1.36	1.38	1.40	1.42	1.44	1.47	1.49	75
16	1.29	1.31	1.33	1.34	1.36	1.38	1.40	1.42	1.44	1.47	1.49	1.51	1.54	1.56	1.59	74
17	1.37	1.39	1.41	1.43	1.45	1.47	1.49	1.51	1.53	1.56	1.58	1.60	1.63	1.66	1.68	73
18	1.45	1.47	1.49	1.51	1.53	1.55	1.57	1.60	1.62	1.64	1.67	1.70	1.72	1.75	1.78	72
19	1.52	1.54	1.57	1.59	1.61	1.63	1.66	1.68	1.71	1.73	1.76	1.79	1.82	1.84	1.87	71
20	1.60	1.62	1.65	1.67	1.69	1.72	1.74	1.77	1.79	1.82	1.85	1.88	1.91	1.94	1.97	70
21	1.68	1.70	1.72	1.75	1.77	1.80	1.82	1.85	1.88	1.91	1.94	1.97	2.00	2.03	2.06	69
22	1.75	1.78	1.80	1.83	1.85	1.88	1.91	1.93	1.96	1.99	2.02	2.06	2.09	2.12	2.16	68
23	1.83	1.85	1.88	1.90	1.93	1.96	1.99	2.02	2.05	2.08	2.11	2.14	2.18	2.21	2.25	67
24	1.90	1.93	1.96	1.98	2.01	2.04	2.07	2.10	2.13	2.16	2.20	2.23	2.27	2.30	2.34	66
25	1.98	2.00	2.03	2.06	2.09	2.12	2.15	2.18	2.22	2.25	2.28	2.32	2.36	2.39	2.43	65
26	2.05	2.08	2.11	2.14	2.17	2.20	2.23	2.26	2.30	2.33	2.37	2.41	2.44	2.48	2.52	64
27	2.12	2.15	2.18	2.21	2.24	2.28	2.31	2.34	2.38	2.42	2.45	2.49	2.53	2.57	2.61	63
28	2.20	2.23	2.26	2.29	2.32	2.36	2.39	2.42	2.46	2.50	2.54	2.58	2.62	2.66	2.70	62
29	2.27	2.30	2.33	2.36	2.40	2.43	2.47	2.50	2.54	2.58	2.62	2.66	2.70	2.75	2.79	61
30	2.34	2.37	2.40	2.44	2.47	2.51	2.54	2.58	2.62	2.66	2.70	2.74	2.79	2.83	2.88	60
31	2.41	2.44	2.48	2.51	2.55	2.58	2.62	2.66	2.70	2.74	2.78	2.83	2.87	2.92	2.97	59
32	2.48	2.51	2.55	2.58	2.62	2.66	2.70	2.74	2.78	2.82	2.86	2.91	2.95	3.00	3.05	58
33	2.55	2.58	2.62	2.66	2.69	2.73	2.77	2.81	2.85	2.90	2.94	2.99	3.04	3.09	3.14	57
34	2.62	2.65	2.69	2.73	2.76	2.80	2.84	2.89	2.93	2.98	3.02	3.07	3.12	3.17	3.22	56
35	2.68	2.72	2.76	2.80	2.84	2.88	2.92	2.96	3.01	3.05	3.10	3.15	3.20	3.25	3.30	55
36	2.75	2.79	2.83	2.87	2.91	2.95	2.99	3.04	3.08	3.13	3.18	3.23	3.28	3.33	3.38	54
37	2.82	2.86	2.90	2.94	2.98	3.02	3.06	3.11	3.15	3.20	3.25	3.30	3.36	3.41	3.47	53
38	2.88	2.92	2.96	3.00	3.04	3.09	3.13	3.18	3.23	3.28	3.33	3.38	3.43	3.49	3.55	52
39	2.95	2.99	3.03	3.07	3.11	3.16	3.20	3.25	3.30	3.35	3.40	3.45	3.51	3.56	3.62	51
40	3.01	3.05	3.09	3.14	3.18	3.22	3.27	3.32	3.37	3.42	3.47	3.53	3.58	3.64	3.70	50
41	3.07	3.11	3.16	3.20	3.24	3.29	3.34	3.39	3.44	3.49	3.54	3.60	3.66	3.72	3.78	49
42	3.13	3.18	3.22	3.26	3.31	3.36	3.41	3.46	3.51	3.56	3.61	3.67	3.73	3.79	3.85	48
43	3.19	3.24	3.28	3.33	3.37	3.42	3.47	3.52	3.57	3.63	3.68	3.74	3.80	3.86	3.93	47
44	3.25	3.30	3.34	3.39	3.43	3.48	3.54	3.59	3.64	3.70	3.75	3.81	3.87	3.94	4.00	46
45	3.31	3.36	3.40	3.45	3.50	3.55	3.60	3.65	3.71	3.76	3.82	3.88	3.94	4.01	4.07	45
	77° 40′	77° 50′	78°	78° 10′	78° 20′	78° 30′	78° 40′	78° 50′	79°	79° 10′	79° 20′	79° 30′	79° 40′	79° 50′	80°	

Table of factors for reduction of transit observations.

TOP ARGUMENT=STAR'S DECLINATION (δ).

SIDE ARGUMENT=STAR'S ZENITH DISTANCE (ζ).

[For factor *A* use left-hand argument. For factor *B* use right-hand argument. For factor *C* use bottom line on *this* page.]

ζ	77° 40′	77° 50′	78°	78° 10′	78° 20′	78° 30′	78° 40′	78° 50′	79°	79° 10′	79° 20′	79° 30′	79° 40′	79° 50′	80°	ζ
°																°
46	3.37	3.41	3.46	3.51	3.56	3.61	3.66	3.71	3.77	3.83	3.89	3.95	4.01	4.08	4.14	44
47	3.42	3.47	3.52	3.57	3.62	3.67	3.72	3.78	3.83	3.89	3.95	4.01	4.08	4.14	4.21	43
48	3.48	3.53	3.57	3.62	3.68	3.73	3.78	3.84	3.89	3.95	4.02	4.08	4.14	4.21	4.28	42
49	3.53	3.58	3.63	3.68	3.73	3.79	3.84	3.90	3.96	4.02	4.08	4.14	4.21	4.28	4.35	41
50	3.59	3.63	3.68	3.74	3.79	3.84	3.90	3.96	4.02	4.08	4.14	4.20	4.27	4.34	4.41	40
51	3.64	3.69	3.74	3.79	3.84	3.90	3.96	4.01	4.07	4.14	4.20	4.26	4.33	4.40	4.48	39
52	3.69	3.74	3.79	3.84	3.90	3.95	4.01	4.07	4.13	4.19	4.26	4.32	4.39	4.46	4.54	38
53	3.74	3.79	3.84	3.89	3.95	4.01	4.06	4.12	4.19	4.25	4.32	4.38	4.45	4.52	4.60	37
54	3.79	3.84	3.89	3.94	4.00	4.06	4.12	4.18	4.24	4.30	4.37	4.44	4.51	4.58	4.66	36
55	3.83	3.89	3.94	3.99	4.05	4.11	4.17	4.23	4.29	4.36	4.43	4.50	4.57	4.64	4.72	35
56	3.88	3.93	3.99	4.04	4.10	4.16	4.22	4.28	4.34	4.41	4.48	4.55	4.62	4.70	4.77	34
57	3.93	3.98	4.04	4.09	4.15	4.21	4.27	4.33	4.39	4.46	4.53	4.60	4.68	4.75	4.83	33
58	3.97	4.02	4.08	4.14	4.19	4.25	4.32	4.38	4.44	4.51	4.58	4.65	4.73	4.80	4.88	32
59	4.01	4.07	4.12	4.18	4.24	4.30	4.36	4.43	4.49	4.56	4.63	4.70	4.78	4.86	4.94	31
60	4.05	4.11	4.17	4.22	4.28	4.34	4.41	4.47	4.54	4.61	4.68	4.75	4.83	4.91	4.99	30
61	4.09	4.15	4.21	4.26	4.32	4.39	4.45	4.52	4.58	4.65	4.72	4.80	4.88	4.96	5.04	29
62	4.13	4.19	4.25	4.31	4.37	4.43	4.49	4.56	4.63	4.70	4.77	4.85	4.92	5.00	5.08	28
63	4.17	4.23	4.29	4.35	4.41	4.47	4.53	4.60	4.67	4.74	4.81	4.89	4.97	5.05	5.13	27
64	4.21	4.26	4.32	4.38	4.44	4.51	4.57	4.64	4.71	4.78	4.86	4.93	5.01	5.09	5.18	26
65	4.24	4.30	4.36	4.42	4.48	4.55	4.61	4.68	4.75	4.82	4.90	4.97	5.05	5.14	5.22	25
66	4.28	4.34	4.40	4.46	4.52	4.58	4.65	4.72	4.79	4.86	4.94	5.01	5.09	5.18	5.26	24
67	4.31	4.37	4.43	4.49	4.55	4.62	4.68	4.75	4.82	4.90	4.97	5.05	5.13	5.22	5.30	23
68	4.34	4.40	4.46	4.52	4.58	4.65	4.72	4.79	4.86	4.93	5.01	5.09	5.17	5.25	5.34	22
69	4.37	4.43	4.49	4.55	4.62	4.68	4.75	4.82	4.89	4.97	5.04	5.12	5.20	5.29	5.38	21
70	4.40	4.46	4.52	4.58	4.65	4.71	4.78	4.85	4.93	5.00	5.08	5.16	5.24	5.32	5.41	20
71	4.43	4.49	4.55	4.61	4.68	4.74	4.81	4.88	4.96	5.03	5.11	5.19	5.27	5.36	5.45	19
72	4.45	4.51	4.57	4.64	4.70	4.77	4.84	4.91	4.98	5.06	5.14	5.22	5.30	5.39	5.48	18
73	4.48	4.54	4.60	4.66	4.73	4.80	4.87	4.94	5.01	5.09	5.17	5.25	5.33	5.42	5.51	17
74	4.50	4.56	4.62	4.69	4.75	4.82	4.89	4.96	5.04	5.11	5.19	5.27	5.36	5.45	5.53	16
75	4.52	4.58	4.65	4.71	4.78	4.85	4.92	4.99	5.06	5.14	5.22	5.30	5.38	5.47	5.56	15
76	4.54	4.60	4.67	4.73	4.80	4.87	4.94	5.01	5.09	5.16	5.24	5.32	5.41	5.50	5.59	14
77	4.56	4.62	4.68	4.75	4.82	4.89	4.96	5.03	5.11	5.18	5.26	5.35	5.43	5.52	5.61	13
78	4.58	4.64	4.70	4.77	4.84	4.91	4.98	5.05	5.13	5.20	5.28	5.37	5.45	5.54	5.63	12
79	4.60	4.06	4.72	4.79	4.85	4.92	5.00	5.07	5.14	5.22	5.30	5.39	5.47	5.56	5.65	11
80	4.61	4.67	4.74	4.80	4.87	4.94	5.01	5.08	5.16	5.24	5.32	5.40	5.49	5.58	5.67	10
81	4.62	4.69	4.75	4.82	4.88	4.95	5.03	5.10	5.18	5.26	5.34	5.42	5.51	5.60	5.69	9
82	4.64	4.70	4.76	4.83	4.90	4.97	5.04	5.11	5.19	5.27	5.35	5.43	5.52	5.61	5.70	8
83	4.65	4.71	4.78	4.84	4.91	4.98	5.05	5.13	5.20	5.28	5.36	5.45	5.53	5.62	5.72	7
84	4.66	4.72	4.79	4.85	4.92	4.99	5.06	5.14	5.21	5.29	5.37	5.46	5.54	5.63	5.73	6
85	4.66	4.73	4.79	4.86	4.93	5.00	5.07	5.14	5.22	5.30	5.38	5.47	5.55	5.64	5.74	5
86	4.67	4.73	4.80	4.86	4.93	5.00	5.08	5.15	5.23	5.31	5.39	5.47	5.56	5.65	5.74	4
87	4.68	4.74	4.81	4.87	4.94	5.01	5.08	5.16	5.23	5.31	5.40	5.48	5.57	5.66	5.75	3
88	4.68	4.74	4.81	4.87	4.94	5.01	5.09	5.16	5.24	5.32	5.40	5.48	5.57	5.66	5.75	2
89	4.68	4.74	4.81	4.88	4.94	5.01	5.09	5.16	5.24	5.32	5.40	5.49	5.57	5.66	5.76	1
90	4.68	4.74	4.81	4.88	4.94	5.02	5.09	5.16	5.24	5.32	5.40	5.49	5.58	5.67	5.76	0
	77° 40′	77° 50′	78°	78° 10′	78° 20′	78° 30′	78° 40′	78° 50′	79°	79° 10′	79° 20′	79° 30′	79° 40′	79° 50′	80°	

PART II.

THE DETERMINATION OF THE DIFFERENCE OF LONGITUDE OF TWO STATIONS.

INTRODUCTORY.

The meridian at Greenwich having been adopted as the initial one to which all longitudes in the United States are to be referred, the determination of the longitude of a new station consists simply in the determination of the difference of longitude of the new station and of Greenwich, or some station of which the longitude reckoned from Greenwich is known. The determination of a difference of astronomic longitude is nothing more nor less than the determination of the difference of the local times of the stations.[1]

There are three general methods of determining longitude now in use, viz, the telegraphic, the chronometric, and the lunar.

In the telegraphic method the error of the local chronometer on local sidereal time is determined at each of the two stations by the methods stated in Part I of this publication, and the two chronometer times are then compared by telegraphic signals sent between the stations.

In the chronometric method certain chronometers which are transported back and forth between the stations take the place of the telegraphic signals and thus serve merely to compare the station chronometers.

In each of the lunar methods the observer at a station of which the longitude is required observes the position of the moon, or at least one coordinate of that position, and notes the *local* time at which his observation was made. He may then consult the Ephemeris and find at what instant of Greenwich time the moon was actually in the position in which he observed it. The difference between this time and the local time of his observation is his longitude reckoned from Greenwich. One coordinate fixing the position of the moon may be determined to serve as a means of deriving a longitude by measuring the right ascension of the moon at a transit across the meridian; by measuring the angular distance between the moon and the sun or one of the four larger planets, or between the moon and one of the brighter stars or by observing the times of disappearance and reappearance (immersion and emersion) of a known star behind the moon—the lunar distance of the star at those instants being the angle subtended by the moon's radius. In each case the Greenwich time at which the moon occupied the position in which it was observed is obtained either from the Ephemeris, from observations at Greenwich at about the time in question, or from similar observations at some station of known longitude.

The determination of longitude by wireless telegraph is not discussed in this publication. This method has been used to a certain extent by some countries with apparently satisfactory results. It will no doubt be used to a considerable extent in the location of islands which have no cable connections. The writer believes that it is much less expensive and more satisfactory at present to use the ordinary telegraph lines for the determination of longitude for geodetic purposes within the United States. These conditions may be reversed in the not distant future.

[1] The times may be either sidereal or mean solar. Usually the sidereal times are compared because the time observations are nearly always made upon stars.

The telegraphic method[1] is the most accurate known method of determining differences of longitude. It is always used in this Survey for all longitude determinations in regions penetrated by telegraph lines, and is therefore set forth fully in this publication.

A method suitable for use in regions not reached by the telegraph,[2] is the chronometric method. As this has been extensively used at coast stations in Alaska and will probably continue to be so used during some years to come, it is also here treated in full.

To use the chronometric method one must be able to travel back and forth carrying chronometers between the two stations. The cost of such a longitude determination increases with increased cost of travel between stations, and its accuracy decreases as the time required to make a round trip increases. These facts cause the chronometric method to give way to lunar methods in certain comparatively rare situations. The points at which the boundary between Alaska and British America (one hundred and forty-first meridian) crosses the Yukon and Porcupine Rivers were determined by lunar methods.[3] Comparatively few such cases have occurred in late years in this Survey in which it was desirable to resort to observations upon the moon to determine important longitudes.[4] To have determined these longitudes by transportation of chronometers would have been exceedingly difficult and costly, and would have given results of a low order of accuracy, for there are more than a thousand miles of slow river navigation between the mouth of the Yukon and either station.

As the lunar methods will probably be used less and less with the lapse of time and the increase of traveling facilities, it does not seem desirable to incorporate details in regard to them in this publication, especially as such details would greatly increase its size. The computations involved are long, complex, and difficult. Those who wish to study the lunar methods are referred for details to Doolittle's Practical Astronomy, to Chauvenet's Astronomy, Volume I, and to the American Ephemeris (aside from the tables), especially to the pages in the back of each volume headed "Use of tables."

PROGRAM AND APPARATUS OF THE TELEGRAPHIC METHOD.

During more than 60 years of its use by the Coast and Geodetic Survey the telegraphic method was gradually modified, but with the adoption of the transit micrometer about 1904 the program of the determination of primary longitudes underwent radical changes. The program and apparatus used at present in the Survey will be described first and then the method formerly used will be briefly explained.

The introduction of the transit micrometer practically eliminated from the time determinations, and consequently from the longitude determinations, the large error which was known as the observer's personal equation. The program of longitude observations was formerly designed to eliminate the personal equation from the results.

GENERAL INSTRUCTIONS FOR LONGITUDE DETERMINATION BY THE COAST AND GEODETIC SURVEY WITH TRANSIT MICROMETERS IN LOW LATITUDES (LESS THAN 50°).

1. The observations upon each star should be given unit weight, regardless of the declination of the star and of whether or not the observation of the transit is complete. If an observed transit is incomplete, only those observations should be used for which the positions of the observing wire are symmetrical with reference to the middle point of the registration interval of the screw; that is, each record is to be rejected for which the symmetrical record is missing.

---

[1] The telegraphic method of determining differences of longitude was originated by the Coast Survey in 1846, two years after the first transmission of telegraphic messages over wires. During the long interval since that time the method has gradually been brought to its present high state of perfection. For a historical note on this subject see Appendix No. 2, Report for 1897, pp. 202-203.

[2] In certain cases in which the telegraph line is wanting, the same principles may be used with the substitution of a flash of light between stations in the place of the electric wave. For example, one might so determine the longitudes of the Aleutian Islands of Alaska, the successive islands being in general intervisible. This method has not, however, been used by this Survey. The cost of determining longitudes by this method will in general be so much greater than by the chronometric method (because of the many intermediate stations which will be required between distant stations), as to more than offset its greater accuracy.

In the final demarcation of the boundary between Alaska and British Columbia, an initial point on the one hundred and forty-first meridian was determined telegraphically, using transits equipped with transit micrometers. The telegraphic longitude came within the range of three determinations by lunar methods. The total range of the several lunar determinations of longitude in different years was 1.1 seconds of time.

[4] A statement of the results of these determinations, which is especially interesting as showing what errors may be expected in such observations, is given in Appendix No. 3 of the Report for 1895.

2. The limit of rejection for an observation upon one star (whether the observed transit is complete or not) is a residual of 0.20 second. No observation corresponding to a residual smaller than this should be rejected unless the rejection is made at the time of observation.

3. Each half set of time observations should consist of observations on from 5 to 7 stars (6 preferred). In rare cases a half set may consist of only four stars. All of these are to be time stars; that is, no azimuth stars are to be observed. For the purpose of this paragraph an azimuth star is defined as one for which the azimuth factor, A, is greater than unity. The algebraic sum of the A factors in each half set should be kept less than unity unless it is found that to secure such a half set considerable delays would be necessary. It is desirable to have the algebraic sum of the A factors as small for each half set as it is possible to make it by the use of good judgment in selecting the stars, but it is not desirable to reduce the number of stars per hour to be observed in order to improve the balancing of the A factors, if said balancing is already within the specified limit.

4. In selecting lists of stars to be observed, one should endeavor to secure the maximum number of stars per hour possible, subject to the conditions of paragraph 3 and to the necessity of securing level readings, reversing the instrument, exchanging signals, et cetera. To observe the same stars at both stations involved in a longitude difference is desirable, but it is of less importance than to secure rapid observations with well-balanced A factors in each half set.

5. The telescope should be placed in the position "illumination west" for the first half set of each night and it should be reversed before the beginning of each of the other half sets.

6. The observations on each night should consist, under normal conditions, of four such half sets as are defined in paragraph 3. In case of interference with the normal progress of the observations by clouds or other causes, a determination on a given night may be allowed to depend upon a smaller number of stars and of half sets at each station. But the determination of the longitude difference on any night is to be rejected if, at either station, there has been no reversal of the instrument, or if less than twelve stars with two reversals are successfully observed at either station, or if the exchange of signals takes place at either station outside the interval covered by the time observations at that station.

7. There is to be no exchange of observers during the determination of any difference of longitude.

8. A determination of a difference of longitude will consist of either three or four such nights of observations as are specified in paragraph 6. If, before an opportunity occurs to take observations upon a fourth night, it becomes known that the result from each of the first three nights of observations agrees with the mean result within $0^s.070$, no observations on a fourth night should be taken. If one or more of the first three nights give results differing by $0^s.070$ or more from the mean, or if observations are secured on a fourth night before the results from the first three nights are all known, then observations on four nights are to constitute a complete determination of a difference of longitude.

9. When referring a longitude station to a triangulation station the angle and distance measurements should be made with a check and with such accuracy that if necessary the longitude station may replace the triangulation station for future surveys.

10. The field computations are to be kept as closely up to date as practicable.

11. In making the computations of time observations in the field, the method shown on pages 21 to 27 of this publication should be followed.

GENERAL INSTRUCTIONS FOR LONGITUDE DETERMINATION BY THE COAST AND GEODETIC SURVEY WITH TRANSIT MICROMETERS IN HIGH LATITUDES (GREATER THAN 50°).

The observing and the field computations for the work in connection with the telegraphic determination of longitude in latitudes greater than 50° should be done in accordance with the instructions for work in latitudes less than 50° except that: (*a*) The stars of a set are given different weights depending upon their positions. (*b*) No rejection limit is fixed for use by the observer; rejections are made, if necessary, in the office after the least square computations have been made. (*c*) It will be impossible, as a rule, to have a half set with all time stars and

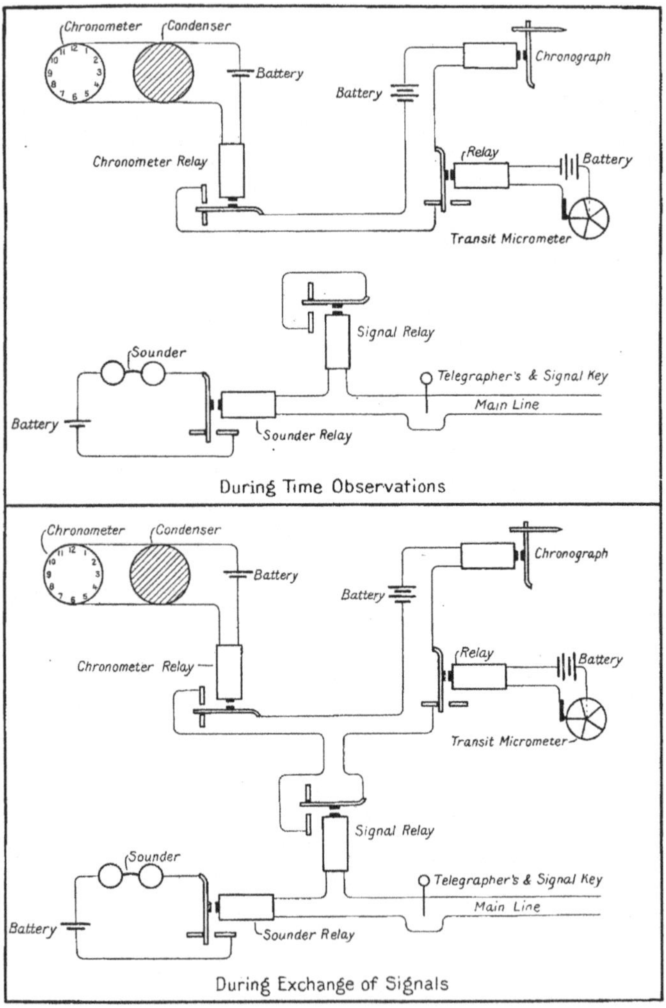

ARRANGEMENT OF ELECTRICAL CONNECTIONS, TELEGRAPHIC LONGITUDE—TRANSIT-MICROMETER METHOD.

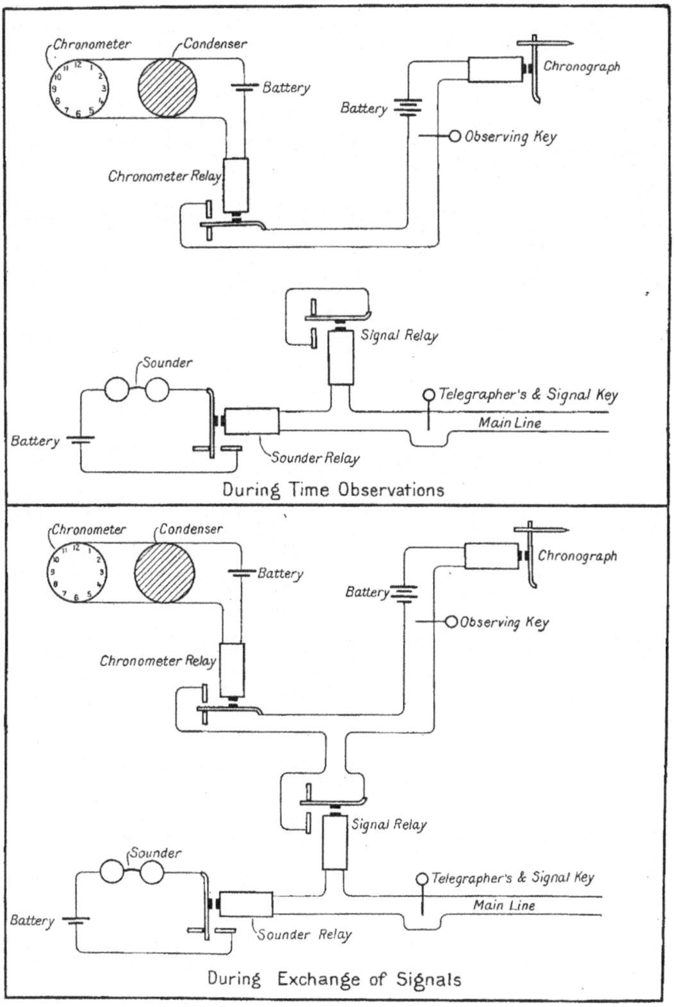

ARRANGEMENT OF ELECTRICAL CONNECTIONS, TELEGRAPHIC LONGITUDE—KEY METHOD.

hence, the half sets are to be made up of time and azimuth stars. (An azimuth star is one having an *A* factor greater than unity.) (*d*) In making the computation of the time observations the observer will use his discretion as to the method to be used, provided it is one of those given in this publication.

USUAL METHOD OF OPERATIONS.

As the personal equation is very small, if it exists at all, it is not considered necessary in determining astronomic longitudes for geodetic or geographic purposes to have an exchange of observers, nor is it necessary that a new station should be in a closed circuit.

The normal determination of longitude between two stations using transit micrometers consists of three nights' observations without exchange of observers. (Under the general instructions a fourth night is sometimes required.) Each night's observations consist of four half-sets of six stars each, the instrument being reversed in its wyes between each two half-sets. Arbitrary signals are usually exchanged between the two stations by telegraph in the interval between the second and third half-sets. This places the arbitrary signals, by which the chronometers at the two stations are compared, as nearly as possible in the middle of the observing period and it makes the longitude determined depend equally on each of the time sets. The two observatories must, of course, be connected by means of a telegraph line. An arrangement is made with the telegraph company for a direct connection between the stations, at the required time, on nights of observation. This is accomplished by running wires from the longitude stations to the switchboards of the local telegraph offices. If possible the line should be without repeaters. The advisability of having the station convenient to the telegraph office should have some weight in determining its location. Occasionally the station may have to be connected directly with a main wire instead of with the telegraph office switchboard.

The general arrangement of the electrical apparatus at each station during star observations and also during exchange of signals is shown in the diagrams of illustrations Nos. 10 and 11. Illustration No. 12 shows the actual switchboard and instruments used in these operations. This board carries an ordinary telegrapher's key, sounder relay, and signal relay, all of which may be included in the telegraph circuit. If desired the signal relay or the sounder relay and key may be cut out by means of plug switches. The sounder is worked by the sounder relay through a separate battery. When the operator is clearing the line or communicating with the operator at the other observatory, the signal relay is cut out, and when signals are being sent it is again cut in, and it operates the pen of the chronograph through a separate battery. Thus, at each station, when the signal relay is on the main line, every break of the telegrapher's key operates the two signal relays and makes records on the chronograph sheets at both stations. The chronometers being placed in the local circuits at both stations continue their records on the chronograph sheets, the circuits being break circuits, and so it is possible to read from the chronograph sheet at each station the chronometer time of sending and receiving the arbitrary signals.

The local circuit, as explained on page 12, consists of one principal circuit, the chronograph circuit, to which the chronometer circuit and the transit circuit are joined through the points of their respective relays. The observing key, when used, replaces the transit circuit. The chronograph circuit, connected with the proper binding posts of the switchboard, includes the points of the signal relay, except when cut out by a plug switch. This plug is kept in during time observations, and taken out only during the exchange of signals.

A few minutes before the time for exchange of signals the telegraph operator secures a clear line between stations, ascertains whether the observations at the other station are proceeding successfully, and telegraphs the exact epoch at which signals will be exchanged. This epoch is arranged, if practicable, not to interfere with the star observations at either station. If at one of the stations floating clouds or other causes are making it difficult to get observations the observer at that station should choose the epoch, for the loss of one or more stars by him might cause the loss of a night's work. When the epoch arrives the points of the signal relay are placed in the local circuit at each station by the removal of a plug of each switchboard. Any break in the main-line circuit will now cause corresponding breaks in the local circuits, and a signal made with the telegraph key[1] will be recorded on both chronographs. The observer at the western station customarily sends signals first, by releasing the telegraph key for an instant between the breaks of his chronometer at an average interval of two seconds. He times these signals so that they will not interfere with his own chronometer record, and he must also be prepared to shift them to another portion of the second, if they are conflicting with the record of the chronometer at the other station. Notice of an interference is given by the other observer by breaking into the circuit and making a succession of quick breaks with the key. After 15 to 20 signals have been sent from the western station, covering a period of over half a minute, double that number of signals are sent by the eastern observer, and then 15 to 20 more are sent by the western observer. This makes a total of 30 to 40 signals each way, with the mean epoch of the signals from the two different directions agreeing closely. The signals, as a rule, cover a total period of less than three minutes. It is well to make a succession of quick breaks at the beginning and end of each series of signals. It is also desirable to vary the position of each of several signals with reference to the chronometer breaks at the beginning of a series or to make several signals at intervals of one second in order to facilitate the identification of corresponding records at the two stations. The number of signals exchanged is arranged to cover a period greater than one minute each way, with a view of eliminating errors in the contact wheel of the chronometer.

A signal sent from one station to the other will be recorded on the chronograph of the sending station slightly before it is on the distant chronograph, and this difference in time of record is called the transmission time. It depends, in fact, both on the retardation of the signal in the telegraph line between the two stations, and on the difference in the time of action of the signal relays at the two stations.[2] Signals sent from west to east will make the difference in longitude too large, and signals from east to west will make it too small by the amount of the transmission time. By taking the mean of the differences as given by the signals in both directions this source of error is eliminated, provided the transmission time is the same in both directions.[3]

During exchange of signals the chronographs are run at double speed, so that the signals may be read to hundredths of seconds. The advantage in sending signals by making arbitrary breaks of the circuit is that they will come at varying parts of the seconds, thus tending to eliminate personal equation in the reading of the fractional parts of the second.[4] If portions of the record are missed, the corresponding signals at the two stations may still be identified by comparing the successive differences between signals.

RECORD OF AN EXCHANGE OF SIGNALS.

The following is one night's record of an actual exchange of signals between two stations, written as read from the chronograph sheet on a special form used for the purpose, on which is also made the computation of the epochs of the signals at the two stations, the computation of the final difference of signals, and the transmission time.

---

[1] It is to be noted that these signals are made by breaking the circuit, which is opposite to the ordinary correspondence use of the key.

[2] The latter is probably a small quantity. Some measurements of the armature time of one of the quick-acting relays used in these longitude determinations showed it to vary from 0.005 to 0.015 second with extreme changes in adjustments and current.

[3] There is always some uncertainty on this score when repeaters are used in the main telegraph line, because of the distinct mechanical arrangements for repeating the signals in the two directions. Repeaters are therefore to be avoided as far as practicable.

[4] Chronometer signals were formerly used—that is, the chronometers were alternately made to send their breaks through the main-line circuit, recording on both chronographs. Some of the objections to this method were liability of damage to the points of the break circuit wheel of the chronometer when put on the main line, possibility of the record of one chronometer interfering with the record of the other, and personal equation in reading a record that always occurred at the same part of a second.

SWITCHBOARD—TELEGRAPHIC LONGITUDE.

Arbitrary signals.

Form 256.

[Station, Key West, Fla. Date, Feb. 14, 1907. Observer, J. S. Hill. Recorder, J. S. Hill.]

From Key West to Miami			From Miami to Key West*		
Miami record	Key West record	Diff.	Miami record	Key West record	Diff.
h m s	*h m s*	*m s*	*h m s*	*h m s*	*m s*
6 33 35.10	6 27 38.28	5 56.82	6 34 54.41	6 28 57.71	5 56.70
36.42	39.63	.79	56.32	59.63	.69
37.50	40.70	.80	58.31	29 01.60	.71
38.50	41.71	.79	35 00.22	03.52	.70
39.45	42.67	.78	02.35	05.64	.71
41.47	44.67	.80	04.30	07.58	.72
43.43	46.63	.80	06.54	09.83	.71
45.50	48.69	.81	08.31	11.60	.71
47.50	50.70	.80	10.26	13.54	.72
49.58	52.77	.81	12.24	15.53	.71
51.60	54.78	.82	14.31	17.61	.70
53.57	56.77	.80	16.22	19.51	.71
55.65	58.85	.80	18.22	21.52	.70
58.18	28 01.37	.81	20.28	23.57	.71
34 00.51	03.71	.80	24.29	27.58	.71
02.52	05.72	.80	26.22	29.51	.71
03.67	06.88	.79	28.23	31.53	.70
04.77	07.95	.82	30.28	33.57	.71
35 42.20	29 45.40	.80	31.24	34.54	.70
43.50	46.72	.78	32.43	35.73	.70
45.08	48.29	.79	34.25	37.54	.71
47.50	50.70	.80			
49.56	52.74	.82			
51.50	54.70	.80			
53.47	56.67	.80			
55.64	58.84	.80			
57.59	30 00.80	.79			
59.57	02.79	.78			
36 01.57	04.77	.80			
03.58	06.80	.78			
05.55	08.76	.79			
07.55	10.76	.79			
09.60	12.80	.80			
11.53	14.74	.79			
12.62	15.85	.77			
13.57	16.77	.80			
14.61	17.81	.80			
15.54	18.73	.81			
Means: 6 34.9	6 29.0	5 56.798	6 35.3	6 29.3	5 56.707
			6 34.9	6 29.0	5 56.798
		Means	6 35.1	6 29.1	5 56.752
					Transmission time=.046

* Complete set of signals from Miami to Key West not obtained.

In the foregoing table the mean epochs are shown for the record of signals on each chronograph sheet, the mean of all the differences of the chronometer records, and the transmission time. It is usually sufficient, in obtaining the mean epochs of signals, where they are symmetrically arranged, to take the mean of the first five and the last five signals.

CHRONOMETER CORRECTIONS AND RATES.

On the following form are tabulated the epochs (T_0) for which chronometer corrections were determined at both stations, the corrections (ΔT) determined, and the rate per minute computed from the two time sets observed on each night. In each case the mean of the epochs and of the corrections is given on the third line for each date. These means furnish a correction for the chronometer very nearly at the epoch of the signals, and they thus reduce the work of computing the chronometer corrections for the epochs of the signals.

Chronometer corrections and rates.

Date	Key West, Fla.		Rate per minute	Miami, Fla.		Rate per minute
	T_0	ΔT		T_0	ΔT	
1907.	h m	s	s	h m	s	s
Feb. 14	5 49.6	+14.691		5 41.4	+45.177	
	7 11.0	+14.726	+0.00043	7 19.1	+45.493	+0.00323
	6 30.3	+14.708		6 30.2	+45.335	
15	5 50.0	+14.327		5 46.9	+50.182	
	7 47.9	+14.220	−0.00091	7 11.0	+50.449	+0.00317
	6 49.0	+14.274		6 29.0	+50.316	
16	5 50.1	+13.479		5 54.7	+55.337	
	7 09.0	+13.460	−0.00024	7 22.4	+55.551	+0.00244
	6 29.5	+13.470		6 38.6	+55.444	

COMPUTATION OF DIFFERENCE OF LONGITUDE.

The next step is the computation of the difference of longitude from the mean of the signals sent in each direction. Each night's observations represents a complete determination of this difference, and a separate and complete computation is accordingly made for each night. The epoch of signals and difference of chronometers are taken from the record of signals for each night, and the chronometer corrections at these epochs are computed for each station and each night, using the rates per minute given in the preceding form. To the difference in chronometers is then applied the difference in chronometer corrections (eastern minus western chronometer), which gives the difference of longitude in time as determined by the night's observations. From this determination the transmission time has already been eliminated by taking the means of eastern and western signals.

The chronometer correction ΔT at the time of exchange T and its probable error r are expressed by

$$\Delta T = \Delta T_1 + \frac{\Delta T_2 - \Delta T_1}{T_2 - T_1}(T - T_1), \text{ and } r = \frac{[(T_2 - T)^2 r_1^{\,2} + (T - T_1)^2 r_2^{\,2}]^{1/2}}{T_2 - T_1}$$

where ΔT_1 and $\pm r_1$ are the chronometer correction and its probable error derived from the first set of time observations at epoch T_1, and ΔT_2 and $\pm r_2$ are the same quantities, respectively, for the second set at epoch T_2.

Computation of difference of longitude.

BETWEEN MIAMI AND KEY WEST, FLA.

Date	T_0		ΔT		Diff: ΔT	Diff. of signals	$\Delta\lambda$	v	Transmission time
	Miami	Key West	Miami	Key West					
1907.	h m	h m	s	s	s	m s	m s	s	s
Feb. 14	6 35.1	6 29.1	+45.351	+14.709	+30.642	5 56.752	6 27.394	−0.031	0.046
15	6 31.7	6 25.8	+50.325	+14.295	+36.030	5 51.285	6 27.315	+ .048	.051
16	6 33.6	6 27.8	+55.432	+13.470	+41.962	5 45.418	6 27.380	− .017	.047
						Mean..	6 27.363		

Reduction to longitude pier of 1896=.97 meter = +0.002 s
Reduction to mean position of pole [1] = 0.000
Miami longitude station east of Key West longitude station = 6 m 27.365 s
= 1° 36′ 50″.525

In the example shown above the second column gives the mean epoch of the exchange of signals as read from the chronograph sheet at the eastern station, Miami, and the fourth column gives the correction to the chronometer at Miami for the mean epoch of the signals, this correction being computed from the corrections to the chronometer and the rate deduced from the time observations. The third and the fifth columns give similar data for the western

[1] See Astronomische Nachrichten No. 4253.

station, Key West. The difference between the chronometer corrections (ΔT) given in the fourth and fifth columns is shown in the sixth column and equals the correction at the eastern station minus the correction at the western station. In the next column is given the difference of signals (eastern minus western). The difference of longitude, $\Delta\lambda$, is then the combination of the difference between the ΔT's at the two stations and the difference of signals. The transmission time is taken from the form on which the record of signals and their reduction is shown, and is placed in the last column, while in the column immediately preceding is placed the difference between each night's determination and the mean of the determinations of all the nights.

The values from the various nights are each given unit weight, and their mean is then considered to be the observed difference of longitude between the transit instruments at the two stations. In the example given this difference has a correction applied to it to reduce it to what it would have been had the transit at the base station, Key West, been placed exactly over the position occupied by the transit in 1896 (adjusted in the longitude net of the United States)[1] instead of at a position 0.97 meters east of it. The particular example given is one of a series of differences of longitude determined in 1907, commencing at Key West and closing on Atlanta. There is also at the latter place an adjusted longitude station of the longitude net of the United States. The longitudes of these two stations, at Key West and Atlanta, being held fixed, a closing discrepancy was developed which was distributed equally among the various differences, each difference being given unit weight. The following table shows the differences of longitude determined between Key West and Atlanta and the distribution of the closing error:

Computation of closing error between Key West and Atlanta.

	Observed difference		Correction to close circuit	Adjusted difference	
	m	*s*	*s*	*m*	*s*
Miami west of Key West	− 6	27. 365	+. 009	− 6	27. 356
Jupiter west of Miami	− 0	27. 404	+. 009	− 0	27. 395
Sebastian west of Jupiter	+ 1	33. 654	+. 009	+ 1	33. 663
Daytona west of Sebastian	+ 2	11. 332	+. 009	+ 2	11. 341
Fernandina west of Daytona	+ 1	46. 878	+. 009	+ 1	46. 887
Atlanta west of Fernandina	+11	42. 609	+. 010	+11	42. 619
Atlanta west of Key West	+10	19. 704	+. 055	+10	19. 759
Atlanta west of Key West (From adjusted longitude net of United States)	+10	19. 759			
Closing error=	+	. 055			

CORRECTION FOR VARIATION OF THE POLE.

A correction is necessary to reduce the observed astronomic longitude to the mean position of the pole. About the middle of each year the Latitude Service of the International Geodetic Association publishes in the Astronomische Nachrichten provisional values of the coordinates of the instantaneous pole for the preceding calendar year, together with tables to reduce observed latitudes, longitudes, and azimuths to the mean position of the pole. The proper correction to the longitude may be computed by means of these tables, knowing the time of observation and the latitude and longitude of the observing station.

DISCUSSION OF ERRORS WHEN TRANSIT MICROMETER IS USED.

Let it be supposed that the regular program for observations with a transit micrometer, three nights' observations without exchange of observers, has been carried out. The computed result, the difference of astronomic longitude of the two places, is subject to the following errors:

[1] See Appendix 2 of the Report for 1897.

First. An accidental error arising from the accidental errors of observations of about 72 stars at each station. If the accidental error of observation of a single star be estimated at $\pm^{s}.07$, which may be considered sufficiently large to cover both the observer's errors and those instrumental errors which belong to the accidental class, then the probable error of the final result arising from this cause would be $\pm^{s}.07 \div \sqrt{36} = \pm^{s}.012$.

Second. An accidental error arising from the accidental errors in the adopted right ascensions of such stars as are observed at one station on a given night but not at the other. It is in such cases only that errors in right ascension have any effect on the computed result. If entirely different stars were observed at the two stations, 24 at each station, and if $\pm^{s}.03$ be accepted as the probable error of a right ascension, then the probable error of the result for one night arising from this source would be $\pm^{s}.03 \div \sqrt{12} = \pm^{s}.009$. In ordinary cases, in which the number of stars not common to both stations is less than 10 per cent, this accidental error is reduced to less than $\pm^{s}.001$.

Third. Errors due to the assumption that the rate of the chronometer is constant during and between the two time sets of a night. As the interval between the mean epochs of the sets is ordinarily only about one hour, these errors are probably exceedingly small. In order to make these errors inappreciable, longitude observers should use chronometers known to show but small variations in rate, and should protect them as thoroughly as is feasible while in use against jars and sudden changes of temperature. The errors from this source will be of about the same value whether the exchange of signals is made at about the mean epoch of the two sets of time observations, or is made at any other epoch within the interval covered by the two sets.

Fourth. The question of the personal equation with the transit micrometer is discussed fully on pages 90 and 91.

Fifth. Errors arising from lateral refraction. The probable minuteness of these errors in time observations has already been commented upon (see p. 48). It is not impossible, however, that small constant errors may arise from this source at stations established in closely built-up portions of great cities, particularly of manufacturing centers.

Sixth. Errors arising from variation of transmission time. By transmission time is meant the interval that elapses from the instant at which the signal relay breaks the local circuit at the sending station to that at which the signal relay breaks the local circuit at the receiving station. This interval is made up of armature time, induction time, and the true transmission time of the electric wave passing along the wire. It is only the variation in transmission time occurring during the exchange of signals on each night that introduces error into the computed result. As this interval is not much over a minute the error is probably insensible if there is a continuous wire connection between stations. If the line between stations passes through a "repeater" the transmission time in one direction through the repeater will be different from that in the other direction unless the two magnets of the repeater are adjusted exactly alike, and half this difference will enter into the computed result as an error. The repeaters used in ordinary telegraph service are not specially designed for quick action, as are the signal relays on the Coast and Geodetic Survey switch board, nor is their adjustment in the control of the longitude observers. Hence the desirability of a continuous wire connection.

Any change in transmission time within the local circuit during the exchange of signals will produce an error in the computed longitude, but such changes are probably insensible. A change at any other time in the local circuit will appear in the observations as a change in the chronometer correction and will probably have no appreciable effect on the final result for the night.

Seventh. The difference of the transmission time through the two signal relays and also the difference in the transmission time through the two transit micrometer relays enter as errors in the final result. These errors are made very small in the present longitude work of the Survey by using relays which are as nearly alike as can be made, and which are specially designed to act very quickly.

If the difference of longitude which is being measured is large, it becomes necessary to abandon the practice of observing the same stars at both stations in order to make the exchange of arbitrary signals come within the period of the night's observations at each station. However, the errors of right ascension thus introduced will not be large.

The combination of the numerical values of the above errors will not fully account for the error of the result as computed from the separate determinations, that is from the residuals, but it may be that some of the above errors for which no numerical values are estimated are much larger than supposed. The discussion of errors of time observations on pages 48–51 of this publication applies to a certain degree to longitude work.

See also Discussion of Errors, when the key method is used, on page 93.

PROGRAM WHERE NO TRANSIT MICROMETER IS USED.

Before the adoption of the transit micrometer for longitude work, when the chronograph and key method was in use, it was necessary in all determinations of differences of longitude to arrange the program of observations so as to eliminate the personal equation of the observers making the time observations. The personal equation was eliminated either directly by exchange of observers, or indirectly by supplementary observations, themselves independent of the longitude observations, but which gave a value for the personal equation to be introduced into the computations. Further on, page 90, the question of personal equation and its determination will be more fully discussed.

In the determination of primary differences of longitude the personal equation was eliminated by the observers exchanging stations when one-half of the observations had been made. One-half the sum of the mean determinations before and after exchange of observers gave a resulting difference of longitude which was independent of the personal equations of the observers provided these personal equations remained constant. Except for this, the program of observations was the same as for observations with a transit micrometer (see p. 81).

The arrangement of the telegraphic apparatus was the same as described on page 81. The observing key took the place of the relay points of the transit micrometer. Illustration No. 11 shows the arrangement of the local and main circuits while time observations were being made, and also while signals were being exchanged. The switchboard is the same as used in transit micrometer observations, and is shown in illustration No. 12. The following records and computations show the various steps in observing and computing an actual difference of longitude.

Record of exchange of signals, and computation of difference of chronometers.

[Station, Atlanta, Ga. Date, Mar. 7, 1896. Observer, G. R. P. Recorder, G. R. P.]

ARBITRARY SIGNALS.

From Atlanta to Key West			From Key West to Atlanta		
Key West record	Atlanta record	Difference of chronometers	Key West record	Atlanta record	Difference of chronometers
h m s	*h m s*	*m s*	*h m s*	*h m s*	*m s*
7 35 59.97	7 25 42.39	10 17.58	7 37 08.76	7 26 51.51	10 17.25
36 01.90	44.30	.60	10.82	53.60	.22
04.03	46.47	.56	12.78	55.52	.26
05.96	48.39	.57	15.28	58.04	.24
*	*	*	*	*	*
*	*	*	*	*	*
56.90	26 39.30	.60	38 38.48	28 21.21	.27
58.91	41.34	.57	40.60	23.33	.27
h m	*h m*	*m s*	*h m*	*h m*	*m s*
Means 7 36.5	7 26.2	10 17.570	7 37.9	7 27.6	10 17.249

SUMMARY OF RESULTS OF TIME DETERMINATIONS AT ATLANTA.

Date	Epoch (by face of chronometer)	Chronometer correction ΔT_W	Rate per minute	Collimation	Azimuth	
					West	East
1896.	h m	s	s	s	s	s
Mar. 7	6 56.4	−13.546	+.00261	+.03	−.154	+.035
7	8 12.6	−13.347		−.01	−.036	−.070
8	6 56.3	− 7.742	+.00310	−.01	+.115	+.089
8	8 12.5	− 7.506		−.06	+.190	+.313
*	*	*	*	*	*	*
*	*	*	*	*	*	*
27	8 12.6	−12.660	+.00043	−.18	+.183	+.155
27	9 22.4	−12.630		−.22	+.378	+.167

SUMMARY OF RESULTS OF TIME DETERMINATIONS AT KEY WEST.

Date	Epoch (by face of chronometer)	Chronometer correction ΔT_E	Rate per minute	Collimation	Azimuth	
					West	East
1896.	h m	s	s	s	s	s
Mar. 7	6 56.4	−11.157	−.00232	−.05	−1.108	−1.236
7	8 12.6	−11.334		−.03	−1.220	−1.108
8	6 56.4	−13.994	−.00227	−.06	−1.649	−1.447
8	8 12.6	−14.167		−.03	−1.644	−1.580
*	*	*	*	*	*	*
*	*	*	*	*	*	*
27	8 12.4	− 4.992	−.00223	−.06	−0.181	−0.256
27	9 21.8	− 5.147		−.00	−0.121	−0.144

$\Delta\lambda$ FROM WESTERN OR ATLANTA SIGNALS.*

Date	Epoch of signals†		Difference of chronometers	Chronometer corrections			$\Delta\lambda_W$ (from western signals)
	Key West T_E	Atlanta T_W		Key West ΔT_E	Atlanta ΔT_W	Difference $\Delta T_E - \Delta T_W$	
1896.	h m	h m	m s	s	s	s	m s
Mar. 7	7 36.5	7 26.2	10 17.570	−11.250	−13.468	+2.218	10 19.788
8	7 36.6	7 26.2	10 26.199	−14.085	− 7.649	−6.436	19.763
*	*	*	*	*	*	*	*
*	*	*	*	*	*	*	*
27	8 51.3	8 41.1	10 12.507	− 5.079	−12.648	+7.569	20.076

* Uncorrected for transmission time and personal equation.
† By face of chronometer.

$\Delta\lambda$ FROM EASTERN OR KEY WEST SIGNALS.*

Date	Epoch of signals†		Difference of chronometers	Chronometer corrections			$\Delta\lambda_E$ (from eastern signals)
	Key West T_E	Atlanta T_W		Key West ΔT_E	Atlanta ΔT_W	Difference $\Delta T_E - \Delta T_W$	
1896.	h m	h m	m s	s	s	s	m s
Mar. 7	7 37.9	7 27.6	10 17.249	−11.253	−13.464	+2.211	10 19.460
8	7 37.6	7 27.2	10 25.881	−14.087	− 7.646	−6.441	.440
*	*	*	*	*	*	*	*
*	*	*	*	*	*	*	*
27	8 53.3	8 43.1	10 12.136	− 5.083	−12.647	+7.564	.700

* Uncorrected for transmission time and personal equation.
† By face of chronometer.

COMBINATION OF LONGITUDE RESULTS.

At one time it was the custom in the Coast and Geodetic Survey to combine the resulting differences of longitude for the various nights' observations by deducing weights and assigning them to the various values. This custom is not now practiced where transit micrometers are used, nor is it followed where an accepted program is carried out even if no micrometers are used. If a regular program is carried out the various nights' determinations are given equal weight, and direct means are taken for the final value of the difference of longitude. However, the following discussion of the combination of longitude results where the different nights' observations are assigned different weights is given here as occasion might arise where the information would be of value.

The following table gives the collection of the results for the different nights and their combination to develop and eliminate the transmission time and personal equation. The mean of the differences of longitude as derived from the western and eastern signals will be free from the transmission time, and their difference is double the transmission time. The relative weights for the resulting differences of longitude for different nights are derived from the expression $p=\frac{p_1 p_2}{p_1+p_2}$, where p_1 and p_2 are the weights of the determinations of the chronometer corrections at the epoch of exchange of signals at the two stations, respectively, or $p_1=\frac{1}{r_1^2}$ and $p_2=\frac{1}{r_2^2}$ in which r_1 and r_2 are the probable errors of the chronometer corrections. To obtain the personal equation the weighted means are taken for each position of the observers, and half their difference is the personal equation to be applied with opposite signs to the two groups. This gives the corrected result for difference of longitude for each night, and the weighted mean of all the nights is the final difference of longitude. The probable error of the latter is $0.674\sqrt{\frac{\Sigma pv^2}{(n-2)\Sigma p}}$ where n is the number of nights of observation and 2 is the number of unknowns (longitude and personal equation). In the table the means in the seventh and ninth columns are weighted means.

The personal equation is one-half the difference in the weighted results for the two positions of the observers, or

$$S-P=\frac{19.884-19.645}{2}=+.120,$$

the sign indicating that S observes later than P. The probable error[1] of the personal equation may be taken as identical with that of the resulting difference of longitude.

The transmission time, as stated, is one-half the difference between the results from western and eastern signals, or in this example, $=\frac{.338}{2}={}^{s}.169$, an unusually large value, due to the marine cable between Key West and the mainland.

Table of resulting difference of longitude between Atlanta, Ga., and Key West, Fla.

Date	Observer at— A	Observer at— K W	From western or Atlanta signals $\Delta\lambda_W$	From eastern or Key West signals $\Delta\lambda_E$	Double transmission time $\Delta\lambda_W-\Delta\lambda_E$	Mean of W. and E. signals	Personal equation	Difference of longitude	Combination weight p	Residuals v
1896.			m s	m s	s	m s	s	m s		s
Mar. 7	P.	S.	10 19.788	10 19.460	0.328	10 19.624	+0.120	10 19.744	8	−.021
8	P.	S.	.763	.440	.323	.602		.722	11	−.043
9	P.	S.	.754	.445	.309	.600		.720	4	−.045
13	P.	S.	.802	.495	.307	.648		.768	13	+.003
14	P.	S.	.842	.522	.320	.682		.802	21	+.037
				Mean	.317	10 19.645				
Mar. 20	S.	P.	10 20.018	10 19.686	.332	10 19.852	−0.120	.732	9	−.033
21	S.	P.	.075	.705	.370	.890		.770	5	+.005
25	S.	P.	.102	.737	.365	.919		.799	4	+.034
26	S.	P.	.074	.721	.353	.897		.777	8	+.012
27	S.	P.	.076	.700	.376	.888		.768	6	+.003
				Mean	0.359	10 19.884		10 19.765	±0.007	

[1] Practically the same result is obtained by deriving separate values for the personal equation by comparing each result in the first position of the observers with the corresponding result in the second position and computing the probable error from the variations in these separate values.

The above formulae and forms are used in the office computation. The field computation differs from that made in the office in that the time computation is made by an approximate field method shown on page 26 or page 34 instead of the least square method given on page 41, and that in the field no probable errors or weights are computed and indiscriminate means are taken instead of weighted means. In the past some of the forms used in the field have been slightly different from those shown above. The office computation will be facilitated by making the field computation as here indicated.

PERSONAL EQUATION.

The absolute personal equation in time observations with a transit is the interval of time from the actual instant of transit of a star image across a line of the diaphragm to the instant to which the transit is assigned by the observer. When the time is observed using a chronograph and an observing key the absolute personal equation is simply the time required for the nerves and the portions of the brain concerned in an observation to perform their functions. In the case of observations by the eye and ear method the mental process becomes more involved, and the personal equation depends on a much more complicated set of physical and psychological conditions than when the observations are made with a key and chronograph.

Although the personal equation has been studied by many persons and for many years, little more can be confidently said in regard to the laws which govern its magnitude than that it is a function of the observer's personality, that probably whatever affects the observer's physical or mental condition affects its value, that it tends to become constant with experience, that it probably differs for slow moving and fast moving stars, and that it is different for very faint stars which the observer sees with difficulty from what it is for stars easily seen.

A systematic error may be present which is due to the tendency of the observer to place the wire always to the right or to the left of the center of the star's image. This tendency is due to the defects in the observer's eye and the error resulting is called the *bisection error*. At some astronomic observatories a reversing prism is used which reverses the image of the star midway in the observations. Thus, during one half of the observations the wire would be placed too far east and during the other half too far west of the center of the star's image (or vice versa) and the mean of all the observations would be free from a bisection error. No numerical values are available for the effect of the *bisection error* but it is known to be so small that it may be neglected in all time and longitude work for the usual geodetic and geographic purposes. (See remarks under the Description of the Zenith Telescope on p. 105.)

There are various mechanical devices for the determination of the absolute personal equation of an observer, but as these are seldom used they will not be discussed here.

The relative personal equation of two observers is the difference of their absolute equations.

When observing time with a transit micrometer the personal equation, if any, may be neglected. The observing does not consist of a series of independent consecutive operations, but rather of a continuous performance, the star's image being bisected by the micrometer wire before the record is begun and kept bisected till after the record is ended.

In Appendix 8 of the Report for 1904, entitled "A Test of the Transit Micrometer," it was shown that if there is an actual personal equation in observing star transits with a transit micrometer it is so small as to be masked by the other errors of observation. Viewed in the light of several years of actual longitude observations with the transit micrometer this conclusion is fully justified. These longitude observations involved four simple or compound loop closures, and one determination with exchange of observers. In observing differences of longitude to close a loop the same observer always kept in front as the work progressed around the loop, thus introducing into the loop closure an accumulation of any relative personal equation that might exist.

In 1906 four differences determined with the transit micrometer between Seattle, Wash., and the point where the one hundred and forty-first meridian boundary of Alaska intersects the Yukon River, were combined with certain Canadian results to form a loop, and the loop closure was reduced to zero by applying a correction of only 0.008 second to each observed difference of longitude.

In Texas in 1906 the three differences of longitude between the three points, Austin, Alice, and Isabel, were determined, using transit micrometers and a program as indicated above. This would introduce into the closure three times any relative personal equation of the observers. The loop closure was 0.038 second, making necessary corrections on the three differences of 0.^{s}013, 0.^{s}013, and 0.^{s}012.

In 1907 a series of longitude differences was determined, using transit micrometers, between Key West and Atlanta, for both of which stations adjusted values are given in the longitude net of the United States,[1] and these adjusted values were held fixed. Six longitude differences between these two stations were determined in such a way as to accumulate any relative personal equation between the two observers. The results are shown on page 85. The correction required to be applied to each observed difference to close the loop was 0.^{s}009. A second loop, closing on one of the links of the first loop or forming with all but the last difference of the first loop a new loop of eight links between the fixed stations, Key West and Atlanta, obtained corrections of only 0.^{s}008 per link to close. The corrections in both loops were of the same sign.

Later in 1907 a series of longitude differences was determined in Minnesota, Dakota, Nebraska, and Iowa, using the transit micrometer. The points held fixed were the stations of the longitude net at Bismarck and Omaha. There were four condition equations and ten unknowns involved in the adjustment of this secondary net. The largest correction to an observed difference of longitude obtained was 0.^{s}038 and the smallest was 0.^{s}003. Four of the corrections obtained were less than 0.^{s}010 and seven were less than 0.^{s}015. Where possible the program of observations was arranged to produce an accumulation of any existing relative personal equation.

In 1908 the difference of longitude between the observatory of the new University of Washington at Seattle and the old longitude station in Seattle was determined, using transit micrometers. Observations were made on six nights, the observers changing stations after each night's observations. The apparent relative personal equation determined by this method of observation amounted to only 0.008 second.

The above evidence justifies the present method of longitude observations with transit micrometers without exchange of observers. The evidence is sufficient to justify the continuation of the present method of carrying on telegraphic longitude work for geographic and geodetic purposes, for the personal equation, if present, is much smaller than the probable errors of the determinations. However, where the greatest accuracy is required, as in the determination of the difference of longitude between two fixed observatories, then an exchange of observers is desirable to eliminate *any* possible personal equation. An exchange of instruments is also required to eliminate differences in the total relay and armature times at the two ends of the line. For a complete elimination of this error the adjustments of the relays and magnets should be the same before and after exchange.

The accuracy of the telegraphic determination of the difference of longitude, where no transit micrometer is used, depends largely upon the accuracy of the determination of the relative personal equation of the two observers, and upon its *constancy*.

The relative personal equation of two observers may be determined in various ways. The method to be selected in a given case depends upon circumstances, involving the question of cost, the difficulty of exchange of observers, and to some degree the desired accuracy of the result.

In primary longitude determinations, where cost and ease of transportation are not prohibitive, the relative personal equation of the observers is eliminated from the result by the observers changing stations after about one-half of the observing has been done. In this way the relative personal equation will enter the resulting differences of longitude before and after exchange of observers with different signs and the mean of such determinations will be the resulting difference of longitude with the effect of personal equation eliminated.

The relative personal equation may be determined independently of the longitude observations by the use of two transits placed in the same observatory or in separate observatories close together, and by having the two observers observe independently the same stars, which should be arranged in time sets. If the two instruments are on the same meridian, or nearly so, and use is made of only one chronometer and chronograph to record both sets of observations,

[1] See Appendix 2, Report for 1897.

it may be necessary to throw one instrument out of adjustment (in collimation) more than the other in order to avoid having the observations overlap. A better arrangement would be to have two chronographs controlled by the same chronometer by means of local relays, and have the chronograph records of the two instruments independent of one another. The difference of the two chronometer corrections thus determined, corrected for the very small longitude difference between the two transit instruments, is the personal equation of the two observers. Sometimes different chronometers are used and compared in the same manner as in actual longitude determinations.

The relative personal equation may also be observed with a single transit instrument as follows: On the first star A observes the transits over the lines of the first half of the diaphragm, then quickly gives place to B who observes the transits across the remainder of the lines, omitting the middle line. On the second star B observes on the first half of the diaphragm and A follows. After observing a series of stars thus, each leading alternately, each observer computes for each star, from the known equatorial intervals of the lines, and from his own observations, the time of transit of the star across the mean line of the diaphragm. The difference of the two deduced times of transit across the mean line is the relative personal equation. If each has led the same number of times in observing, the result is independent of any error in the assumed equatorial intervals of the lines. No readings of the striding level need be taken, and the result is less affected by the instability of the instrument than in the other method. If the stars observed by this method are so selected as to form time sets, and the chronometer corrections are computed from each observer's observations independently, the difference of these chronometer corrections will be the relative personal equation.

As the accuracy of the telegraphic determination of longitude without the use of the transit micrometer depends also upon the constancy of the relative personal equation of the two observers concerned, there is shown below a table which gives some values of the relative personal equation as derived from telegraphic longitude observations (key and chronograph method). The values in this table indicate to what extent the relative personal equation may be expected to vary from month to month and year to year. The plus sign indicates that the observer first named observes later (slower) than the other.

Relative personal equation (not reduced to equator).

C. H. Sinclair—E. Smith [14 years]			C. H. Sinclair—R. A. Marr [4 years]			C. H. Sinclair—G. R. Putnam [5 years]		
	s	s		s	s		s	s
1881 Aug. and Sept.	−0.123	±0.008	1886 Sept. and Oct.	+0.288	±0.008	1891 May and June	+0.184	±0.011
1881 Nov. and Dec.	− .085	06	1888 Sept.	+ .210	09	1891 June and July	+ .140	08
1885 Apr. and May	− .047	08	1888 Oct. and Nov.	+ .144	11	1891 July	+ .172	06
1885 May and June	− .131	03	1888-9 Dec. and Jan.	+ .214	10	1891 Aug.	+ .161	10
1885 July and Aug.	− .110	10	1889 Jan.	+ .233	05	1891 Aug. and Sept.	+ .176	11
1886 May and June	− .062	08	1889 Jan. and Feb.	+ .225	07	1892 Feb. and Mar.	+ .160	06
1886 June and July	+ .010	06	1889 Feb. and Mar.	+ .267	07	1892 Mar.	+ .192	04
1886 July and Aug.	− .023	12	1889 Mar. and Apr.	+ .278	12	1892 Mar. and Apr.	+ .140	02
1886 Aug. and Sept.	+ .056	04	1889 Apr. and May	+ .217	12	1892 Apr.	+ .150	05
1887 May and June	+ .038	10	1889 May and June	+ .282	18	1892 Apr. and May	+ .126	04
1887 June, July, and Aug.	+ .109	13	1889 June and July	+ .246	07	1892 June and July	+ .109	10
			1889 July	+ .275	08	1893 Feb. and Mar.	+ .082	10
1887 Sept.	+ .111	13	1889 July	+ .265	05	1896 Feb. and Mar.	+ .155	03
1887 Sept. and Oct.	+ .160	09	1889 July and Aug.	+ .228	15	1896 Mar.	+ .129	07
1895 Feb. and Mar.	+ .093	11	1889 Aug.	+ .284	08	1896 Apr.	+ .122	05
1895 Mar.	+ .075	11	1889 Aug. and Sept.	+ .226	06	1896 Apr. and May	+ .181	05
1895 Apr.	+0.086	±0.005	1889 Sept.	+ .258	07	1896 May and June	+ .142	13
			1890 May and June	+ .166	14	1896 June and July	+0.124	±0.008
The relative personal equation of these two observers seems to be a function of the time and a mean of the above values would therefore have but little meaning.			1890 July	+ .238	10			
			1890 July and Aug.	+ .237	14	Mean S.—P.=	+0.147	
			1890 Aug.	+0.278	±0.006	Prob. error* of a single value		±0.020
			Mean S.—M.=	+0.241				
			Prob. error* of a single value		±0.026			

* This value may be taken as a measure of the variability of the personal equation.

Each value in the table depends upon 8 or 10 nights of observation, 4 or 5 nights each before and after the exchange of observers, and may therefore be considered to be a mean value covering a period of from two weeks to a month or more. It is improbable that the variation of the relative personal equation from night to night is as small as would be inferred directly from the above table. The error due to personal equation, remaining in the deduced longitude after the exchange of observers, is one-half the difference between the mean value of the relative personal equation before the exchange of observers and its mean value after the exchange.

DISCUSSION OF ERRORS WHEN KEY AND CHRONOGRAPH ARE USED.

This discussion is based upon the supposition that the regular program for longitude observations when using an observing key and chronograph, consisting of 5 nights each before and after exchange of observers, has been carried out, and also that the method of selection of stars is the one formerly in use on primary longitude work in this Survey, in which a time set consisted of 10 stars, 5 before and 5 after reversal of the horizontal axis.

These sources of error are given the same order as those shown on pages 85–87 under the heading: Discussion of Errors when Transit Micrometer is Used.

First. An accidental error arising from the accidental errors of observations of 200 stars at each station. If the accidental error of observation of a single star be estimated at $\pm 0.^s10$, and this is surely a sufficiently large estimate to cover both the observer's errors and those instrumental errors which belong to the accidental class, the probable error of the final result arising from this cause would be $\pm 0.^s10 \div \sqrt{100} = \pm 0.^s010$.

Second. The statement on page 86 regarding the accidental error arising from the accidental errors in the adopted right ascensions of the stars used, is applicable to all methods of observing.

Third. For a statement regarding the errors due to the variation of the rate of the chronometer see page 86.

Fourth. Errors arising from the variation of the relative personal equation from night to night. These are probably among the largest errors involved in longitude determinations. A constant error, not eliminated by the exchange of observers, may possibly arise from this source if the temperature, altitude, moisture conditions, etc., are very different at the two stations. Other than this, the errors arising from this source belong to the accidental class when considered with reference to the computed difference of longitude and are exhibited in the residuals corresponding to the separate nights of observation.

Fifth. The statement concerning errors due to lateral refraction on page 86 is equally applicable here.

Sixth. No change is necessary in the statement on page 86 regarding the errors due to variation in the transmission time.

Seventh. The difference of the transmission time through the two signal relays enters as an error in the final result. This error is made very small in the present work of the Survey by the use of fast-acting signal relays which are as nearly alike as possible. It might be further reduced if each observer carried his own switchboard with him when exchange of stations is made.

As stated on page 87, if the difference in longitude which is being measured is large, say more than 30 minutes of time, it is well to abandon the practice of endeavoring to observe the same stars at both stations to such an extent as will bring the exchange of time signals near the middle of the time observations at each station. The error of right ascension thus introduced will be more than offset by the accuracy gained by the proper placing of the exchange.

Are there appreciable errors which are constant for the night in the time determinations or in the other operations involved in the determination of a longitude difference by the telegraphic method; and if so, what is the average magnitude of such errors? The excess of the probable error of a longitude difference computed as indicated on page 89 over its value as derived from the computed probable errors of the chronometer corrections at exchange is due to errors which are constant for and peculiar to each night. Using this principle[1] the error peculiar

[1] For the formulæ used in applying a similar principle to latitude observations, see pp. 119–123.

to a night has been computed from fifteen longitude determinations made since 1890. It was found that the error peculiar to each night, and therefore not capable of elimination by increasing the number of observations per night, expressed as a probable error, was ± 0.^{s}022, while the probable error in the result for a night arising from accidental errors of observation, and therefore capable of further elimination by increased observation, was ± 0.^{s}013. It should be noted that the errors discussed under all but the first heading above are each capable of contributing to the error peculiar to a night. It is likely that variation in the personal equation is the most potent cause of such errors. It is evident from the probable errors given above that very little is lost in ultimate accuracy if clouds interfere so as to cut off a part, say one-fourth, of the regular program of time observations (two sets of ten stars each), and that almost no gain in accuracy would result from lengthening the program.

Are there appreciable errors in a telegraphic determination of a difference of longitude which are constant for the interval of several days over which the determination extends; and, if so, what is the average magnitude of such errors? We may obtain an answer to this question by comparing the probable errors of longitude difference computed as on page 89 with the same probable errors as computed from the residuals developed in adjusting such a longitude net as that given in Appendix No. 2 of the Report for 1897. The excess of the last-named probable errors over the first-named is due to errors which are constant for the station during the time of occupation. From the published adjustment of the great longitude net referred to above (see pp. 246, 247, 255, of Report for 1897), after omitting the first eleven determinations (all made not later than 1872, and several involving trans-Atlantic cables) and the fifty-eighth determination (publication incomplete), it follows that the constant error peculiar to each longitude determination and not capable of elimination by increasing the number of nights per station, expressed as a probable error, is ±0.^{s}022, while the accidental error of the deduced difference of longitude, which is capable of further reduction by increasing the number of nights per station (beyond the standard number, ten), is ± 0.^{s}011. It follows that a reduction of the number of nights per station to six, or even four, would result in but a slight decrease in accuracy—about 10 per cent. Three sources of errors peculiar to a station in the order of their probable magnitude are those mentioned under the fourth, sixth, seventh, and fifth headings above, namely: Variation in personal equation, variation in transmission time (especially when a repeater interrupts a circuit), the difference of the two signal relay times, and possibly lateral refraction in some cases.

REDUCTION TO MEAN POSITION OF POLE.

This correction will be applied in the office in accordance with the Preliminary Results published annually by the International Geodetic Association (see p. 85).

A STATEMENT OF COSTS.

Since 1906 forty-two differences in longitude have been determined in the United States, using the transit micrometer. Forty-one were determined in four seasons. The average cost for the field work and preparing for the field, including all expenses and salaries, was $440. The average cost per difference for the various seasons varied from $360 to $550. The cost of a difference of longitude between two places will vary according to the conditions under which work is done, and consequently it should be planned to have the parties in the field when the weather may be expected to be most favorable. The work should be localized for any season as much as is possible. The longer the season the more economically should the work be done. If possible, the stations should be located near the line of the telegraph in order to avoid the delay and the expense of building a long line to the observatory. The determination of longitude differences telegraphically in remote regions, such as Alaska, may cost from three to six or more times the average cost of a difference in the United States.

No data are readily available showing the cost of the determination of longitudes telegraphically, using the key and chronograph. But owing to the necessity of exchanging observers for each difference of longitude and of observing over more nights than when the transit micrometer is used, it is probable that the cost would be from 25 to 50 per cent more than the costs stated above.

LONGITUDE BY THE CHRONOMETRIC METHOD.

The equipment, program of observations, and methods of computation pertaining to a determination of a difference of longitude by the chronometric method, in which chronometers transported back and forth between stations take the place of the telegraphic signals, may be most conveniently explained by giving a concrete example.

The longitude of a station at Anchorage Point, Chilkat Inlet, Alaska, was determined in 1894 by transporting chronometers between that station and Sitka, of which the longitude had previously been determined. At Anchorage Point observations were taken on every possible night from May 15 to August 12, namely on fifty-three nights, by the eye and ear method, using a meridan telescope. The hack or observing chronometer kept sidereal time, and there were also four other chronometers at the station, two keeping mean time and two sidereal. These four chronometers were never removed during the season from the padded double-walled box in which they were kept for protection against sudden changes of temperature and in which the hack chronometer was also kept when not in use. The instrumental equipment and procedure at Sitka was similar to that just described. A sidereal chronometer was the hack, and two other chronometers, one sidereal and one mean time, were used in addition. Nine chronometers, eight keeping mean time and one sidereal, were carried back and forth between the stations on the steamer *Hassler*.

Aside from the time observations, the programme of operations was as follows: Just before beginning the time observations at Anchorage Point, and again as soon as they were finished on each night, the hack chronometer was compared with the two mean time chronometers by the method of coincidence of beats (described on p. 96). These two were then compared with each of the two remaining (sidereal) chronometers at the station. These comparisons, together with the transit time observations, served to determine the correction of each chronometer to local time at the epoch of the transit observations. Whenever the steamer first arrived at the station, and again when it was about to leave, the hack chronometer was compared with the other station chronometers, as indicated above, was carried on board the steamer and compared with the nine traveling chronometers, and then immediately returned to the station and again compared with the other four station chronometers. On board the steamer the hack was compared by coincidence of beats with each of the eight mean time chronometers, and the remaining (sidereal) chronometer was then compared with some of the eight. The comparisons on shore before and after the trip to the steamer served to determine the correction of the hack at the epoch of the steamer comparisons. The steamer comparisons [1] determined the corrections of each of the traveling chronometers to Anchorage Point time. Similar operations at Sitka determined the corrections of the nine traveling chronometers to Sitka time as soon as they arrived and again just before they departed from Sitka. During the season the steamer made seven and a half round trips between the stations.

CARE OF CHRONOMETERS.

To secure the greatest possible uniformity of rate a chronometer should be kept running continuously, both when in use and when out of use between seasons of work. When it is allowed to remain stopped for a considerable time, the oil in the bearings tends to become gummy. When started again, the chronometer will tend to have a varying rate for some time until the effects of the stoppage have been worn off.

If a chronometer is to be shipped (by express, for example), and therefore is to be subjected presumably to comparatively violent handling and jarring, it should always be stopped and the balance wheel locked by gently inserting small wedge-shaped pieces of clean cork under it.

[1] In addition to the chronometer comparisons referred to in this paragraph the steamer chronometers and the station chronometers were each intercompared daily. This was done merely as a check upon their performance.

A running chronometer should always be protected as carefully as possible against jars, and especially against such sharp quick jars as result from setting it down upon a hard surface. Either the surface upon which it is set should be padded or a cushion should be carried with the chronometer. When it becomes necessary to carry a chronometer in the hand—as, for example, when a hack chronometer is carried back and forth between an observatory and a steamer in connection with chronometric longitudes—the gimbals should be locked to prevent the chronometer from swinging. It is important that the locking should be done in such a way that there will be no looseness and the corresponding tendency to a chucking motion. While the chronometer is being carried, swinging of the arm should be avoided as much as possible. Any swinging of the chronometer in azimuth is especially objectionable, as it tends to make it skip seconds and to damage it. Chronometers have been known to skip seconds, probably from this cause, even in the hands of an experienced and careful officer. On shipboard chronometers should be left free to swing in their gimbals, which should be so adjusted that the face of the chronometer will be approximately horizontal. Any change in this adjustment is apt to produce a change of rate.

COMPARISON OF CHRONOMETERS BY COINCIDENCE OF BEATS.

The process of comparing a sidereal and a mean time chronometer is analogous to that of reading a vernier. The sidereal chronometer gains gradually on the mean time chronometer, and once in about three minutes the two chronometers tick exactly together (one beat $=0^s.5$). As one looks along a vernier to find a coincidence, so one listens to this audible vernier and waits for a coincidence. As in reading a vernier one should look at lines on each side of the supposed coincidence to check, and perhaps correct the reading by observing the symmetry of adjacent lines, so here one listens for the approaching coincidence, hears the ticks nearly together, apparently hears them exactly together for a few seconds, and then hears them begin to separate, and notes the real coincidence as being at the instant of symmetry. The time of coincidence is noted by the face of one of the chronometers. Just before or just after the observation of the coincidence the difference of the seconds readings of the two chronometers is noted to the nearest half second (either mentally or on paper). This difference serves to give the seconds reading of the second chronometer at the instant of coincidence. The hours and minutes of both chronometers are observed directly. When a number of chronometers are to be intercompared, the experienced observer is able to pick out from among them two that are about to coincide. He compares those, selects two more that are about to coincide and compares them, and so on; and thus to a certain extent avoids the waits, of a minute and a half on an average, which would otherwise be necessary to secure an observation on a pair of chronometers selected arbitrarily.

At Sitka on July 13, 1894, it was observed that $18^h\ 30^m\ 08^s.00$ on chronometer No. 194 (sidereal) $=11^h\ 52^m\ 30^s.00$ on chronometer No. 208 (mean time); and that $11^h\ 15^m\ 35^s.50$ on chronometer No. 1510 (mean time) $=14^h\ 48^m\ 10^s.00$ on chronometer No. 387 (sidereal). It was known that at the epoch of the comparisons the correction of No. 194 to Sitka sidereal time was $-1^m\ 54^s.01$, and of No. 1510 to Sitka mean time was $-6^m\ 26^s.34$. The required corrections to No. 208 and No. 387 were computed as follows:

	h	m	s		h	m	s
Time by 194	=18	30	08.00	Time by 1510	= 11	15	35.50
Correction to 194	=	−01	54.01	Correction to 1510	=	− 6	26.34
Sidereal time	=18	28	13.99	Mean time	= 11	09	09.16
Sidereal time of mean noon	= 7	26	53.66	Correction mean to sidereal	=	+01	49.93
Sidereal interval	=11	01	20.33	Sidereal interval	= 11	10	59.09
Correction, sidereal to mean	=	−01	48.34	Sidereal time of mean noon	= 7	26	53.66
Mean time	=10	59	31.99	Sidereal time	= 18	37	52.75
Time by 208	=11	52	30.00	Time by 387	= 14	48	10.00
Correction to 208	=	−52	28.01	Correction to 387	=+3	49	42.75

The correction to reduce a sidereal to a mean time interval, or *vice versa*, may be taken from the tables in the back part of the American Ephemeris. The sidereal time of mean noon may be taken from that part of the Ephemeris headed "Solar ephemeris," and it should not be overlooked that it is the sidereal time of *local* mean noon that is required, and that, therefore, the longitude (approximate) of the station must be taken into account. The correction to be applied to Washington sidereal time of mean noon to obtain that for the station is the same as the correction to reduce a mean time interval equal to the longitude of the station from Washington to a sidereal interval.

COMPUTATION OF LONGITUDE FROM A SINGLE ROUND TRIP.

From the operations at Anchorage Point the correction of each station chronometer at the epoch of each set of time observations became known. The intercomparisons on shore before leaving for the steamer and after returning, together with the assumption that each station chronometer runs at a uniform rate between time sets, gave five separate determinations of the correction to the hack at the epoch of the steamer comparisons.

Thus, on June 18, 1894, at $3^h.45$ by its own face, the middle epoch of the steamer comparisons, the correction to the hack (No. 380), was

	m	s
By its own rate	−2	38.16 (weight ½).
By No. 4969 rated		38.30
By No. 2490 rated		38.26
By No. 207 rated		38.16
By No. 2637 rated		38.62 (weight ¼).
Mean	= −2	38.30
Weighted mean	= −2	38.25

The comparisons of No. 380 with No. 4969 at the station on this date, computed upon the supposition that No. 4969 ran at a uniform rate between preceding and following time observations, showed that the correction to No. 380 at $2^h.64$ by its face was $-2^m\ 38^s.34$, and at $4^h.36$ was $-2^m\ 38^s.25$. Assuming it to run uniformly between these epochs, its correction was $-2^m\ 38^s.30$ at $3^h.45$, as shown above.

An examination of the daily rates of the five chronometers showed that No. 2637 ran very irregularly, and that No. 380 did not run as regularly as the other three. Hence these chronometers were assigned less weight than the others, as indicated above.[1]

Using the weighted mean value for the correction to No. 380 at the epoch of the steamer comparisons these comparisons give the correction of each traveling chronometer on Anchorage Point time.

Similar operations at Sitka gave the correction to each traveling chronometer on Sitka time on each arrival at and departure from Sitka.

Computation of difference of longitude of Sitka and Anchorage Point.

FIRST TRIP STARTING FROM ANCHORAGE POINT.

Chronometers M. T. or Sid.	Anchorage Point, May 15		Sitka, May 17		Sitka, May 20		Anchorage Point, May 23	
	Chr. epoch	Correction	Chr. epoch	Correction	Chr. epoch	Correction	Chr. epoch	Correction
	h	h m s	h	h m s	h	h m s	h	h m s
M. T. 231	11.83	−0 03 31.39	7.54	−0 03 02.93	7.55	−0 03 02.14	7.65	−0 03 29.46
1 507	11.84	−0 01 03.88	7.81	−0 00 34.93	7.67	−0 00 33.73	7.65	−0 01 01.34
1 510	12.15	−0 03 42.50	7.75	−0 03 19.43	7.52	−0 03 28.22	7.75	−0 04 05.90
196	9.49	+2 26 28.51	5.20	+2 26 53.00	5.19	+2 26 46.08	5.29	+2 26 16.72
1 542	11.92	−0 02 55.84	7.53	−0 02 29.37	7.72	−0 02 31.83	7.81	−0 03 02.63
1 728	9.38	+2 34 40.23	5.08	+2 34 59.90	4.91	+2 34 46.00	5.23	+2 34 02.46
208	12.71	−0 42 08.24	8.17	−0 42 01.19	8.48	−0 42 35.76	8.56	−0 43 38.37
2 167	8.73	+3 18 39.99	4.39	+3 19 09.98	4.15	+3 19 12.09	4.59	+3 18 47.44
Sid. 387	11.97	+3 46 50.04	7.65	+3 47 22.97	7.78	+3 47 29.67	8.29	+3 47 09.31

[1] If considered desirable, the relative weights to be assigned to the station chronometers may be determined more accurately by the method outlined in the footnote on p. 100.

Computation of difference of longitude of Sitka and Anchorage Point—Continued.

FIRST TRIP STARTING FROM ANCHORAGE POINT—Continued.

Chronometers M. T. or Sid.	Total		At Sitka		Traveling		Daily rate	From Anchorage Point to Sitka		Correction at Sitka on Anchorage Point time	Difference of longitude
	Time	Rate	Time	Rate	Time	Rate		Time	Rate		
	d h	*s*	*d h*	*s*	*d h*	*s*	*s*	*d h*	*s*	*h m s*	*m s*
M.T. 23f	7 19.82	+ 1.93	2 24.01	+ 0.79	4 19.81	+ 1.14	+ 0.24	1 19.71	+ 0.43	−0 03 30.96	0 28.08
1 507	.81	+ 2.54	23.86	+ 1.20	19.95	+ 1.34	+ 0.28	19.97	+ 0.51	−0 01 03.37	28.44
1 510	.60	−23.40	23.77	− 8.79	19.83	−14.61	− 3.03	19.60	− 5.50	−0 03 48.00	28.57
196	.80	−17.79	23.99	− 6.92	19.81	−10.87	− 2.25	19.71	− 4.10	+2 26 24.41	28.59
1 542	.89	− 6.79	24.19	− 2.46	19.70	− 4.33	− 0.90	19.61	− 1.64	−0 02 57.48	28.11
1 728	.85	−37.77	23.83	−13.90	20.02	−23.87	− 4.94	19.70	− 8.99	+2 34 31.24	28.66
208	.85	−90.13	24.31	−34.57	19.54	−55.56	−11.54	19.46	−20.90	−0 42 29.14	27.95
2 167	.86	+ 7.45	23.76	+ 2.71	20.10	+ 4.74	+ 0.98	19.66	+ 1.78	+3 18 41.77	28.21
Sid. 387	20.32	+19.27	24.13	+ 6.70	20.19	+12.57	+ 2.60	19.68	+ 4.73	+3 46 54.77	28.20

In the form on page 97 the column headed "Chr. epoch" gives the face reading of the chronometer, expressed in hours and hundredths (rather than minutes and seconds) for convenience in computation. The corrections at Anchorage Point are to the local time of that station, and at Sitka to Sitka local time.

In the form above, the second and third columns give the elapsed chronometer time and the accumulated rate between the Anchorage Point steamer comparisons, and the fourth and fifth columns give the same quantities between the Sitka steamer comparisons. The second column minus the fourth, and the third minus the fifth are the traveling time (both ways) and the accumulated rate while traveling, from which the daily traveling rate as given in the eighth column becomes known. The ninth column gives the traveling time between steamer comparisons from Anchorage Point to Sitka, and the tenth column gives the accumulated rate during this interval computed by the use of the eighth column. This accumulated rate being applied as a correction to the chronometer correction on Anchorage Point time at the beginning of the trip gives the correction on Anchorage Point time on arrival at Sitka. This difference subtracted from the directly observed correction on Sitka time at that epoch, shown in the upper form, gives the required difference of longitude.

It should be noted that in this computation the traveling rate is supposed to be a constant during the round trip, but is not assumed to be the same as the rate while in port.

The longitude difference if computed from the return half of the trip, from Sitka to Anchorage Point, would necessarily by this process of computation be identical with that shown above.

If the steamer had stopped so short a time at Sitka that only one set of steamer comparisons had been made while there, as was frequently the case, the above computation would have been simplified in an obvious manner.

COMBINATION OF RESULTS.

The difference of longitude was thus computed from each traveling chronometer for each round trip, starting from Anchorage Point, the last half trip (7½ round trips being made) from Anchorage Point to Sitka, being omitted. A similar computation was also made for each round trip, starting from Sitka, the first half trip, Anchorage Point to Sitka, now being omitted.[1] Each of these computations would be subject to a constant error if the traveling chronometers had uniformly accelerated or uniformly retarded rates, but their mean is free from this error. One half of the computation also serves as a check on the other half.

[1] If the steamer had returned again to Anchorage Point, so as to complete eight round trips, all of the eight would have been used in the first computation; and in the second computation (round trips, starting from Sitka) the *last* trip from Sitka to Anchorage Point, combined with the *first* trip in the opposite direction, would have been used as the eighth round trip. This principle of computing the difference of longitude from the round trips starting from each station in turn, and combining the two results was used for the first time by Assistant C. A. Schott in 1857 in deriving the difference of longitude of Savannah, Ga., and Fernandina, Fla. (See Coast Survey Report for 1857, pp. 314–324.)

The method of combining these separate results is shown in the following form:

Difference of longitude between Sitka and Anchorage Point, Chilkat Inlet, Alaska.

SUMMARY OF RESULTS FROM SEVEN ROUND TRIPS, STARTING FROM ANCHORAGE POINT.

Chronometers, M. T. or Sid.		1st	2d	3d	4th	5th	6th	7th	Means $\Delta\lambda$	Weights
		s	s	s	s	s	s	s	s	
M. T.	231	28.03	26.36	28.36	28.19	28.45	28.19	28.18	27.97	3
	1507	28.44	29.06	29.18	28.26	28.27	28.20	28.54	28.56	4
	1510	28.57	29.25	29.00	28.52	28.63	28.06	28.58	28.66	7
	196	28.59	29.09	29.54	28.59	28.43	28.51	28.92	28.81	3
	1542	28.11	28.11	28.66	28.23	28.47	28.38	28.37	28.33	22
	1728	28.66	28.94	29.16	28.63	28.58	28.43	28.59	28.71	6
	208	27.95	27.40	28.21	28.19	28.42	28.42	28.09	28.10	6
	2167	28.21	28.56	28.90	28.55	28.68	28.27	28.64	28.54	17
Sid.	387	28.20	28.44	28.91	27.93	28.41	27.93	28.59	28.34	6
Mean		28.31	28.36	28.88	28.34	28.48	28.27	28.50	28.45	
Weighted mean		28.25	28.38	28.82	28.35	28.52	28.28	28.49	28.44	
Weight		3	1	2	2	2	1	2		

Weighted mean 0^h 0^m $28^s.44 \pm 0^s.05$

SUMMARY OF RESULTS FROM SEVEN ROUND TRIPS, STARTING FROM SITKA.

Chronometers, M. T. or Sid.		1st	2d	3d	4th	5th	6th	7th	Means $\Delta\lambda$	Weights
		s	s	s	s	s	s	s	s	
M. T.	231	28.87	28.78	28.74	28.39	28.37	28.71	28.11	28.57	3
	1507	27.69	29.08	29.11	27.76	28.78	27.93	28.64	28.43	4
	1510	28.37	28.88	28.82	27.91	28.83	28.10	28.58	28.50	7
	196	28.59	29.07	28.95	27.66	28.03	29.56	29.20	28.72	3
	1542	28.93	28.57	28.59	28.22	28.50	28.50	28.32	28.52	22
	1728	27.59	28.90	28.75	27.99	29.01	28.09	28.75	28.44	6
	208	27.71	28.03	28.52	28.58	27.88	28.76	27.65	28.16	6
	2167	28.24	28.71	28.80	28.27	28.77	28.31	28.49	28.51	17
Sid.	387	28.68	28.80	28.43	27.69	28.97	27.98	28.73	28.47	6
Mean		28.30	28.76	28.75	28.05	28.57	28.44	28.50	28.48	
Weighted mean		28.41	28.69	28.70	28.13	28.61	28.38	28.44	28.48	
Weight		1	2	2	2	2	2	2		

Weighted mean 0^h 0^m $28^s.48 \pm 0^s.05$

Final mean $\Delta\lambda$ $= +0^h\ 00^m\ 28^s.46 \pm 0^s.05$

Let N be the number of days during which the chronometers are depended upon to carry the time during each round trip, reckoned by adding to the "traveling time," as given in the sixth column of the form on page 98, the interval between each comparison of the hack chronometer with the traveling chronometers and the nearest (either before or after) time observation made at that station. The weight assigned to each trip is proportional to the reciprocal of N. This weighting depends upon the assumptions that errors in the computed longitude arising from the time determinations and from the chronometer comparisons are small as compared with those arising from variations in chronometer rates; that the time is carried by the combined station chronometers over the intervals during which they are depended upon with about the same degree of accuracy (due regard being paid to the length of the interval) as the combined traveling chronometers carry the time during the trip, and, finally, that the errors arising from the variations in the chronometer rates belong to the accidental class and are proportional to the square root of the length of the interval over which the time is carried.

WEIGHTS ASSIGNED TO SEPARATE CHRONOMETERS.

Even a cursory examination of such a table as that given on the preceding page shows that some chronometers run much more uniformly than others, and therefore furnish determinations of the longitude difference which are entitled to greater weight. Let $l_1, l_2, l_3, \ldots l_n$ be the derived values of the difference of longitude as given by one chronometer on the different trips, and let l be their mean. Let n be the number of trips. Then, by the ordinary laws of least squares, assigning equal weights to the separate trips, the probable error of any one of these l's is

$$\pm\sqrt{\frac{(0.455)\,[(l-l_1)^2+(l-l_2)^2+\quad\ldots\quad(l-l_n)^2]}{n-1}}$$

The weight p, inversely proportional to the square of this probable error to be assigned to a chronometer, is proportional to

$$\frac{n-1}{[(l-l_1)^2+(l-l_2)^2+\quad\ldots\quad(l-l_n)^2]}$$

The computation of weights may be put in the following convenient tabular form:

COMPUTATION OF WEIGHTS.

From the seven round trips starting from Anchorage Point.

Chronometer l	231 27s.97	1507 28s.56	1510 28s.66	196 28s.81	1542 28s.33	1728 28s.71	208 28s.10	2167 28s.54	387 28s.34
$l-l_1$	−.06	+.12	+.09	+.22	+.22	+.05	+.15	+.33	+.14
$l-l_2$	+1.61	−.50	−.59	−.28	+.22	−.23	+.70	−.02	−.10
$l-l_3$	−.39	−.62	−.34	−.73	−.33	−.45	−.11	−.36	−.57
$l-l_4$	−.22	+.30	+.14	+.22	+.10	+.08	−.09	−.01	+.41
$l-l_5$	−.48	+.29	+.03	+.38	−.14	+.13	−.32	−.14	−.07
$l-l_6$	−.22	+.36	+.60	+.30	−.05	+.28	−.32	+.27	+.41
$l-l_7$	−.21	+.02	+.08	−.11	−.04	+.12	+.01	−.10	−.25
$(l-l_1)^2$	.00	.01	.01	.05	.05	.00	.02	.11	.02
$(l-l_2)^2$	2.59	.25	.35	.08	.05	.05	.49	.00	.01
$(l-l_3)^2$	.15	.38	.12	.53	.11	.20	.01	.13	.32
$(l-l_4)^2$	.05	.09	.02	.05	.01	.01	.01	.00	.17
$(l-l_5)^2$	.23	.08	.00	.14	.02	.02	.10	.02	.00
$(l-l_6)^2$	.05	.13	.36	.09	.00	.08	.10	.07	.17
$(l-l_7)^2$	.04	.00	.01	.01	.00	.01	.00	.01	.06
$\Sigma(l-l_n)^2$	3.11	.94	.87	.95	.24	.37	.73	.34	.75
By 2d comb.*	.47	2.30	.89	2.73	.30	1.78	1.22	.36	1.32
Mean of 2	1.79	1.62	0.88	1.84	0.27	1.08	0.98	0.35	1.04
n−1	6	6	6	6	6	6	6	6	6
$\frac{n-1}{\Sigma(l-l_n)^2}$	3.3	3.7	6.8	3.3	22.2	5.6	6.1	17.0	5.8

* From similar results from seven round trips starting from Sitka.

A similar computation was made using the seven round trips starting from Sitka, the results of which are shown in the line marked "by 2d combination," and the weights were derived from the mean results of the two computations.[1]

DISCUSSION OF ERRORS.

The error in a difference of longitude observed and computed as indicated in the preceding sections depends upon the errors in the transit time observations, errors in the comparison of chronometers, errors arising from variations in the rates of chronometers, and, finally, the relative personal equation of the two observers concerned.

[1] The relative weights to be assigned to the station chronometers when they are used to determine the correction of the hack at the epoch of the steamer comparisons might be computed by an analogous process. Let O be the correction to a chronometer at the epoch of transit time observations as determined from those observations. Let I be its correction at that same epoch interpolated between its observed corrections at the last preceding and first following transit time observations on the assumption that its rate during that interval is constant. For a group of chronometers whose corrections are all determined a number of times in succession by the same transit observations, the relative weights are evidently proportional to $\frac{1}{\Sigma(I-O)^2}$.

The errors in the time observations will in general be very small in comparison with the other errors affecting the result. For the probable magnitude of the time errors see the first part of this publication. In Appendix No. 3 of the Report for 1894 and in No. 3 of 1895 may be found detailed statements of the results of several determinations of longitude by the chronometric method which will serve to give a concrete idea of the magnitude of the errors involved in such determinations. The relative magnitude of the errors arising from the time determinations increases as the time, N (see p. 99), required for a round trip decreases.

The errors made in comparing chronometers by the method of coincidences are negligible in their effect upon the final result. The checks obtained during the intercomparisons of chronometers show that the probable error in a single comparison is about $\pm 0^s.01$, corresponding to a probable error of about $\pm 4^s$ in estimating the time of coincidence of ticks.

The errors arising from variations in the rates of chronometers are by far the most serious class of errors involved in chronometric determinations of longitude. The table of results given on page 99 gives a fair indication of the magnitude of the errors to be expected from this source.

The various traveling chronometers are subjected to variations of temperature, humidity, and barometric pressure, and to disturbances arising from the motion of the ship, which are *common to them all.* Do these common conditions produce variations in rate which are common to all the chronometers, and therefore introduce a common error into the various values of the longitude difference resulting from any one trip? An examination of the results of six chronometric determinations of longitude in Alaska, printed in the 1894 and 1895 Reports, indicates that such errors in the deduced longitudes, common to all the chronometers on a given trip, are exceedingly small upon an average—so small that they are concealed by the accidental errors.

Chronometers are compensated for temperature as well as possible by the maker, but such compensation is necessarily somewhat imperfect. In general, however, this compensation is so nearly perfect that little or nothing is gained in accuracy by deriving and using temperature coefficients connecting the temperature and the rate. There are occasional exceptions; for example, the Hutton chronometer No. 194 (see pp. 77–78 of the Report for 1894) shows a very large variation in rate due to change of temperature.

In considering the errors due to variations in chronometer rates it should not be overlooked that the station chronometers are depended upon to carry the time over the interval from the nearest time observations to the steamer comparisons in precisely the same manner in which the traveling chronometers are depended upon during the trip. It is because of this fact that it may be desirable during periods of very bad weather to supplement the transit observations upon stars by transit observations upon the sun, as indicated on page 51, or in low latitudes by theodolite or vertical circle observations for time, or even by sextant observations for time.

Unless the relative personal equation is eliminated from the computed longitude it is apt to be one of the largest errors affecting the mean result, except when the round trips are very long or very few chronometers are carried. It may be eliminated by any of the methods suggested on pages 90–93.

Assuming that the relative personal equation is eliminated by direct determination or otherwise, the error of the mean result of a chronometric longitude determination will be nearly inversely proportional to the square root of the number of chronometers carried (provided the stations are supplied with a sufficient number of good chronometers to make the shore errors small), to the square root of the number of round trips, and the square root of the average value of N (the interval over which the time is carried by the chronometers). It will depend very intimately upon the quality of the chronometers and upon the care with which they are protected from temperature changes and jars. It will be affected very little by an increase in the errors of the time observations proper, resulting from very fragmentary observations on cloudy nights or from substituting some more approximate method for transit observations upon stars.

From the above principles and the numerical values given in Appendix No. 3 of the 1894 Report and in No. 3 of the 1895 Report, one may make an estimate of the errors to be expected if the above elaborate plan of operations can be carried out only in part, as, for example, when an observer determines the longitude of a new station by making a single trip to it, carrying a few chronometers only and making all time observations at both ends of the trip himself.

In connection with any plan of operations which involves long intervals between the arrival at and the departure from a given station, it should be kept in mind that the computation usually involves the assumption that the rates of the traveling chronometers are the same on the trip to the station as on the return trip, and therefore a long stay at the station is apt to increase the error of the final result by giving the chronometers a long time to acquire new rates. Under extreme conditions it may sometimes be well to avoid this assumption and to use a separate traveling rate for each half trip derived from observations just preceding or following that half trip.

INTRODUCTORY.

A measurement of the meridional zenith distance of a known star, or other celestial object, furnishes a determination of the latitude of the station of observation. In the zenith telescope, or Horrebow-Talcott,[1] method of determining the latitude, there is substituted for the measurement of the absolute zenith distance of a star the measurement of the small difference of meridional zenith distances of two stars culminating at about the same time, and on opposite sides of the zenith. The effect of this substitution is the attainment of a much higher degree of precision, arising from the increased accuracy of a differential measurement, in general, over the corresponding absolute measurement; from the elimination of the use of a graduated circle from the essential part of the measurement; and from the fact that the computed result is affected, not by the error in estimating the absolute value of the astronomic refraction, but simply by the error in estimating the very small difference of refraction of two stars at nearly the same altitude.

Because of its great accuracy, combined with convenience and rapidity, the Horrebow-Talcott method has become the only standard method of this Survey. For other methods of determining the latitude, involving in most cases absolute measurements of zenith distance or altitude, the reader is referred to treatises on astronomy.

The method of determining the latitude by observing the time of transit of a star across the prime vertical, is one which is capable of a very high degree of accuracy and is well adapted to field use, as the effects of instrumental errors may be readily eliminated. To determine the latitude of a station by this method, the times of transit of various stars (of positive declination less than the latitude) across the plane of a transit placed approximately in the prime vertical are observed. The inclination of the transverse axis is determined accurately with a striding level. The effects of error of collimation and pivot inequality are eliminated by reversal of the axis. The effects of azimuth error (deviation of the instrument from the prime vertical) and of constant errors in the observed times (personal equation) are eliminated by observing some stars to the eastward of the zenith and others to the westward. The declinations of the stars observed must be accurately known, as the declination errors enter directly into the latitude at about their full value, but the right ascensions need be known but approximately.

This method has been little used by this Survey; perhaps because more time is required to prepare an extended observing list than in the zenith telescope method, but it may be found useful in the future. If the only instrument available is a theodolite having a good striding level, but not equipped for observations by the zenith telescope method, observations in the prime vertical will give the best possible determination of the latitude. (For details in regard to this method, see Chauvenet's Astronomy, Vol. II, pp. 238–271, and Doolittle's Practical Astronomy, pp. 348–377. For an interesting early test of the method [1827] by Bessel, with a very small portable instrument, see Astronomische Nachrichten, Vol. 9, pp. 413–436.)

GENERAL INSTRUCTIONS FOR LATITUDE WORK.

1. In order that the records and computations of the latitude work of this Survey may be uniform in character and that there may be approximately the same accuracy in the results, some general directions are given here which should be carried out by all observers of this Survey,

[1] See p. 245 of Appendix 14, Report for 1880, for some general remarks on Talcott's method.

engaged upon this class of work, unless they are directed otherwise by special instructions or unless exceptional circumstances are encountered which make changes necessary or desirable.

2. The Horrebow-Talcott method should be followed, using the zenith telescope or the meridian telescope. (See p. 8 for description of the latter instrument. The zenith telescope is described below.)

3. A pair of stars should be observed only once at a given station, unless some gross error is discovered, in which case the pair may be reobserved. Not more than two stars should be observed at one setting of the instrument. A star may be observed on more than one night, if paired with a different star on each night.

4. A sufficient number of pairs should be observed at a station to make it reasonably certain that the probable error of the mean result is not greater than $\pm 0''.10$ (see directions for procedure in making the office computation). No additional expenditure of time or money should be made in trying to reduce the probable error below this limit. In no case, however, should the number of pairs observed at a station be less than 10.

5. No determination of the micrometer value should be made in the field, as this value is computed at the office from the regular observations for latitude.

6. The pairs observed should be so selected that the algebraic sum of the measured micrometer differences in turns at a station is less than the total number of pairs. This sum should be made small, in order that the computed latitude may be nearly free from any effect of error in the mean value of the micrometer screw.

7. The stars observed upon should be taken from "The Preliminary General Catalogue of 6188 Stars for the Epoch 1900" by Lewis Boss, which was published by the Carnegie Institution of Washington in 1910.

8. Duplicates of the latitude records, in the form of entries in the latitude computation sheets, should be made and checked as the work progresses. Only such portions of the latitude computations should be made in the field as are necessary to ascertain the degree of accuracy secured.

9. The duplicates and computations, both complete and incomplete, for each station should be sent to the office by registered mail, as soon as practicable after the completion of the occupation of the station. Each book of original records should be sent to the office by registered mail soon after the last of the corresponding duplicates and computations have been forwarded, but not so soon as to arrive in Washington by the same mail. It is desirable to have the records and computations sent to the office promptly, in order to avoid their possible loss.

10. Original descriptions of stations should be inserted in the original record of latitude observations and a duplicate description of each station should be written in a volume kept especially for the purpose. This volume should be sent to the office at the close of a season's work.

11. The form of record of observations and of field and office computations of results should conform to those shown in this publication.

These General Instructions will be referred to from time to time in the succeeding text.

DESCRIPTION OF THE ZENITH TELESCOPE.

Illustration No. 13 shows one of the best zenith telescopes now in use in this Survey. This instrument, Zenith Telescope No. 4, was originally made by Troughton & Simms, of London, in 1849, and was remodeled at the Coast and Geodetic Survey Office in 1891. It carries a telescope with a clear aperture of about 76mm (3 inches), and a focal length of about 116.6cm (46 inches). The magnifying power with the eyepiece ordinarily used is 100 diameters. Two latitude levels are used instead of one, to secure increased accuracy. Each of these levels carries a graduation which is numbered continuously from one end to the other (instead of each way from the middle), the numbering of the upper one running from 0 to 50 and of the lower from 60 to 110. A 2mm division on the upper level has a value of about $1''.6$ and on the lower about $1''.4$. The vertical axis of the instrument is in the vertical plane in which the telescope swings. The clamp arm, perforated for the sake of lightness, gives the telescope a

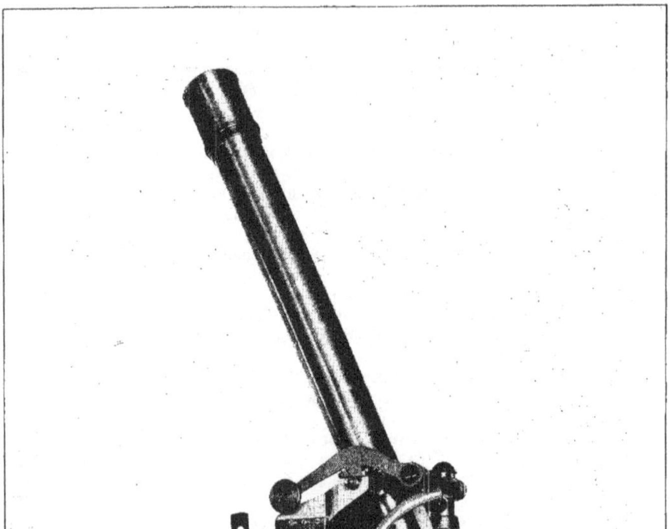

ZENITH TELESCOPE.

marked degree of stability in so far as changes of inclination are concerned. The eyepiece micrometer, arranged to measure zenith distance, has a value of about 45″ per turn, and the micrometer head is graduated to hundredths of a turn.

The better known type of zenith telescope, in which the telescope is mounted eccentrically on one side of the vertical axis instead of in front of it, is also in use in the Survey. The meridian telescopes described on page 8 are extensively used for latitude determinations, as well as for time.

In latitude work with the meridian circle at astronomic observatories the instrument is usually fitted with a reversing prism. By rotating this prism the apparent motion of the star is changed from the direction right to left to the direction left to right or vice versa. A pointing is made on the star before it transits, the prism is reversed, and a second pointing is made after the transit. The observer may always place the wire above the center of the star's image (or below) but as the image is reversed by the prism, one of the pointings is made on the south side of the center of the star and the other pointing on the north side. The mean of the two pointings will be free from any constant or systematic error in the bisection of the star. It is believed that the systematic error of bisection does not affect the results of latitude observations made by the Talcott method, except to a small degree due to the fact that an observer's systematic error of bisection may be slightly different for stars of different magnitude. A pair may be composed of stars of very different magnitudes. The reversing prism need not be used in any latitude observations by the Talcott method which are made for the usual geodetic or geographic purposes.

SUPPORT FOR THE INSTRUMENT.

The support for the latitude instrument most frequently used in this survey is a wooden tripod made of lumber about 6 inches square in cross-section, well braced and set firmly in the ground to a depth of from 1 to 3 feet, depending on the nature of the soil. Piers made of brick, of cement blocks, or of concrete are also used. The concrete pier is not as satisfactory as the other types, if it is used very soon after it is constructed. When latitude and azimuth are both observed at a station the same pier may be used for mounting both the latitude instrument and the theodolite. A type of pier used by some of the parties of this Survey is shown in illustration No. 24 and is described on page 140.

OBSERVATORIES AND OBSERVING TENTS.

At the field stations only a temporary structure to protect the instrument from wind during the observations and from rain during the stay at the station is needed. The observer is seldom at a station more than a week after everything has been made ready for the observing, and an observatory such as is shown in illustration No. 14, built of rough lumber, answers every purpose. It is advisable to have 2 doors in the observatory to insure the free circulation of air. No part of the building should touch the ground except at the corners. The roof may be made water-tight by boards or a covering of felt or tar paper. A canvas sheet is sometimes carried with the outfit and the roof is made by stretching this sheet over the rafters and tying it to the sides of the observatory. The canvas may be removed during the observations, thus leaving the whole top of the observatory open to the sky.

When a station is located in a town, although for only a short time, the observatory should as a rule be made neatly, of smooth lumber, as shown in illustration No. 15. Buildings at permanent latitude stations need not be discussed here, as this publication deals only with observations made for geodetic or geographic purposes.

An observing tent such as is shown in illustration No. 16 or in illustration No. 17 is more frequently used on latitude work than the wooden observatory, and it has the great advantage that it is easily transported and quickly set up. Except on mountain peaks or at other places where transportation is difficult the tent has a floor similar to that used with an observatory.

Where a floor or platform is not used, the observer must be extremely careful not to shift his weight during the interval between the pointing on a star and the reading of the levels, and in this case the bubble readings must be made by an attendant who must also stand in one place without shifting his weight from the time the observation is made until the level is read.

ADJUSTMENTS.

When setting up the instrument place two of the foot screws in an east and west line. The level correction may then be kept small during the progress of the observations by using one foot screw only.

The vertical axis may be made approximately vertical by use of the plate level, if there is one on the instrument, and the final adjustment made by using the latitude level. The position of the horizontal axis may then be tested by readings of the striding level. If the horizontal axis is found to be inclined, it must be made horizontal by using the screws which change the angle between the horizontal and vertical axes, if the instrument is of the old form. With the new form of instrument (illustration No. 13), or with a meridian telescope, the two axes will always remain so nearly at right angles that no means for making this adjustment is needed. With these instruments the vertical axis may be made vertical by using both the striding level and the latitude level at the same time.

The eyepiece and objective should be carefully focused as indicated on pages 14 and 15. It is important that the focus of the objective should be kept constant during the stay at a station, since the angular value of one turn of the eyepiece micrometer is depended upon to remain constant for the station. However, the results of the determination of the value of a turn of the micrometer vary in some cases as much as 0''.13, corresponding to a range of about 3.3 millimeters in the distance between the objective and the micrometer lines (see p. 129). In connection with the common habit of carefully keeping the draw tube clamped for the purpose of holding the micrometer value constant, it is interesting to note that while in the field in 1905 Assistant W. H. Burger focused zenith telescope No. 2 five times in rapid succession with a range of only 0.1 millimeter in the position of the sliding tube.

The movable micrometer thread with which all pointings are to be made must be truly horizontal. This adjustment may be made, at least approximately, in daylight after the other adjustments. Point, with the movable thread, upon a distant well-defined object, with the image of that object near the apparent right-hand side of the field of the eyepiece, and with the telescope clamped in zenith distance. Shift the image to the apparent left-hand side of the field by turning the instrument about its vertical axis. If the bisection is not still perfect, half the correction should be made with the micrometer and half with the slow-motion screws which rotate the whole eyepiece and reticle about the axis of figure of the telescope. Repeat, if necessary. The adjustment should be carefully tested at night after setting the stops by taking a series of pointings upon a slow-moving star as it crosses the field with the telescope in the meridian. If the adjustment is perfect, the mean reading of the micrometer before the star reaches the middle of the field should agree with the mean reading after passing the middle, except for the accidental errors of pointing. It is especially important to make this adjustment carefully, for the tendency of any inclination is to introduce a *constant* error into the computed values of the latitude.

The line of collimation (see p. 13) as defined by the middle vertical line of the reticle must be very nearly perpendicular to the horizontal axis. If the instrument is a meridian telescope, or of the form shown in illustration No. 13, this adjustment may be made as for a transit (p. 15) by reversing the horizontal axis in the wyes. If the instrument is of the form in which the telescope is to one side of the vertical axis, the method of making the test must be modified accordingly. It may be made by using two collimating telescopes which are pointed upon one another in such positions that the zenith telescope may be pointed first upon one and then upon the other with no intermediate motion except a rotation of 180° about the horizontal axis. It may be made as for an engineer's transit, but using two fore and two back points, the distance apart of each pair of points being made double the distance between the vertical axis and the axis of collimation of the telescope. A single pair of points at that distance apart may be used and the horizintal circle trusted to determine when the instrument has been turned

No. 14.

OBSERVATORY.

No. 15.

OBSERVATORY.

180° in azimuth. Or a single point at an approximately known distance may be used and the horizontal circle trusted as before, and a computed allowance made on the horizontal circle for the parallax of the point when the telescope is changed from one of its positions to the other. Thus, let d = the distance of the vertical axis from the axis of collimation of the telescope, D = the distance to the point, and p = the parallax for which correction is to be made; then, in seconds of arc:

$$p = \frac{2d}{D \sin 1''}$$

If one considers the allowable limit of error in this adjustment (see p. 134) it is evident that refined tests are not necessary, and that a telegraph pole or small tree, if *sufficiently distant* from the instrument, may be assumed to be of radius = d, and the adjustment made accordingly.

The stops on the horizontal circle must be set so that when the abutting piece is in contact with either of them the line of collimation is in the meridian. For this purpose the chronometer correction must be known roughly—within one second, say. Set the telescope for an Ephemeris star which culminates well to the northward of the zenith, and look up the apparent right ascension for the date. Follow the star with the middle vertical line of the reticle, at first with the azimuth motion free and afterwards using the tangent screw on the horizontal circle, until the chronometer, corrected for its error, indicates that the star is on the meridian. Then clamp a stop in place against the abutting piece. Repeat for the other stop, using a star which culminates far to the southward of the zenith. It is well, if time permits, to test the setting of each stop by an observation of another star before commencing latitude observations.

The correction to the chronometer may be obtained by observations on the sun or stars with a sextant or a vertical circle (see pp. 52–56), by observing the time of transit of stars with a theodolite, or by using the zenith telescope as a transit instrument. With the zenith telescope in good adjustment and approximately in the meridian and the sidereal time known within several minutes, the chronometer time of transit of a star near the zenith is noted. This observation gives a close approximation to the chronometer error. Then a north star of high declination is used and the telescope is put more nearly in the meridian by the method explained above. Next the chronometer time of transit of a second zenith star is observed, which will usually give the chronometer correction within a second. With this value of the chronometer correction the telescope may be put closely enough in the meridian for observing.

The finder circle must be adjusted to read zenith distances (see p. 16).

THE OBSERVING LIST.

The Boss catalogue[1] of 6188 stars is now available, and is at present the best list from which to select pairs of stars. (See paragraph 7 of General Instructions, p. 104.) The latitude of the station should be obtained to the nearest minute from a map, a triangulation station, or from preliminary observations on the sun or stars. In the Boss catalogue the declinations of the stars are given and the observing list may be made out like the form shown below. Any other arrangement of the data may be used. To find *all* available pairs in a given list one may, for each star in succession within the zone of observation, 45° each way from the zenith, subtract the declination from twice the latitude and then compare this difference with the declination of each star in the list within the *following* 20^m of right ascension. Any star whose declination[2] is within 20′ of the above difference will combine with the star under consideration to make a pair, provided the other conditions stated below are fulfilled. By proceeding thus every available pair will be found.[3]

[1] Preliminary general catalogue of 6188 stars for the epoch 1900. Lewis Boss, Carnegie Institution of Washington, 1910.

[2] Or $180° - \delta$ for subpolars.

[3] At stations in Alaska there are but few stars in the zone extending 45° northward from the zenith as compared with the corresponding zone to the southward, and the above process may be improved by taking in succession only stars to the north of the zenith and comparing each with stars in *both the preceding and the following* 20^m. To make the search with a subpolar star subtract $180° - \delta$ from twice the latitude and pair with any star whose declination is within 20′ of this difference, provided its right ascension differs from that of the subpolar anywhere from $11^h\,40^m$ to $12^h\,20^m$.

Observing list (Form 1).

[St. Anne, Ill., June 25, 1908. Zenith telescope No. 4. $\phi=41° 01'.3$. Search factor $=2\phi=82° 03'$.]

Star No. Boss catalogue	Mag.	Right ascension	Declination δ	Difference between δ's	$\Sigma\delta$= sum of declinations	$\Sigma\delta-2\phi$	N−S= a* ($\Sigma\delta-2\phi$)	Star north or south	Setting = ½ difference of δ's	Turns
		h m s	° ′	° ′	° ′	′			° ′	
4327	4.5	16 55 22	82 11					N		12
4379	4.9	17 11 53	−0 21	82 32	81 50	−13	−17	S	41 16	28
4441	5.9	17 28 13	28 28					S		10
4494	5.8	17 42 04	53 50	25 22	82 18	+15	+20	N	12 41	30
4623	5.1	18 13 22	64 22					N		24
4651	5.4	18 18 45	17 47	46 35	82 09	+ 6	+ 8	S	23 18	16
4669	5.9	18 22 26	29 47					S		20
4711	5.5	18 31 52	52 17	22 30	82 04	+ 1	+ 1	N	11 15	20

* a = number of turns of the micrometer screw in one minute of arc = 1.34. The value of one turn of the micrometer screw = 44″.650.

The approximate mean right ascensions and declinations for the observing list are obtained for the time of the observations by multiplying the annual variation by the number of years elapsed since the epoch of the catalogue and combining the products algebraically with the right ascension and declination given in the catalogue used.

In the above form there is no column for zenith distances. The setting for a pair is one-half the difference between the declinations of the two stars of a pair. To get the values in the column N − S subtract double the latitude (for station St. Anne, 82° 03′) from the sum of the declinations of the two stars and multiply the result in minutes of arc by the number of turns of the micrometer screw in a minute of arc. N − S is positive if the north star has the greater zenith distance and is negative if the south star has the greater zenith distance. The center of the comb in the micrometer eyepiece is called 20, and increasing readings on the graduated head go with increasing zenith distances. Then the setting of the micrometer wire for any north star is $20+\frac{N-S}{2}$ and for any south star $20-\frac{N-S}{2}$. These settings are given in the last column of the above table.

When one star of the pair is a subpolar, the finder circle setting is $90°-\frac{1}{2}\Sigma\delta$. N − S in this case is a (180° − difference of δ's -2ϕ) and is positive or negative according as the north star has the greater or lesser zenith distance. The setting of the micrometer wire will be given by the same general expression as above.

For the purposes of the observing list it is sufficiently accurate to know the mean right ascensions to within one second and the declinations and derived quantities to the nearest minute of arc. The approximate reading of the turns is given to facilitate identification of the stars and to enable the observer to put the micrometer line approximately in position before the star enters the field of view. The middle reading of the micrometer comb is called 20 to avoid negative readings.

If the Ten Year Catalogues for 1880 and 1890 and the Nine Year Catalogue for 1900, by the Royal Observatory at Greenwich, are used, then the form of the observing list could be made to advantage in a manner somewhat different from that shown above, for in those publications the north polar distances are given instead of the declinations. The list may be similar to that shown below, where the settings, etc., are derived from the north polar distances of the stars. In the first column of the example are given the Boss catalogue numbers, though the stars are also in the lists of the Greenwich catalogues mentioned above. They are the same stars as those in the first form of star list.

No. 16.

OBSERVING TENT.

No. 17.

OBSERVING TENT.

Observing List (Form 2).

[St. Anne, Ill., June 25, 1908. Zenith Telescope No. 4. $\phi = 41° 01'.3$. Search factor $= 180° - 2\phi = 97° 57'$.]

Star No., Boss catalogue	Mag.	α	North polar distances and difference	Sum of N. P. D.'s; and search factor minus sum of N. P. D.'s	N−S*	Star north or south	Setting = ½ dif. of N. P. D.'s	Turns
		h m s	° ′	° ′			° ′	
4327	4.5	16 55 22	7 49			N		12
4379	4.9	17 11 53	90 21	98 10		S	41 16	28
			82 32	−13	−17			
4441	5.9	17 28 13	61 32			S		10
4494	5.8	17 42 04	36 10	97 42		N	12 41	30
			25 22	+15	+20			
4623	5.1	18 13 22	25 38			N		24
4651	5.4	18 18 45	72 13	97 51		S	23 18	16
			46 35	+ 6	+ 8			
4669	5.9	18 22 26	60 13			S		20
4711	5.5	18 31 52	37 43	97 56		N	11 15	20
			22 30	+ 1	+ 1			

* N−S = c (search factor − sum of N. P. D.'s), where c = number of turns of the micrometer screw in one minute of arc = 1.34. The value of one turn of the micrometer screw = 44″.650.

When a subpolar star is used slight changes will be necessary, similar to those described for the case where the observing list is prepared in terms of the declinations.

Among the requisites for a pair of stars for an observing list, are, that their right ascensions shall not differ by more than 20^m, or $12^h \pm 20^m$ when a subpolar is used, to avoid too great errors arising from instability in the relative positions of different parts of the instrument; nor by less than about 1^m, that interval being required to take the readings upon the first star and prepare for the second star of a pair; that their difference of zenith distances shall not exceed the half length of the micrometer comb, 20′ for many instruments; that each star shall be bright enough to be seen distinctly, not fainter than the seventh magnitude for the larger instruments; and that no zenith distance shall exceed 45°, to guard against too great an uncertainty in the refraction. The third of the above conditions may be used more conveniently in this form; the sum of the two declinations must not differ from twice the latitude by more than 20′. The total range of the list in right ascension is governed by the hours of darkness on the proposed dates of observation.

In the list of pairs resulting directly from the search there will be many pairs which overlap in time. A feasible observing list may be formed by omitting such pairs that among the remainder the shortest interval between the last star of one pair and the first star of the next is not less than 2^m. In that interval a rapid observer can finish the readings upon one pair and set for the next, under favorable circumstances. The omitted pairs may be included in a list prepared for the second or third night of observation. It will frequently be found that the same star occurs in two or more different pairs. Such pairs may be treated like those which overlap in time.[1]

DIRECTIONS FOR OBSERVING.

All adjustments having previously been made, set for the first star and await it with the bubble of the latitude level nearly in the middle of the tube, and with the micrometer line at that part of the comb at which the star is expected, as shown by the observing list. Watch the chronometer so as to know when to expect the star. When the star enters the field, place the micrometer line approximately upon it. As soon as the star comes within the safe observing limits of the field bisect it carefully. As the star moves along watch the bisection and correct

[1] Past records furnish abundant evidence that observations made by pointing twice upon a close zenith star, once in each position of the instrument, give results of a low order of accuracy, probably because of the hurry with which the observations must be made, and of the fact that one or both of the observations must be made out of the meridian. It is therefore not advisable to make such observations.

it if any error is detected. Because of momentary changes in the refraction, the star will usually be seen to move along the line with an irregular motion, now partly above it and now partly below. The mean position of the star is to be covered by the line.[1] It is possible, but not advisable, to make several bisections of the star while it is passing across the field. As soon as the star reaches the middle vertical line of the diaphragm read off promptly from the comb the whole turns of the micrometer, read the level, and then the fraction of a micrometer turn, in divisions, from the micrometer head. Set promptly for the next star, even though it is not expected soon. In setting for the second star of a pair all that is necessary is to reverse the instrument in azimuth and set the micrometer line to a new position. The abutting piece must be brought gently against the stop and the circle securely clamped in that position.

Especial care should be taken in handling the micrometer screw, as any longitudinal force applied to it produces a flexure of the telescope which tends to enter the result directly as an error. The last motion of the micrometer head in making a bisection should always be in one direction (preferably that in which the screw acts positively against its opposing spring), to insure that any lost motion is always taken up in one direction. The bubble should be read promptly, so as to give it as little time as possible to change its position after the bisection. The desired reading is that at which it stood at the instant of bisection. Avoid carefully any heating of the level by putting the reading lamp, warm breath, or face any nearer to it than necessary. During the observation of a pair the tangent screw of the setting circle must not be touched, for the angle between the telescope and the level must be kept constant. If it is necessary to relevel, to keep the bubble within reading limits, use the tangent screw which changes the inclination of the telescope. Even this may introduce an error, due to a change in the flexure of the telescope, and should be avoided if possible. It is desirable to relevel the instrument from time to time between pairs, so as to keep the level correction small, less than one division of the level if possible.

Occasionally the approximate time should be noted at which the star being observed crosses the middle vertical line of the diaphragm, so as to make sure that the adjustment of the stops in azimuth remains satisfactory. It is desirable (though not necessary) to have a recorder. He should be a man above the average in intelligence, and should be able to prepare an observing list after a little practice and to assist in computing the results. It is not economical to take a man from place to place unless he can assist in the computations. The recorder may count seconds aloud from the face of the chronometer in such a way as to indicate when the star is to culminate. Such counting aloud serves a double purpose. It is a warning to the observer to be ready and it indicates where to look for the star if it is faint and difficult to find. It also gives for each star a rough check upon the position of the azimuth stops. It is only a rough check, because the observing list gives mean right ascensions instead of apparent right ascensions for the date, but it is sufficiently accurate (see p. 119). The observer, or recorder, can easily make allowance for the fact that all stars (except circumpolars) will appear to be too early or too late, according to the observing list, by about the same interval, 0^s to 5^s, the difference between the mean and apparent right ascension. If a star can not be observed upon the middle line, on account of temporary interference by clouds or tardiness in preparing for the observation, it may be observed anywhere within the safe limits of the field (often indicated by vertical lines on the diaphragm) and the chronometer time of observation recorded. In practice a star is seldom observed off the meridian.

It is desirable to make all settings with such accuracy that the mean of the two micrometer readings on a pair shall not differ from 20 turns by more than 1 turn. It is not infrequently true that the value of a micrometer screw increases slightly but steadily from one end to the other. In such cases the correction to each observed value of the latitude, due to this irregularity of the screw, will be insensible if the settings are made with the indicated accuracy, but not otherwise.

[1] This wording must be modified to correspond if, in accordance with the considerations stated on p. 141, two close parallel lines are used instead of a single line.

EXAMPLE OF RECORD AND COMPUTATIONS.

Zenith telescope record for latitude.

Form 255.

[Station, St. Anne. Date, June 25, 1908. Chronometer, 2637. Observer, W. Bowie.]

No. of pair	Star number Boss Cat.	N. or S.	Micrometer Turns	Micrometer Div's.	Level North	Level South	Chronometer time of culmination	Chronometer time of observation	Meridian distance	Remarks
					d	d				
9	4327	N.	11	69.0	6.0 67.8	39.1 99.5	(*)	16 55 24	(*)	+30† +46
	4379	S	27	34.4	40.2 100.5	7.2 68.7		17 11 47		Struck instrument
10	4441	S	9	61.0	40.3 101.2	7.2 69.4		17 28 07		+24
	4494	N	31	47.0	7.1 69.4	40.4 101.8		17 41 58		
11	4623	N	24	88.2	9.2 71.6	42.6 103.8		18 13 18		+16
	4651	S	16	66.0	42.2 103.2	8.7 71.0		18 18 39		
12	4669	S	19	62.5	44.2 106.0	10.9 73.8		18 22 20		+15
	4711	N	20	55.4	11.2 74.4	44.7 106.5		18 31 45		Mean of double star

* These columns are only used when a star is observed off the meridian.

† This is the continuous sum, up to this pair, of the south minus the north micrometer turns.

Reduction, mean to apparent declination, with Cape tables.

Form 32a.

[Station, St. Anne triangulation latitude station.]

Order								
1	Date	June 25, 1908						
2	Star No.	4327	4379	4441	4494	4623	4651	4669
3	α_0	16 55.4	17 11.9	17 28.2	17 42.1	18 13.4	18 18.7	18 22.4
7	$G+\alpha_0$	18 33.8	18 50.3	19 06.6	19 20.5	19 51.8	19 57.1	20 00.8
11	$H+\alpha_0$	4 40.0	4 56.5	5 12.8	5 26.7	5 58.0	6 03.3	6 07.0
4	δ_0	82 11	−0 20	28 28	53 50	64 21	17 46	29 46
8	P'	+2.95	+4.36	+5.74	+6.90	+9.40	+9.81	+10.08
9	$P'x$	−2.49	−3.68	−4.84	−5.82	−7.92	−8.27	− 8.50
12	Q'_0	+6.26	0.00	+1.78	+2.14	+0.14	−0.08	− 0.27
13	$\delta Q'$	0.00	−0.03	+0.03	+0.02	0.00	0.00	0.00
14	Q'	+6.26	−0.03	+1.81	+2.16	+0.14	−0.08	− 0.27
15	$Q'y$	+0.66	0.00	+0.19	+0.23	+0.01	−0.01	− 0.03
5	$\delta_0('')$	23.30	−30.91	24.93	23.54	57.64	46.61	31.42
10	$P'+P'x$	+0.46	+ 0.68	+0.90	+1.08	+1.48	+1.54	+ 1.58
16	$Q'+Q'y$	+6.92	− 0.03	+2.00	+2.39	+0.15	−0.09	− 0.30
17	I	+0.08	+ 0.58	+0.52	+0.35	+0.25	+0.56	+ 0.51
18	μ' $\tau\mu'$	−.001 .00	−.059 −.03	+.024 +.01	−.035 −.02	+.029 +.01	+.007 .00	−.033 −.02
19	$\delta ('')$	30.76	29.71	28.36	27.34	59.53	48.62	33.19

6

$\log g = 0.49813$ $\qquad$ $\log h = 1.31041$ $\qquad$ $G = 1^h\ 38.4^m$

$\log g_0 = 1.30216$ $\qquad$ $\log h_0 = 1.26717$ $\qquad$ $H = 11^h\ 44.6^m$

$\log \frac{g}{g_0} = \log (1+x) = 9.19597$ $\qquad$ $\log \frac{h}{h_0} = \log (1+y) = 0.04324$ $\qquad$ $i = +0.585$

$1+x = +0.157$ $\qquad$ $1+y = +1.105$ $\qquad$ $\tau = 0.483$

Make computation by horizontal lines in the order indicated. For explanation of Q_0' and $\delta Q'$, see pages (2) and (5) of Cape tables. Opposite δ_0 in the sixth line place the degrees and minutes, and opposite δ_0 ('') the seconds of the mean declination. The quantities x, y, i, and τ may be assumed constant for a night, and should be taken for an epoch midway between the first and last stars. The quantities G and H may be assumed constant for periods not exceeding four hours each, and should be taken for the midway epoch of each such period. Use α_0, G, H, $G+\alpha_0$, and $H+\alpha_0$, to tenths of minutes of time; x, y, and τ to three significant figures; and all other quantities to two decimal places.

Form 33.

Latitude

[Station, St. Anne. State, Illinois.

Date	Catalogue		Micrometer		Level			Meridian distance	Declination
	Star No.	N or S	Reading	Diff. Z. D.	n	s	Diff.		
			t *d*	*t* *d*	*d*	*d*	*d*	*S*	° ′ ″
1908.									
June 25	4327	N	11 69.0		06.0	39.1			82 11 30.76
				+15 65.4	67.8	99.5			
	4379	S	27 34.4		40.2	07.2	+2.1		−0 20 29.71
					100.5	68.7			
	4441	S	9 61.0		40.3	07.2			28 28 28.36
				−21 86.0	101.2	69.4			
	4494	N	31 47.0		07.1	40.4	−0.05		53 50 27.34
					69.4	101.3			
	4623	N	24 88.2		09.2	42.6			64 21 59.53
				− 8 22.2	71.6	103.8			
	4651	S	16 66.0		42.2	08.7	−1.05		17 46 48.62
					103.2	71.0			
	4669	S	19 62.5		44.2	10.9			29 46 33.19
				− 92.9	106.0	73.8			
	4711	N	20 55.4		11.2	44.7	−0.95		52 16 49.44
					74.4	106.5			

computation.

Observer, W. Bowie. Instrument, zenith telescope No. 4.]

Sum and half sum	Corrections				Latitude	Remarks
	Micrometer	Level	Refraction	Meridian		
° ′ ″	′ ″	″	″	″	° ′ ″	
81 51 01. 05 40 55 30. 52	+5 49. 48	+0. 78	+0. 18		41 01 20. 96	Struck instrument
82 18 55. 70 41 09 27. 85	−8 08. 02	−0. 02	−0. 14		41 01 19. 67	
82 08 48. 15 41 04 24. 08	−3 03. 56	−0. 39	−0. 06		41 01 20. 07	
82 03 22. 63 41 01 41. 32	−0 20. 74	−0. 35	−0. 01		41 01 20. 22	

Value of one division of latitude level: Upper −1″. 600
Lower −1. 364
Mean −1. 482

Summary of latitude computation.

[St. Anne, Ill., June 25, 1908.]

Star No. Boss catalogue		Mic. diff. M_1	ϕ 41° 01′	$\Delta\phi$	$\overline{\Delta\phi}^2$	M_1r_1	Corrected ϕ	$\Delta\phi$	$\overline{\Delta\phi}^2$
			″				″		
3667	3729	+ 8.3	20.26	−0.03	0.00	−0.11	20.15	+0.09	0.01
*(2265)	3803	+ 0.1	19.77	+0.46	0.21	0.00	19.77	+0.47	0.22
3842	3856	−14.3	20.02	+0.21	0.04	+0.20	20.22	+0.02	0.00
3949	3979	+14.1	20.40	−0.17	0.03	−0.19	20.21	+0.03	0.00
4063	4072	− 9.1	20.24	−0.01	0.00	+0.12	20.36	−0.12	0.01
4081	4090	+13.8	20.54	−0.31	0.10	−0.19	20.35	−0.11	0.01
4112	4129	+16.3	20.15	+0.08	0.01	−0.22	19.93	+0.31	0.10
4161	4201	+ 2.3	19.80	+0.43	0.18	−0.03	19.77	+0.47	0.22
4327	4379	+15.7	20.96	−0.73	0.53	−0.22	20.74	−0.50	0.25
4441	4494	−21.9	19.67	+0.56	0.31	+0.30	19.97	+0.27	0.07
4623	4651	− 8.2	20.07	+0.16	0.03	+0.11	20.18	+0.06	0.00
4669	4711	− 0.9	20.22	+0.01	0.00	+0.01	20.23	+0.01	0.00
4745	4758	+ 2.8	20.77	−0.54	0.29	−0.04	20.73	−0.49	0.24
*(3019)	4799	−16.4	20.53	−0.30	0.09	+0.22	20.75	−0.51	0.26
4824	4892	−16.1	20.06	+0.17	0.03	+0.22	20.28	−0.04	0.00
+ sum		73.4		2.08	1.85			1.73	1.39
− sum		86.9		2.09				1.77	
Algebraic sum		−13.5		−0.01					
Mean		− 0.9	20.23				20.24		

* 2265 and 3019 are ten-year 1880 numbers. The mean declinations for these stars were obtained from several sources.

$$e_p = \pm\sqrt{\frac{0.455\times1.85}{14}} = \pm0''.25$$

The value of one-half turn of the micrometer as used in the field = 22″.325.

Mean ϕ, 8 pairs with plus micrometer difference = 41° 01′ 20″.33.

Mean ϕ, 7 pairs with minus micrometer difference = 41° 01′ 20″.12.

The mean of 7 pairs with minus micrometer differences minus the mean of 8 pairs with plus micrometer differences = −0″.21.

Observation equations

$c - 8.3r_1 - 0.03 = 0$
$c - 0.1r_1 + 0.46 = 0$
$c + 14.3r_1 + 0.21 = 0$
$c - 14.1r_1 - 0.17 = 0$
$c + 9.1r_1 - 0.01 = 0$
$c - 13.8r_1 - 0.31 = 0$
$c - 16.3r_1 + 0.08 = 0$
$c - 2.3r_1 + 0.43 = 0$
$c - 15.7r_1 - 0.73 = 0$
$c + 21.9r_1 + 0.56 = 0$
$c + 8.2r_1 + 0.16 = 0$
$c + 0.9r_1 + 0.01 = 0$
$c - 2.8r_1 - 0.54 = 0$
$c + 16.4r_1 - 0.30 = 0$
$c + 16.1r_1 + 0.17 = 0$

Normal equations

$15c + 13.5r_1 - 0.01 = 0$

$13.5c + 2346.59r_1 + 31.872 = 0$

$r_1 = -0''.0137$

$c = +0''.0130$

$$r_1 = \pm\sqrt{\frac{0.455\times1.39}{2334\times13}} = \pm0''.0010$$

Corrected value of one-half turn of micrometer screw $= 22''.3113 \pm 0''.0046$

$$e_p = \pm\sqrt{\frac{0.455\times1.39}{13}} = \pm0''.22$$

$$e_\phi = \pm\sqrt{\frac{0.455\times1.39}{13\times14.92}} = \pm0''.06$$

Latitude of St. Anne latitude station	= 41° 01′ 20″.24 ± 0″.06
Reduction to sea level, elevation of station, 206 meters	= −0.03
Reduction to mean position of pole [1]	= +0.07
Latitude of St. Anne latitude station, reduced to sea level and the mean position of the pole	= 41° 01′ 20″.28 ± 0″.06

For an explanation of the above adjustment see page 130.

[1] See Astronomische Nachrichten No. 4414.

GENERAL NOTES ON COMPUTATIONS OF LATITUDE IN THE UNITED STATES COAST AND GEODETIC SURVEY.

The result from each pair of stars is given equal weight. This is done upon the supposition that the theoretical weights are so nearly equal that, if they were used, the final value for the latitude of a station would seldom be changed by more than 0''.01.

A first rejection limit of 3''.00 from the mean value of the latitude is used. After the 3''.00 rejection limit has been applied the probable error of a result from a single pair, e_p, is computed from all the remaining values, and then $5e_p$ is used as an absolute rejection limit, and $3.5e_p$ is used as a doubtful limit beyond which rejection is to be made if strong evidence in favor of rejection is found other than the residual itself. Such evidence may consist of positive notes indicating bad conditions during the observation of the particular pair concerned, contradictions in the record indicating a probable misreading, or a mean declination of a star with a probable error so large that it might account for the large residual.

A new value of one-half turn of the micrometer is to be derived from the latitude observations only in those cases in which the mean latitude from pairs with plus micrometer differences differs by more than 0''.20 from the mean latitude from pairs with minus micrometer differences. It is believed that, when the agreement is within 0''.20, a new value of one-half turn, if derived from the observations, would differ from the old by less than 0''.01 and the final latitude would ordinarily be changed by less than 0''.01. It is also believed that the derived correction to the old value would, in these cases, be but little, if any, larger than its own probable error.

The formulæ used in computing the probable errors, if a correction to the micrometer value is derived from the latitude observations, are:

$$e_p = \sqrt{\frac{(0.455)\Sigma\overline{\Delta\phi}^2}{(p-2)}}$$

$$e_{\frac{1}{2}} = \sqrt{\frac{(0.455)\Sigma\overline{\Delta\phi}^2}{(p-2)\left(p-\frac{(\Sigma M_1)^2}{\Sigma M_1^2}\right)}}$$

$$e_{r_1} = \text{probable error of } r_1 = \sqrt{\frac{(0.455)\Sigma\overline{\Delta\phi}^2}{(p-2)\left(\Sigma M_1^2-\frac{(\Sigma M_1)^2}{p}\right)}}$$

The correction for elevation to reduce the mean latitude to sea level is always applied. (See p. 130.)

The reduction to a triangulation station or to other points is also applied on the latitude computation and the relation of the latitude station to such point or points is there indicated. Unless the latitude station is within a few meters of the triangulation station and due east or west of it, the latitude computation should show the latitude of both the latitude station and the triangulation station.

EXPLANATION OF COMPUTATION.

Let ζ and ζ' equal the true meridional zenith distances of the southern and northern stars, and δ and δ' the apparent declinations of the same, respectively; then the expression for the latitude is

$$\varphi = \tfrac{1}{2}(\delta+\delta') + \tfrac{1}{2}(\zeta-\zeta')$$

Now, if z, z' denote the observed zenith distances of the south and the north stars; n, s the north and the south readings of the level for the south star, and n', s' the same for the north star; d the value of one division of level; r and r' the refraction corrections and m and m' the reductions of the measured zenith distances to the meridian for the south and the north stars, respectively, then

$$\varphi = \tfrac{1}{2}(\delta + \delta') + \tfrac{1}{2}(z - z') + \frac{d}{4}\Big\{(n + n') - (s + s')\Big\}^{1} + \tfrac{1}{2}(r - r') + \tfrac{1}{2}(m + m')$$

and if M and M' be the micrometer readings of the south and the north stars, increased micrometer readings corresponding to increased zenith distances, and R the value of one turn, then

$$\tfrac{1}{2}(z - z') = (M - M')\frac{R}{2}.$$

The details of the computation of the second and third terms in the above formula are sufficiently indicated in the computation shown above. The first, fourth, and fifth terms are explained more fully on the following pages (117–119).

Tenths of divisions of the micrometer head are usually estimated.

COMPUTATION OF APPARENT PLACES.

The data given in the Boss preliminary general catalogue of stars for 1900 in regard to a star, from which its apparent place at the time of observation is to be computed, are the mean right ascension and declination, α_m and δ_m for the year 1900, t_m; the annual variation in right ascension, $\frac{d\alpha_m}{dt}$; the annual variation in declination $\frac{d\delta_m}{dt}$, (the annual precession and proper motion together constitute the annual variation); and the secular variation of the precession in declination, given for 100 years, which, by moving the decimal point, becomes $\frac{d^2\delta_m}{dt^2}$. There are also given the proper motion in declination, μ'; the mean epoch E; the probable error of the declination at the mean epoch $e_{\delta Ep}$; $e_{\mu'}$, the probable error of 100 μ'; and the probable error of the declination for 1910, e_δ. The probable error of the declination for any date, T, is

$$\sqrt{(e_{\delta Ep})^2 + \left(\frac{T - E}{100}e_{\mu'}\right)^2}$$

The reduction to the apparent place at observation is made in two steps; first, the given mean place is reduced to the mean place at the beginning of the year of observation, and upon that as a basis the apparent place computation is then made.

Let the mean right ascension and declination at the beginning of the year of observation be called α_o and δ_o

Then

$$\alpha_o = \alpha_m + (t_o - t_m)\frac{d\alpha_m}{dt}$$

$$\delta_o = \delta_m + (t_o - t_m)\frac{d\delta_m}{dt} + \tfrac{1}{2}(t_o - t_m)^2\frac{d^2\delta_m}{dt^2}$$

The Boss catalogue shows that for the star 4327, $\alpha_m = \alpha_{1900} = 16^h\ 56^m\ 12^s$, with an annual variation $\frac{d\alpha_m}{dt} = -6^s.304$. Also $\delta_m = \delta_{1900} = 82°\ 12'\ 07''.66$. The annual variation, $\frac{d\delta_m}{dt} = -5''.510$, the secular variation, $\frac{d^2\delta_m}{dt^2} = -''.00880$, the proper motion, $\mu' = -''.001$; the mean epoch, E, $= 1875.5$, and the probable error, $e_{\delta Ep} = \pm 0''.03$; $e_{\mu'}$, the probable error of $100\mu'$ $= \pm 0''.13$, and the probable error of the declination for $1910 = \pm 0''.05$.

---

[1] The correction for inclination as here given is for a level of which the graduation is numbered in both directions from the middle. If the graduation is numbered continuously from one end to the other with numbers increasing *toward the objective*, the level correction is

$$\frac{d}{4}\Big\{(n' - n) + (s' - s)\Big\}$$

(Compare this with the similar formula for a striding level on page 23.) If the numbering on the level graduation increases *toward the eyepiece* this formula becomes

$$\frac{d}{4}\Big\{(n - n') + (s - s')\Big\}$$

This star was observed for latitude in June, 1908, at St. Anne, Ill., $0^h\,43^m$ west of Washington.

$\alpha_0 = 16^h\,56^m\,12^s - 8\,(6^s.304) = 16^h\,55^m\,22^s$, which is sufficiently close to the apparent right ascension for use in connection with latitude observations.

$\delta_0 = 82°\ 12'\ 07''.66 + 8\,[-5''.510 + \frac{1}{2}(8)(-''.00880)] = 82°\,11'\,23''.30$. The probable error of the declination for $1908 = \sqrt{(0''.03)^2 + \{.325(0''.13)\}^2} = \pm 0''.05$.

The apparent declination,[1] δ, at the instant of observation may now be computed by the formula given on page 526 of the American Ephemeris for 1908, namely,

$$\delta = \delta_0 + \tau\mu' + g\cos(G + \alpha_0) + h\cos(H + \alpha_0)\sin\delta_0 + i\cos\delta_0,$$

in which g, G, h, H, and i are quantities called independent star numbers which are functions of the time only and are given in the Ephemeris (pp. 532 to 539, 1908) for every Washington mean midnight during the year. τ is the elapsed decimal fraction of the fictitious year and is given in the Ephemeris with the independent star numbers.

This formula has been put in a more convenient form, conducive to more rapid computation, and adapted to the use of natural numbers and Crelle's Rechentafeln, in an appendix to the Cape Meridian Observations, 1890–91, entitled "Star-Correction Tables," by W. H. Finlay, M. A.

The formula is

$$\delta = \delta_0 + P'\,(1 + x) + Q'\,(1 + y) + I + \tau\mu'.$$

in which I, P', and Q' are tabulated in the Finlay tables.

$P' = g_0\cos(G + \alpha_0)$ and is tabulated with respect to the argument $G + \alpha_0$ and can be obtained from one opening of the tables for all stars and dates.

$Q' = h_0\cos(H + \alpha_0)\sin\delta_0$ and is tabulated with respect to the arguments $(H + \alpha_0)$ and δ_0.

$I = i\cos\delta_0$ and is tabulated with respect to i and δ_0. Q' and I can be obtained from the same opening of the tables for any given star and date, and all interpolations involve such small tabular differences that they may be made mentally.

$$1 + x = \frac{g}{g_0} \text{ and } 1 + y = \frac{h}{h_0}.$$

The values chosen for g_0 and h_0 are 20''.0521 and 18''.500, respectively, so that x is generally negative and never greater numerically than unity, while y is always positive and never greater than 0.11; thus the multiplications by x and y can be easily effected by Crelle's Rechentafeln. x and y are functions of the time only, and with sufficient accuracy may usually be considered constant for a single night.

If the period over which the observations extend on any night is not more than four hours long, the quantities g, h, G, H, i, and τ may be taken from the Ephemeris for the middle of the observing period and assumed to be constant for the night. The errors from this assumption will be small and of both algebraic signs.

The computation of the apparent places of seven stars observed at the St. Anne latitude station is shown on page 111.

When a given star is observed on several nights in succession it is not necessary to compute the apparent place for every night of observation. The apparent place may be computed for certain nights at intervals of not more than three days and the declination for intermediate nights may be obtained by interpolation.

CORRECTION FOR DIFFERENTIAL REFRACTION.

The difference of refraction for any pair of stars is so small that we may neglect the variation in the state of the atmosphere at the time of the observation from that mean state supposed in the refraction tables, except for stations at high altitudes. The refraction being nearly proportional to the tangent of the zenith distance, the difference of refraction for the two stars will be given by

$$r - r' = 57''.7\sin(z - z')\sec^2 z,$$

---

[1] In the comparatively rare cases in which it is necessary to compute the apparent right ascension of a star it may be done by the use of the formula given in Finlay's tables,

$$\alpha = \alpha_0 + P(1+x) + Q(1+y) + f + \tau\mu.$$

and since the half difference of zenith distances, as measured by the micrometer, is the quantity applied in the computation, the following table of corrections to the latitude for differential refraction has been prepared with the argument one-half difference of zenith distance at the side, and the argument zenith distance at the top.

If the station is so far above sea level that the mean barometric pressure at the station is less than 90 per cent of the mean barometric pressure at sea level (760^{mm}) it may be desirable to take this fact into account by diminishing the values given in the following table (computed for sea level) to correspond to the reduced pressure. That is, if the mean pressure is 10 per cent less than at sea level diminish each value taken from the table by 10 per cent of itself, if 20 per cent less diminish tabular values by 20 per cent, and so on. This need only be done roughly, since the tabular values are small.

Correction to latitude for differential refraction $=\frac{1}{2}(r-r')$.

[The sign of the correction is the same as that of the micrometer difference.]

One-half diff. of zenith distances	Zenith distance							
	0°	10°	20°	25°	30°	35°	40°	45°
′	″	″	″	″	″	″	″	″
0.0	0.00	0.00	0.00	0.00	0.00	0.00	0.00	0.00
0.5	0.01	0.01	0.01	0.01	0.01	0.01	0.01	0.02
1.0	0.02	0.02	0.02	0.02	0.02	0.03	0.03	0.03
1.5	0.03	0.03	0.03	0.03	0.03	0.04	0.04	0.05
2.0	0.03	0.03	0.04	0.04	0.04	0.05	0.06	0.07
2.5	0.04	0.04	0.05	0.05	0.06	0.06	0.07	0.08
3.0	0.05	0.05	0.06	0.06	0.07	0.08	0.09	0.10
3.5	0.06	0.06	0.07	0.07	0.08	0.09	0.10	0.12
4.0	0.07	0.07	0.08	0.08	0.09	0.10	0.11	0.13
4.5	0.08	0.08	0.09	0.09	0.10	0.11	0.13	0.15
5.0	0.08	0.09	0.10	0.10	0.11	0.13	0.14	0.17
5.5	0.09	0.10	0.10	0.11	0.12	0.14	0.16	0.18
6.0	0.10	0.10	0.11	0.12	0.13	0.15	0.17	0.20
6.5	0.11	0.11	0.12	0.13	0.14	0.16	0.19	0.22
7.0	0.12	0.12	0.13	0.14	0.16	0.18	0.20	0.23
7.5	0.13	0.13	0.14	0.15	0.17	0.19	0.21	0.25
8.0	0.13	0.14	0.15	0.16	0.18	0.20	0.23	0.27
8.5	0.14	0.15	0.16	0.17	0.19	0.21	0.24	0.29
9.0	0.15	0.16	0.17	0.18	0.20	0.23	0.26	0.30
9.5	0.16	0.16	0.18	0.19	0.21	0.24	0.27	0.32
10.0	0.17	0.17	0.19	0.20	0.22	0.25	0.29	0.34
10.5	0.18	0.18	0.20	0.21	[illegible]	[illegible]	[illegible]	[illegible]
11.0	0.18	0.19	0.21	0.22	0.25	0.28	0.31	0.37
11.5	0.19	0.20	0.22	0.23	0.26	0.29	0.33	0.39
12.0	0.20	0.21	0.23	0.25	0.27	0.30	0.34	0.40
12.5	0.21	0.22	0.24	0.26	0.28	0.31	0.36	0.42
13.0	0.22	0.22	0.25	0.27	0.29	0.33	0.37	0.44
13.5	0.23	0.23	0.26	0.28	0.30	0.34	0.39	0.45
14.0	0.23	0.24	0.27	0.29	0.31	0.35	0.40	0.47
14.5	0.24	0.25	0.28	0.30	0.32	0.36	0.41	0.49
15.0	0.25	0.26	0.29	0.31	0.34	0.38	0.43	0.50
15.5	0.26	0.27	0.29	0.32	0.35	0.39	0.44	0.52
16.0	0.27	0.28	0.30	0.33	0.36	0.40	0.46	0.54
16.5	0.28	0.29	0.31	0.34	0.37	0.41	0.47	0.55
17.0	0.29	0.29	0.32	0.35	0.38	0.43	0.49	0.57
17.5	0.29	0.30	0.33	0.36	0.39	0.44	0.50	0.59
18.0	0.30	0.31	0.34	0.37	0.40	0.45	0.51	0.60
18.5	0.31	0.32	0.35	0.38	0.41	0.46	0.53	0.62
19.0	0.32	0.33	0.36	0.39	0.43	0.48	0.54	0.64
19.5	0.33	0.34	0.37	0.40	0.44	0.49	0.56	0.65
20.0	0.34	0.35	0.38	0.41	0.45	0.50	0.57	0.67

REDUCTION TO THE MERIDIAN.

If a star is observed off the meridian while the line of collimation of the telescope remains in the meridian, the measured zenith distance is in error on account of the curvature of the apparent path of the star. Let m be the correction to reduce the measured zenith distance to what it would have been if the star had been observed upon the meridian.

Then,

$$m = \frac{\sin^2 \frac{1}{2}\tau}{\sin 1''} \sin 2\delta$$

in which τ is the hour-angle of the star. The signs are such that the *correction to the latitude* $(= \frac{m}{2})$ is always plus for the stars of positive declination and minus for stars of negative declination (below the equator), *regardless of whether the star is to the northward or to the southward of the zenith.* $\frac{m}{2}$ or $\frac{m'}{2}$ is, then, always applied as a correction to the latitude, with the sign of the right-hand member of the above equation. For a subpolar $180° - \delta$ must be substituted for δ, making the correction negative in this case just as for stars of southern declination. The following table gives the corrections to the latitude computed from the above formula. If both stars of a pair are observed off the meridian, two such corrections must be applied to the computed latitude.

Correction to latitude for reduction to meridian.

[Star off the meridian but instrument in the meridian. The sign of the correction to the latitude is positive except for stars south of the equator and subpolars.]

δ	10^m	15^m	20^m	22^m	24^m	26^m	28^m	30^m	32^m	34^m	36^m	38^m	40^m	42^m	44^m	46^m	48^m	50^m	52^m	54^m	56^m	58^m	60^m	δ
°	″	″	″	″	″	″	″	″	″	″	″	″	″	″	″	″	″	″	″	″	″	″	″	°
1										.01	.01	.01	.01	.01	.01	.01	.01	.01	.01	.01	.01	.02	.02	89
2					.01	.01	.01	.01	.01	.01	.01	.01	.02	.02	.02	.02	.02	.02	.03	.03	.03	.03	.03	88
3			.01	.01	.01	.01	.01	.01	.01	.02	.02	.02	.02	.03	.03	.03	.03	.04	.04	.04	.04	.05	.05	87
4			.01	.01	.01	.01	.01	.02	.02	.02	.02	.03	.03	.03	.04	.04	.04	.05	.05	.06	.06	.06	.07	86
5		.01	.01	.01	.01	.02	.02	.02	.02	.03	.03	.03	.04	.04	.05	.05	.05	.06	.06	.07	.07	.08	.09	85
6		.01	.01	.01	.02	.02	.02	.03	.03	.03	.04	.04	.05	.05	.06	.06	.07	.07	.08	.08	.09	.10	.10	84
7		.01	.01	.02	.02	.02	.03	.03	.03	.04	.04	.05	.05	.06	.06	.07	.08	.08	.09	.10	.10	.11	.12	83
8		.01	.02	.02	.02	.03	.03	.03	.04	.04	.05	.05	.06	.07	.07	.08	.09	.09	.10	.11	.12	.13	.14	82
9		.01	.02	.02	.02	.03	.03	.04	.04	.05	.05	.06	.07	.07	.08	.09	.10	.11	.11	.12	.13	.14	.15	81
10		.01	.02	.02	.03	.03	.04	.04	.05	.05	.06	.07	.07	.08	.09	.10	.11	.12	.13	.14	.15	.16	.17	80
12	.01	.01	.02	.03	.03	.04	.05	.05	.06	.06	.07	.08	.09	.10	.11	.12	.13	.14	.15	.16	.17	.19	.20	78
14	.01	.01	.03	.03	.04	.04	.05	.06	.07	.07	.08	.09	.10	.11	.12	.14	.15	.16	.17	.19	.20	.22	.23	76
16	.01	.02	.03	.03	.04	.05	.06	.07	.07	.08	.09	.10	.12	.13	.14	.15	.17	.18	.20	.21	.23	.24	.26	74
18	.01	.02	.03	.04	.05	.05	.06	.07	.08	.09	.10	.12	.13	.14	.16	.17	.18	.20	.22	.23	.25	.27	.29	72
20	.01	.02	.04	.04	.05	.06	.07	.08	.09	.10	.11	.13	.14	.15	.17	.19	.20	.22	.24	.26	.28	.29	.32	70
22	.01	.02	.04	.05	.05	.06	.07	.09	.10	.11	.12	.14	.15	.17	.18	.20	.22	.24	.26	.28	.30	.32	.34	68
24	.01	.02	.04	.05	.06	.07	.08	.09	.10	.12	.13	.15	.16	.18	.20	.21	.23	.25	.27	.29	.32	.34	.36	66
26	.01	.02	.04	.05	.06	.07	.08	.10	.11	.12	.14	.15	.17	.19	.21	.23	.25	.27	.29	.31	.34	.36	.39	64
28	.01	.03	.05	.05	.07	.08	.09	.10	.12	.13	.15	.16	.18	.20	.22	.24	.26	.28	.31	.33	.35	.38	.41	62
30	.01	.03	.05	.06	.07	.08	.09	.11	.12	.14	.15	.17	.19	.21	.23	.25	.27	.30	.32	.34	.37	.40	.42	60
32	.01	.03	.05	.06	.07	.08	.10	.11	.13	.14	.16	.18	.20	.22	.24	.26	.28	.31	.33	.36	.39	.41	.44	58
34	.01	.03	.05	.06	.07	.09	.10	.11	.13	.15	.16	.18	.20	.22	.24	.27	.29	.32	.34	.37	.40	.42	.45	56
36	.01	.03	.05	.06	.07	.09	.10	.12	.13	.15	.17	.19	.21	.23	.25	.28	.30	.32	.35	.38	.41	.44	.47	54
38	.01	.03	.05	.06	.08	.09	.10	.12	.13	.15	.17	.19	.21	.23	.26	.28	.30	.33	.36	.39	.41	.44	.48	52
40	.01	.03	.05	.07	.08	.09	.11	.12	.14	.16	.17	.19	.21	.24	.26	.28	.31	.34	.36	.39	.42	.45	.48	50
45	.01	.03	.05	.07	.08	.09	.11	.12	.14	.16	.18	.20	.22	.24	.26	.29	.31	.34	.37	.40	.43	.46	.49	45

The catalogues now available contain so many stars which may be observed for latitude that it is not desirable to move the instrument out of the meridian to observe a star which is missed as it crosses the meridian.

COMBINATION OF RESULTS, EACH PAIR OBSERVED MORE THAN ONCE.

Separate values of the latitude being computed from each observation upon each pair, it remains to combine these in such a way as to obtain the most probable value of the latitude and to obtain certain probable errors.

Let p be the total number of pairs observed. Let the number of observations upon pair No. 1 be n_1, upon pair No. 2, n_2, and so on, and let the total number of observations at the station be $n_0 = n_1 + n_2 + n_3 \quad . \quad . \quad .$ Let Δ be a residual obtained by subtracting the result from a single observation on a certain pair from the mean result from all the observations upon that pair. Let e be the probable error of a single observation of the latitude, excluding the error arising from defective adopted declinations.

The various values of Δ depend upon and are a measure of the probable error of observation, but are independent of the errors of the adopted declinations. According to the principles of least squares,

$$e^2 = \frac{0.455 \Sigma \Delta^2}{\text{No. obs.} - \text{No. unknowns}} = \frac{0.455 \Sigma \Delta^2}{n_0 - p}.$$

Let φ_1 be the mean latitude from observations on pair No. 1, φ_2 from pair No. 2, and so on. Let v be the residual obtained by subtracting $\varphi_1, \varphi_2 \quad . \quad . \quad .$ in turn from the indiscriminate mean for the station of $\varphi_1, \varphi_2, \varphi_3 \quad . \quad . \quad .$ There will be p such residuals, and they are a measure of the probable error of the mean result from a pair, which will be called e_p, arising from both errors of observation and errors of declination.

$$e^2_p = \frac{0.455 \Sigma v^2}{p-1}.$$

Let $e_{p1}, e_{p2} \quad . \quad . \quad .$ be the probable errors, respectively, of $\varphi_1, \varphi_2, \varphi_3 \quad . \quad . \quad .$ Let $e_{\frac{\delta\delta}{2}}$ be the probable error of the mean of two declinations. Then

$$e^2_{p1} = e^2_{\frac{\delta\delta}{2}} + \frac{e^2}{n_1}, \quad e^2_{p2} = e^2_{\frac{\delta\delta}{2}} + \frac{e^2}{n_2} \quad . \quad . \quad . \quad . \quad .$$

These various values $e^2_{p1}, e^2_{p2}, \quad . \quad . \quad .$ differ from each other because of the various values of $n_1, n_2, \quad . \quad . \quad .$ even though $e^2_{\frac{\delta\delta}{2}}$ and e^2 are assumed to be constant, and the value derived above for e^2_p is their mean value. Adding these various equations, p in number, and taking the mean, member by member, there is obtained

$$e^2_p = e^2_{\frac{\delta\delta}{2}} + \frac{\frac{e^2}{n_1} + \frac{e^2}{n_2} + \frac{e^2}{n_3} \quad . \quad . \quad .}{p} = e^2_{\frac{\delta\delta}{2}} + \frac{e^2}{p}\left[\frac{1}{n_1} + \frac{1}{n_2} + \frac{1}{n_3} \quad . \quad . \quad .\right]$$

Placing

$$\frac{e^2}{p}\left[\frac{1}{n_1} + \frac{1}{n_2} + \frac{1}{n_3} \quad . \quad . \quad .\right] = \varepsilon^2$$

to abbreviate the notation, and solving for $e^2_{\frac{\delta\delta}{2}}$ there is obtained

$$e^2_{\frac{\delta\delta}{2}} = e^2_p - \varepsilon^2$$

Having determined the values of $e^2_{\frac{\delta\delta}{2}}$ and e^2, the proper relative weights, w_1, w_2, inversely proportional to the squares of their probable errors, may now be assigned to $\varphi_1, \varphi_2, \varphi_3, \quad . \quad . \quad .$

$$w_1 = \frac{1}{e^2_{p1}} \qquad w_2 = \frac{1}{e^2_{p2}} \quad . \quad . \quad .$$

or

$$w_1 = \left(e^2_{\frac{\delta\delta}{2}} + \frac{e^2}{n_1}\right)^{-1} \qquad w_2 = \left(e^2_{\frac{\delta\delta}{2}} + \frac{e^2}{n_2}\right)^{-1} \quad . \quad . \quad .$$

An exception to the above weights arises when two or more north stars are observed at one setting of the telescope in connection with the same south star, or *vice versa*, and the computation is made as if two or more independent pairs had been observed. The results of the component pairs in such a combination are not independent, since they involve in common the error of observation and the error of declination of the common star. The weight to be assigned to each component pair in a doublet is on this account but two-thirds of that given above,[1] and to each component pair in a triplet is one-half. The combination of two stars on one side of the zenith with one on the other side is called a doublet, and three stars on one side of the zenith with one on the other side is called a triplet. The present practice in the United States Coast and Geodetic Survey is not to observe doublets or triplets. (See paragraph 3 of General Instructions, p. 104.)

If a combination observed at one setting of the telescope includes two or more stars on each side of the zenith, it may be broken up in the computation into two or more independent doublets or triplets, each of which may be treated as indicated above.

If a given star on one side of the zenith is observed in connection with a certain star on the other side of the zenith on a certain night (or nights), and on a certain *other* night (or nights) is observed in connection with some *other* star, the two results are independent in so far as the observations are concerned, but involve a common adopted declination for one of the two stars of each pair. The proper weight to be assigned depends in this case upon the relative magnitude of $e_{\frac{s}{2}}$ and e, but is for their ordinary values so nearly equal to the weight for an independent pair that it may, with little error, be assumed to be such without going to the trouble of evaluating it.

The weight to be assigned to a zenith star observed in both positions of the telescope is $\left(2e^2_{\frac{s}{2}}+\frac{e^2}{N_n}\right)^{-1}$ in which N_n is the number of nights' observations upon it.

The most probable value φ_0 for the latitude of the station is the weighted mean of the mean results from the various pairs, or

$$\varphi_0=\frac{w_1\varphi_1+w_2\varphi_2+w_3\varphi_3\quad.\quad.\quad.}{w_1+w_2+w_3\quad.\quad.\quad.}=\frac{\Sigma w\varphi}{\Sigma w}$$

The probable error of φ_0 is

$$e_\varphi=\sqrt{\frac{0.455\Sigma w\Delta\varphi}{(p-1)\Sigma w}}$$

in which $\Delta\varphi$ is the residual obtained by subtracting φ_1, φ_2, φ_3 . . . in turn from φ_0.

A concrete illustration of the processes indicated by the above formulæ is furnished by the following reproduction of certain parts of the computation of the latitude of the New Naval Observatory from observations made in 1897 with a zenith telescope.

[1] This may be made evident as follows: Let a_1 and a_2 be respectively the declination plus the measured zenith distance of a first and second south star, and a_3 the declination minus the measured zenith distance of a north star observed in combination with them. Let the probable errors of a_1, a_2, a_3 be e_1, e_2, e_3, respectively. Note that e_1, e_2, e_3 each include errors both of declination and observation. If the two component pairs are computed separately and the mean taken, the result is of the form $\left\{\frac{a_1+a_3}{2}+\frac{a_2+a_3}{2}\right\}\frac{1}{2}=\frac{a_1}{4}+\frac{a_2}{4}+\frac{a_3}{2}$ and its probable error squared is $\left(\frac{e_1}{4}\right)^2+\left(\frac{e_2}{4}\right)^2+\left(\frac{e_3}{2}\right)^2$. Assuming that $e_1=e_2=e_3$, this becomes $\frac{3}{8}e_1^2$, the square of the probable error of the mean result from the combination. By the same reasoning it may be shown that the square of the probable error of the result from a single independent pair is $\left(\frac{e_1}{2}\right)^2+\left(\frac{e_2}{2}\right)^2=\frac{1}{2}e_1^2$. The weights to be assigned to the combination and to an independent pair are then in the ratio of $(\frac{3}{8}e_1^2)^{-1}$ and $(\frac{1}{2}e_1^2)^{-1}$, or of $\frac{4}{3}$ to 1. If the weight for an independent pair is unity the weight of *each component* of a doublet is therefore two-thirds.

Pairs Star Nos.	ϕ 38° 55′+	Δ	Δ^2
	″		
(2058)	09. 81	−. 01	. 00
4440	09. 80	. 00	. 00
	09. 80		
4513	08. 07	−. 07	. 00
4550	07. 92	+. 08	. 01
	08. 00		
4513	08. 12	. 00	. 00
4555	08. 13	−. 01	. 00
	08. 12		
4526	09. 31	−. 34	. 12
4550	08. 40	+. 57	. 32
	08. 80	+. 17	. 03
	09. 44	−. 47	. 22
	09. 01	−. 04	. 00
	08. 85	+. 12	. 01
	08. 97		
4526	09. 36	−. 26	. 07
4555	08. 62	+. 48	. 23
	09. 51	−. 41	. 17
	08. 91	+. 19	. 04
	09. 12	−. 02	. 00
	09. 11	−. 01	. 00
	09. 10		
....	....	....	....
....	....	....	....
....	....	....	....
Sum........................			6. 69

$$e=\sqrt{\frac{0.455\ \Sigma\Delta^2}{n_0-p}}=\sqrt{\frac{(0.455)\ (6.69)}{100-21}}=\sqrt{0.039}=\pm 0''.20$$

Pair, Star Nos. B. A. C. (10yr.)	[C. S.]	ϕ	v	v^2	n	w	$w\phi$	$\Delta\phi$	$\overline{\Delta\phi}^2$	$w\overline{\Delta\phi}^2$
(2058)	4440	38° 55′ 09″. 80	−1. 00	1. 00	2	11	19. 80	−0. 99	0. 98	10. 78
4513	4550	08 . 00	+ . 80	. 64	2	5	0. 00	+ . 81	. 66	3. 30
4513	4555	08 . 12	+ . 68	. 46	2	5	0. 60	+ . 69	. 48	2. 40
4526	4550	08 . 97	− . 17	. 03	6	4*	3. 88	− . 16	. 03	0. 12
4526	4555	09 . 10	− . 30	. 09	6	4*	4. 40	− . 29	. 08	0. 32
4577	(2158)	08 . 83	− . 03	. 00	6	8	6. 64	− . 02	. 00	0. 00
(2158)	4646	08 . 72	+ . 08	. 01	6	8	5. 76	+ . 09	. 01	0. 08
(2195)	4688	09 . 11	− . 31	. 10	5	12	13. 32	− . 29	. 08	0. 96
4706	4726	08 . 25	+ . 55	. 30	5	12	3. 00	+ . 56	. 31	3. 72
4742	(2233)	08 . 50	+ . 30	. 09	5	12	6. 00	+ . 31	. 10	1. 20
(2254)	4847	08 . 93	− . 13	. 02	5	12	11. 16	− . 12	. 01	0. 12
4876	4937	08 . 92	− . 12	. 01	5	12	11. 04	− . 11	. 01	0. 12
4958	(2341)	08 . 83	− . 03	. 00	5	12	9. 96	− . 02	. 00	0. 00
(2350)	5026	09 . 15	− . 35	. 12	5	9*	10. 35	− . 34	. 12	1. 08
[1259]	(2365)	09 . 35	− . 55	. 30	5	5*	6. 75	− . 54	. 29	1. 45
5076	5084	08 . 64	+ . 16	. 03	5	12	7. 68	+ . 17	. 03	0. 36
5115	5153	08 . 87	− . 07	. 00	5	12	10. 44	− . 06	. 00	0. 00
5168	5178	08 . 62	+ . 18	. 03	5	12	7. 44	+ . 19	. 04	0. 48
5249	5293	08 . 50	+ . 30	. 09	5	12	6. 00	+ . 31	. 10	1. 20
5313	5322	09 . 22	− . 42	. 18	5	12	14. 64	− . 41	. 17	2. 04
5344	(2537)	08 . 44	+ . 36	. 13	5	12	5. 28	+ . 37	. 14	1. 68
Sums		16 . 87	+3. 41 −3. 48	3. 63		203	164 . 14			31. 41
Means		38° 55′ 08″. 80					08″. 81			

* For explanation of these four weights, see p. 123.

$$e_p{}^2 = \frac{(0.455)\ (3.63)}{20} = 0.083 \qquad \varepsilon^2 = \frac{0.039}{21}(4.97) = 0.009$$

$$e^2_{\frac{*}{2}} = 0.083 - 0.009 = 0.074 \qquad e_\phi = \sqrt{\frac{(0.455)\ (31.41)}{(20)\ (203)}} = \pm 0.06.$$

Latitude $= 38^\circ\ 55'\ 08''.81 \pm 0''.06$.

In computing the values of $w\phi$, $38^\circ\ 55'\ 08''.00$ was first dropped from each value of ϕ.

An independent determination of $e_{\frac{*}{2}}$ may be obtained from the probable errors of the mean declinations of the stars observed, as given in the Boss catalogue.

For the stars observed at a station the mean value of the probable error of the mean of two declinations is

$$e_{\frac{*}{2}} = \sqrt{\frac{\Sigma e^2_*}{2N_s}},$$

in which N_s is the total number of stars observed.

For a particular pair

$$e^2_{\frac{*}{2}} = \frac{\Sigma e^2_*}{4},$$

in which only the two stars of the pair are included in the summation in the numerator. From this formula and from that given on page 120 (viz, $e^2_{\frac{*}{2}} = e^2_p - \varepsilon^2$) two separate values for $e_{\frac{*}{2}}$ for each pair may be computed. Which should be used in the formula

$$w_n = \left(e^2_{\frac{*}{2}} + \frac{e^2}{n_n}\right)^{-1},$$

fixing the weight to be assigned to the mean result from a pair? There are two objections to the rigid use in all cases of the second value (from the latitude computation). That value is a mean for all the pairs of a list, and in using it the fact that some declinations have very much larger probable errors than others in the same list is ignored. Moreover, in practice, the formula $e^2_{\frac{*}{2}} = e^2_p - \varepsilon^2$ is sometimes found to give a value for $e_{\frac{*}{2}}$ which is so small as to be evidently erroneous, and sometimes $e^2_{\frac{*}{2}}$ is even *negative*, which is an absurdity. On the other hand, whenever the value $e^2_{\frac{*}{2}} = \frac{\Sigma e^2_*}{2N_s}$ is smaller than $e^2_{\frac{*}{2}} = e^2 - \varepsilon^2_p$, and that is usually the case, it indicates that there is in the observations some error peculiar to each star, which combines with the declination error, and so apparently increases it. When such errors exist, the weights should be correspondingly reduced, and therefore the values of $e^2_{\frac{*}{2}} = e^2_p - \varepsilon^2$ should be used in the weighting.

The following method of weighting, therefore, seems to be the best for use in the office computation. *In the weight formula* (see page 120), $w_n = \left(e^2_{\frac{*}{2}} + \frac{e^2}{n_n}\right)^{-1}$, *use for each pair the larger of the two available values of* $e^2_{\frac{*}{2}}$, namely, $e^2_{\frac{*}{2}} = \frac{\Sigma e^2_*}{4}$ and $e^2_{\frac{*}{2}} = e^2_p - \varepsilon^2$. By so doing all the disadvantages of each of the two methods discussed in the preceding paragraph are avoided. To find quickly which of the values of $e^2_{\frac{*}{2}}$ from the mean place computation are greater than $e^2_{\frac{*}{2}} = e^2_p - \varepsilon^2$ one may first note on the list of mean places for what stars e^2_* exceeds $2\ (e^2_p - \varepsilon^2)$. Only pairs involving such stars need be examined further. To illustrate, of the pairs involved in the latitude computation shown on page 122, there were only four for which the mean place computation gave values of $e^2_{\frac{*}{2}}$ exceeding 0.074. The stars involved in these four pairs were 4526, 4550, 4555, (2350), 5026, [1259], (2365), and the corresponding values of e^2_* were 0.37, 0.08, 0.10, 0.18, 0.24, 0.08, 0.73. The weights assigned to these four pairs therefore depend upon $e^2_{\frac{*}{2}} = \frac{\Sigma e^2_*}{4}$ in each case.

COMBINATION OF RESULTS WHEN EACH PAIR IS OBSERVED BUT ONCE.

It is the present practice of this Survey to observe a pair of stars only once at a station, and in the final computations the resulting latitude from each pair observed is given unit weight. (See the first paragraph under the heading "General Notes on Computations of Latitude in the U. S. Coast and Geodetic Survey" on p. 115.)

Whenever the plan of observing each pair but once at a station is carried out the method of combining results and computing probable errors outlined in the preceding pages fails, and for it must be substituted the following procedure, for which little additional explanation is needed:

$$e^2_p = \frac{0.455\ \Sigma v^2}{p-1}$$

in which e_p is the probable error of the result from a pair, including both the error of observation and the declination errors, v is the residual obtained by substracting the latitude from a single pair from the indiscriminate mean of all the pairs, and p is the number of pairs. In the field computation and also in the final computation this indiscriminate mean is considered to be the final value of the latitude. Its probable error is

$$e_\phi = \sqrt{\frac{0.455\ \Sigma v^2}{p(p-1)}}$$

No value of the probable error of observation not involving the declination error is available from such a field computation. But the computed values of e_p and e_ϕ give sufficiently good indications of the accuracy of the observations to enable the observer to decide in the field whether the instrument is in good condition and whether more observations are needed and that is all that is necessary. (See p. 104.)

If desired, the office computation may be carried further as the probable error of the declination of a star e_* may be obtained from the catalogue.

The probable error of a single observation is given by the formula $e^2 = e^2_p - \frac{\Sigma e^2_*}{2 N_s}$, in which N_s is the total number of stars observed.

If weights were given each pair (not the present practice in this Survey), the weight to be assigned to a pair would be

$$w = (e^2_{\frac{*}{2}} + e^2)^{-1}$$

in which for each pair $e^2_{\frac{*}{2}} = \frac{\Sigma e^2_*}{4}$, the summation covering the two stars of that pair only.

DETERMINATION OF LEVEL AND MICROMETER VALUES.

For methods of determining the level value see page 46.

Until recently the method most frequently used in this Survey for determining the micrometer value is as follows:[1] The time is observed that is required for a close circumpolar star, near elongation, to pass over the angular interval measured by the screw. Near elongation the apparent motion of the star is nearly vertical and nearly uniform. That one of the four close circumpolars given in the Ephemeris, namely, α, δ, and λ Ursæ Minoris and 51 Cephei, may be selected which reaches elongation at the most convenient hour. In selecting the star it may be assumed with sufficient accuracy that the elongations occur when the hour-angle is six hours on either side of the meridian. In planning the observations and in making the computation it is necessary to know the time of elongation more accurately, and it may be computed from the formula

$$\cos t_E = \cot \delta \tan \phi$$

[1] See Appendix No. 8, United States Coast and Geodetic Survey, Report for 1900, for a full discussion of the determination of micrometer value.

Chronometer time of elongation $=\alpha-\Delta T\pm t_E$, the plus sign being used for western elongation and the minus for eastern elongation. t_E is the hour-angle at elongation reckoned eastward or westward from upper culmination, and ΔT is the chronometer correction.

If desired ζ_E, the zenith distance of the star at elongation may be computed from the formula

$$\cos \zeta_E = \operatorname{cosec} \delta \sin \phi$$

It is advisable to have the middle of the series of observations about elongation. The observer may obtain an approximate estimate of the rate at which the star moves along the micrometer by a rough observation or from previous record, and time the beginning of his observations accordingly.

To begin observations the star is brought into the field of the telescope and to the proper position, the telescope is clamped both in zenith distance and azimuth, the micrometer is made to read an integral number of turns, and the bubble is brought approximately to the middle of the level tube. The chronometer time of transit of the star across the thread is observed and the level read. The micrometer thread is then moved one whole turn in the direction of the apparent motion of the star, the time of transit again observed and the level read, and the process repeated until a sufficiently large portion of the middle of the screw has been covered by the observations to correspond with what is actually used in the latitude observations. If desired, an observation may be made at every half turn, or even at every quarter turn, by allowing an assistant to read the level. It is well to note the temperature.

The form of record and computation is shown below, the first four columns being the record, and the remainder the computation, of the value of one turn of micrometer from observations made at the New Naval Observatory June 18, 1897.

$\phi = 38° 55' 08''.8$.

For the star B. A. C. 8213 at the time of observation $\alpha = 23^h 27^m 45^s.6$ and $\delta = 86° 44' 13''.4$. The chronometer correction at the time of the observations was known to be $+2^s.3$. Whence the chronometer time of eastern elongation was computed to be $17^h 38^m 16^s.5$ and the zenith distance $51° 00'.5$.

Computation of value of micrometer.

Station, New Naval Observatory, Washington, D. C. Observer, O. B. F. Star, B. A. C. 8213 E. E. Date, June 18, 1897. Instrument, Zenith telescope, No. 4.]

Micrometer reading	Chronometer time of observation	Level		Time from elongation	Reduction to mean state of level	Corrections		Reduced time	Time at 20 turns	Δ	Δ'
		n	s			Time	Level				
t	h m s	d	d	m	d	s	s	h m s	h m s	s	s
35	17 15 08.5	13.1 68.4	39.9 101.9	23.1	−0.10	+2.3	−0.1	17 15 10.7	17 28 10.7	+4.7	−0.3
34	16 02.0			22.2		+2.1	−0.1	16 04.0	12.0	+3.4	−1.2
33	16 53.5			21.4		+1.9	−0.1	16 55.3	11.3	+4.1	−0.2
32	17 45.0	13.2 68.6	40.0 102.0	20.5	+0.15	+1.6	+0.1	17 46.7	10.7	+4.7	+0.7
31	18 37.5			19.6		+1.4	+0.1	18 39.0	11.0	+4.4	+0.7
30	19 30.0			18.8		+1.3	+0.1	19 31.4	11.4	+4.0	+0.7
29	20 22.5			17.9		+1.1	+0.1	20 23.7	11.7	+3.7	+0.7
28	21 16.0			17.0		+0.9	+0.1	21 17.0	13.0	+2.4	−0.3
27	22 07.5			16.1		+0.8	+0.1	22 08.4	12.4	+3.0	+0.7
26	23 00.0			15.2		+0.7	+0.1	23 00.8	12.8	+2.6	+0.6
25	23 53.0			14.4		+0.6	+0.1	23 53.7	13.7	+1.7	0.0
24	24 45.5			13.5		+0.5	+0.1	24 46.1	14.1	+1.3	0.0
23	25 37.5			12.6		+0.4	+0.1	25 38.0	14.0	+1.4	+0.4
22	26 30.5			11.8		+0.3	+0.1	26 30.9	14.9	+0.5	−0.2
21	27 23.0	13.2 68.3	40.0 101.8	10.9	−0.10	+0.2	−0.1	27 23.1	15.1	+0.3	0.0
20	28 16.0			10.0		+0.2	−0.1	28 16.1	16.1	−0.7	−0.7
19	29 08.0	13.2 68.5	40.0 102.1	9.1	+0.15	+0.1	+0.1	29 08.2	16.2	−0.8	−0.5
18	30 00.5			8.2		+0.1	+0.1	30 00.7	16.7	−1.3	−0.6
17	30 53.0			7.4		+0.1	+0.1	30 53.2	17.2	−1.8	−0.8
16	31 44.5			6.5		+0.1	+0.1	31 44.7	16.7	−1.3	0.0
15	32 37.0			5.7		0.0	+0.1	32 37.1	17.1	−1.7	0.0
14	33 29.5			4.8		0.0	+0.1	33 29.6	17.6	−2.2	−0.2
13	34 22.0	13.2 68.3	40.1 101.9	3.9	0.00	0.0	0.0	34 22.0	18.0	−2.6	−0.3
12	35 14.0			3.0		0.0	0.0	35 14.0	18.0	−2.6	+0.1
11	36 06.5			2.2		0.0	0.0	36 06.5	18.5	−3.1	−0.1
10	36 58.5			1.3		0.0	0.0	36 58.5	18.5	−3.1	+0.2
9	37 50.5			0.4		0.0	0.0	37 50.5	18.5	−3.1	+0.6
8	38 43.5			−0.4		0.0	0.0	38 43.5	19.5	−4.1	−0.1
7	39 35.5			−1.3		0.0	0.0	39 35.5	19.5	−4.1	+0.2
6	40 28.0			−2.2		0.0	0.0	40 28.0	20.0	−4.6	0.0
5	41 19.5			−3.0		0.0	0.0	41 19.5	19.5	−4.1	+0.9
								Mean	17 28 15.4		

Assumed value of $R_1 = 52^s$.

$2480\ r_1 = +820.3$

$r_1 = +\ 0.3308$

$R_1 + r_1 = 52^s.3308$

$\log (R_1 + r_1) = 1.7187573$

$\log 15 = 1.1760913$

$\log \cos \delta = 8.7552522$

$\{ 1.6501008$

$\{ 44''.679$

Corr. for refraction $-\ 0\ .030$

One turn $44''.649$

For explanation of notation, see page 128.

Because of the curvature of the apparent path of the star its rate of change of zenith distance is not constant, even near elongation. The rate of change *at elongation* may readily be computed. It is at that instant in seconds of arc 15 cos δ per second of sidereal time. The table of curvature corrections given below enables one to correct the observed times to what they would have been if in the place of the actual star there were substituted an ideal star whose motion was vertical at a constant rate 15 cos δ and which coincided with the actual star at the instant of elongation.

Correction for curvature of apparent path of star, in computation of micrometer value.

[The correction tabulated is $\frac{1}{6}(15 \sin 1'')^2 \tau^3 - \frac{1}{120}(15 \sin 1'')^4 \tau^5$ in which τ is the time from elongation. Apply the corrections given in the table to the observed chronometer times, adding them before either elongation, and subtracting them after either elongation.]

τ	Corr.	τ	Corr.	τ	Corr.	τ	Corr.	τ	Corr.	τ	Corr.
m	*s*	*m*	*s*	*m*	*s*	*m*	*s*	*m*	*s*	*m*	*s*
6	0.0	16	0.8	26	3.3	36	8.9	46	18.5	56	33.3
7	0.1	17	0.9	27	3.7	37	9.6	47	19.7	57	35.1
8	0.1	18	1.1	28	4.2	38	10.4	48	21.0	58	37.0
9	0.1	19	1.3	29	4.6	39	11.3	49	22.3	59	39.0
10	0.2	20	1.5	30	5.1	40	12.2	50	23.7	60	41.0
11	0.2	21	1.8	31	5.7	41	13.1	51	25.2	61	43.1
12	0.3	22	2.0	32	6.2	42	14.1	52	26.7	62	45.2
13	0.4	23	2.3	33	6.8	43	15.1	53	28.3	63	47.4
14	0.5	24	2.6	34	7.5	44	16.2	54	29.9	64	49.7
15	0.6	25	3.0	35	8.2	45	17.3	55	31.6	65	52.1

In the computation the fifth column shows the values of τ, and the seventh column the resulting curvature corrections.

When the reading of the level changes, it indicates, upon the usual assumption that the relation between the level vial and the telescope remains constant, that the inclination of the telescope has changed. The effect of the movement of the telescope may be eliminated in the computation by applying to each observed time the correction in seconds of time,

$$\pm \{(n-s)-(n'-s')\}\frac{d}{30 \cos \delta}$$

to reduce it to what it would have been if the readings of the north and south end of the bubble had been n' and s', respectively.

If, as in the present case, the level graduation is numbered continuously from one end to the other with the numbers increasing toward the eye end, instead of being numbered in both directions from the middle, the required correction becomes

$$\pm \{(n'+s')-(n+s)\}\frac{d}{30 \cos \delta}.$$

In each of these formulæ the plus sign is to be used for western elongation and the minus sign for eastern elongation. It is convenient to take for the assumed n' and s' the actual readings at some one moment during the set of observations.

Zenith telescope No. 4 had two latitude levels, and the correction was computed by taking the mean of the two and using the mean value of d $(=1''.482)$. The sixth column shows the mean values of $(n'+s')-(n+s)$ and the eighth column the resulting corrections, the factor $\frac{d}{30 \cos \delta}$ being 0.87.

Let R_1 be an assumed approximate value of one turn in time and let r_1 be a required correction to R_1. Let T_0 be an approximate value of the chronometer time of transit of the star across the micrometer line set at 20 turns (the middle of the screw) and t_0 a required correction to T_0. Then, upon the assumption that the screw has a uniform value throughout the part observed upon and that the star moves in the direction of *increasing* readings (western elongation), for each observed time an observation equation may be written of the form

$$t+(20-R_0)\ (R_1+r_1)-(T_0+t_0)=0$$

in which t is the observed time of transit across the line set at the reading R_0 after correction for curvature and level. After transposition this may be written

$$(20-R_0)r_1-t_0=\Delta$$

in which

$$\Delta=T_0-[t+(20-R_0)R_1]$$

whence the normal equations become

$$\Sigma(20-R_0)^2r_1-\Sigma(20-R_0)t_0=\Sigma(20-R_0)\Delta$$

$$-\Sigma(20-R_0)\ r_1+nt_0=-\Sigma\Delta.$$

If the turns observed upon are symmetrical about 20, $\Sigma(20-R_0)$ becomes zero. If, moreover, as in the numerical case here shown, T_0 is purposely taken equal to the mean value of $t+(20-R_0)R_1$, $\Sigma\Delta$ is zero and t_0 derived from the second normal equation is necessarily zero. Also the first normal equation reduces to the working form

$$\Sigma(20-R_0)^2\ r_1=\Sigma(20-R_0)\Delta$$

If the star is observed at eastern elongation it moves in the direction indicated by *decreasing* micrometer readings and throughout the preceding formulæ R_0-20 must be substituted for $20-R_0$.

In the computation form printed above, the values of $t+(R_0-20)R_1$ are shown in the column headed "Time at 20 turns," R_1 being assumed $=52^s$. T_0 was assumed $=17^h\ 28^m\ 15.^s4$, the mean for this column, and the Δ's written accordingly.

The equation $\Sigma(R_0-20)^2r_1=\Sigma(R_0-20)\Delta$ reduces numerically to $2480r_1$[1] $=820.3$.

Δ' is the residual obtained by substituting the derived value r_1 in each observation equation, or $\Delta'=\Delta-(R_0-20)r_1$.

The remainder of the computation needs no explanation except that the correction for refraction to be applied to the value of one turn is the change of refraction for a change of zenith distance equal to one turn, or in the most convenient form for use, it is the value of one turn in minutes of arc times the difference of refraction for 1′ at the altitude at which the star was observed (approximately $=\phi$). The difference of refraction for 1′ may be obtained from any table of mean refractions with sufficient accuracy. The correction for refraction is always negative, since the change of refraction is always such as to make a star appear to move slower than it really does.

It will sometimes be necessary to apply a correction for rate. This correction, to be applied to the computed value of one turn, is in seconds of arc

$$\frac{(\text{rate of chronometer in seconds per day})\ (\text{value of one turn in seconds of arc})}{86400}$$

The correction is negative if the chronometer runs too fast.

The micrometer value is sometimes determined by turning the micrometer box 90° and observing upon a close circumpolar near culmination. There are two serious objections to this

[1] In this computation it becomes necessary to find the sum of the series $1^2+2^2+3^2+4^2\ .\ .\ .\ .\ .\ +15^2$. It is convenient for this purpose to use the formula $1^2+2^2+3^2+4^2\ .\ .\ .\ +x^2=\frac{x^3}{3}+\frac{x^2}{2}+\frac{x}{6}$. Occasionally in least square computations it becomes necessary to compute the sum of a similar series of fourth powers. One may then use the formula $1^4+2^4+3^4+4^4\ .\ .\ +x^4=\frac{x^5}{5}+\frac{x^4}{2}+\frac{x^3}{3}-\frac{x}{30}$. To obtain the sum of the series $(\frac{1}{4})^4+(\frac{1}{2})^4+(\frac{3}{4})^4+(1)^4+(1\frac{1}{4})^4\ .\ .\ .\ +x^4$, apply the formula to the series $1^4+2^4+3^4+4^4\ .\ .\ +(4x)^4$ and divide by $256=4^4$. See *Sammlung von Formeln der reinen und angewandten Mathematik von Dr. W. Láska, p. 88 (Braunschweig, 1888–1894).*

procedure. The focal adjustment is liable to be disturbed more or less when the micrometer box is turned, and a corresponding constant error introduced into the result. In observing at elongation the telescope is depended upon to be stable in zenith distance, the direction in which it is designed to be stable, and the level readings furnish a means of correcting in large part for small movements in that direction. But when the observations are made at culmination the instrument is depended upon to remain fixed in azimuth, the direction in which, because of its peculiar design, it is weakest, and there is no check upon changes in azimuth corresponding to the level readings. Hence, it is not advisable to observe for micrometer value at culmination. The only modifications in the computations are that there are no corrections for level or refraction, and that in computing the curvature correction τ is now the hour-angle. The curvature correction is additive before either culmination, and subtractive after it.

It is decidedly questionable whether it is advisable to determine the mean value of the micrometer screw by observations upon close circumpolars either at culmination or elongation. Such observations consume a great deal of time both in observation and in the subsequent computation, and experience shows that they are subject to unexpectedly large and unexplained errors. For example, during the observations for variation of latitude at Waikiki, Hawaiian Islands, in 1891–92, the micrometer value was thus determined twelve times. The results show a range of about 0''.13 or one three-hundred-and-thirtieth of the mean value, corresponding to a range of about 3.3 millimeters in the distance between the objective and the micrometer line, though the draw tube was kept clamped continuously, and the range of temperature during the entire year was only about 11° C. (Coast and Geodetic Survey Report, 1892, Part II, p. 61.) Similarly, sixteen determinations of the value of a micrometer used at fifteen stations on the Mexican Boundary Survey of 1892–93 showed a range of 0''.33 or one one-hundred-and-ninetieth of the mean value.[1] In this case the draw tube was unclamped and the telescope refocused at the beginning of the observations at each station. The observed value was apparently not a function of the temperature. The San Francisco series of observations for variation of latitude also show a similar large range in the observed micrometer value (viz: 0''.17). (Coast and Geodetic Survey Report, 1893, Part II, p. 447.) In general, whenever the micrometer value is determined repeatedly by the circumpolar method so large a range of results is developed as to force one to suspect that large constant errors are inherent in this method of observation. It can hardly be urged that the differences between the results represent actual changes in the micrometer value, for such differences are developed even when successive determinations are made during a single evening. Moreover, whenever the mean micrometer value is determined from the latitude observations themselves it is frequently found to differ radically from that derived from circumpolar observations on the same nights. So marked and so frequent has the latter form of disagreement been, that many of the office latitude computations have actually been made during the last few years by rejecting the micrometer value from circumpolar observations, when there is a marked difference between it and the value computed from the latitude observations as indicated below, and using the latter value in the latitude computation.

DETERMINATION OF MICROMETER VALUE FROM LATITUDE OBSERVATIONS.

After considering the above facts and conclusions this Survey decided to adopt the method of computing the micrometer value from the latitude observations, and since the beginning of the year 1905 no observations have been made on close circumpolar stars for that purpose.

The total range in the values of one turn of the micrometer screw of zenith telescope No. 2, as determined from the latitude observations for 36 of the 63 stations established by Assistant W. H. Burger, from 1905 to 1908, is 0''.17. This is one two hundred and seventy-third of the mean value.

As to the accuracy of the micrometer value determined from the latitude observations, it may be noted that if it be assumed that the probable error of a single observation of latitude

[1] Report of the International Boundary Commission, United States and Mexico, 1891-1896 (Washington, 1898), p. 103.

is $\pm 0''.40$, of the mean of two declinations is $\pm 0''.16$ (see p. 133) and of the latitude as derived from independent pairs is $\pm 0''.10$, the probable error of the micrometer value, as determined from a *single observation* upon a pair having a difference of zenith distance of ten turns would be

$$\frac{1}{10}\sqrt{(0.40)^2+4(0.16)^2+(0.10)^2}=\pm 0''.05.$$

There can be little doubt, therefore, that the mean micrometer value determined from all the latitude observations at a station is more accurate than that determined from even three or four sets of circumpolar observations each requiring an hour or more of time.

It has been urged that to determine an instrumental constant from the observations in the computation of which it is to be used is a questionable procedure; that it "smooths out" the results, but probably does not give real accuracy. The force of this objection disappears when one contrasts the proposed practice of deriving a *single* instrumental constant from observations on twelve or more pairs with the usual and unquestioned practice in transit time computations of deriving *three* instrumental constants (two azimuth and one collimation constant) from only *ten to twelve observations* on as many stars.

It should be noted that the form of the computation of circumpolar micrometer observations given on page 126 is especially adapted to the detection of irregularities and periodic errors, as they will at once become evident from an inspection of the values of Δ'. One common form of irregularity in screws is a continuous increase in the value from one end to the other, in which case Δ' tends to have the same sign at the two ends of the set and the opposite sign in the middle.

To derive the mean micrometer value from the latitude observations let M_1 be the difference, in turns, of the micrometer readings on the two stars of a pair, taken with the same sign as in the latitude computation, let r_1 be the required correction to the assumed value of *one-half* turn with which the computation of the latitude was made, let p be the number of pairs, and let c be the correction to the mean latitude ϕ_0. Let $\Delta\phi$ have the same meaning as before, viz, $\phi_0-\phi_1$, $\phi_0-\phi_2$, etc. (See computation on p. 114.) For each pair an observation equation of the form $c-M_1r_1+\Delta\phi=0$ may be written. The resulting normal equations, from which r_1 may be derived, are

$$pc-\Sigma M_1r_1+\Sigma\Delta\phi=0$$
$$-\Sigma M_1c+\Sigma M^2{}_1r_1-\Sigma M_1\Delta\phi=0$$

The computation will be sufficiently accurate if M_1 is carried to tenths of turns only, and as here indicated without assigning weights to the separate pairs.

To the preliminary values of ϕ_1, ϕ_2 [illegible] the results from the separate pairs, may now be applied the corrections M_1r_1 and the latitude computation completed as before.

REDUCTION TO SEA LEVEL.

The reduction of the observed latitude to sea level is given by the expression

$$\Delta\phi=-0.000171\ h\ \sin 2\phi$$

in which $\Delta\phi$ is the correction in seconds of arc to be applied to the observed latitude, h is the elevation of the station above sea level in meters, and ϕ is the latitude of the station. This correction may be gotten from the following table:

Reduction of latitude to sea level.

[The correction is negative in every case.]

h \ ϕ		5° 85°	10° 80°	15° 75°	20° 70°	25° 65°	30° 60°	35° 55°	40° 50°	45°
Feet	*Meters*	″	″	″	″	″	″	″	″	″
100	30	0.00	0.00	0.00	0.00	0.00	0.00	0.00	0.01	0.01
200	61	.00	.00	.01	.01	.01	.01	.01	.01	.01
300	91	.00	.01	.01	.01	.01	.01	.01	.02	.02
400	122	.00	.01	.01	.01	.02	.02	.02	.02	.02
500	152	.00	.01	.01	.02	.02	.02	.02	.03	.03
600	183	.01	.01	.02	.02	.02	.03	.03	.03	.03
700	213	.01	.01	.02	.02	.03	.03	.03	.04	.04
800	244	.01	.01	.02	.03	.03	.04	.04	.04	.04
900	274	.01	.02	.02	.03	.04	.04	.04	.05	.05
1000	305	.01	.02	.03	.03	.04	.05	.05	.05	.05
1100	335	.01	.02	.03	.04	.04	.05	.05	.06	.06
1200	366	.01	.02	.03	.04	.05	.05	.06	.06	.06
1300	396	.01	.02	.03	.04	.05	.06	.06	.07	.07
1400	427	.01	.02	.04	.05	.06	.06	.07	.07	.07
1500	457	.01	.03	.04	.05	.06	.07	.07	.08	.08
1600	488	.01	.03	.04	.05	.06	.07	.08	.08	.08
1700	518	.02	.03	.04	.06	.07	.08	.08	.09	.09
1800	549	.02	.03	.05	.06	.07	.08	.09	.09	.09
1900	579	.02	.03	.05	.06	.08	.09	.09	.10	.10
2000	610	.02	.04	.05	.07	.08	.09	.10	.10	.10
2100	640	.02	.04	.05	.07	.08	.09	.10	.11	.11
2200	671	.02	.04	.06	.07	.09	.10	.11	.11	.11
2300	701	.02	.04	.06	.08	.09	.10	.11	.12	.12
2400	732	.02	.04	.06	.08	.10	.11	.12	.12	.13
2500	762	.02	.04	.07	.08	.10	.11	.12	.13	.13
2600	792	.02	.05	.07	.09	.10	.12	.13	.13	.14
2700	823	.02	.05	.07	.09	.11	.12	.13	.14	.14
2800	853	.03	.05	.07	.09	.11	.13	.14	.14	.15
2900	884	.03	.05	.08	.10	.12	.13	.14	.15	.15
3000	914	.03	.05	.08	.10	.12	.14	.15	.15	.16
3100	945	.03	.06	.08	.10	.12	.14	.15	.16	.16
3200	975	.03	.06	.08	.11	.13	.14	.16	.16	.17
3300	1006	.03	.06	.09	.11	.13	.15	.16	.17	.17
3400	1036	.03	.06	.09	.11	.14	.15	.17	.17	.18
3500	1067	.03	.06	.09	.12	.14	.16	.17	.18	.18
3600	1097	.03	.06	.09	.12	.14	.16	.18	.18	.19
3700	1128	.03	.07	.10	.12	.15	.17	.18	.19	.19
3800	1158	.03	.07	.10	.13	.15	.17	.19	.20	.20
3900	1189	.04	.07	.10	.13	.16	.18	.19	.20	.20
4000	1219	.04	.07	.10	.13	.16	.18	.20	.21	.21
4100	1250	.04	.07	.11	.14	.16	.19	.20	.21	.21
4200	1280	.04	.07	.11	.14	.17	.19	.21	.22	.22
4300	1311	.04	.08	.11	.14	.17	.19	.21	.22	.22
4400	1341	.04	.08	.11	.15	.18	.20	.22	.23	.23
4500	1372	.04	.08	.12	.15	.18	.20	.22	.23	.23
4600	1402	.04	.08	.12	.15	.18	.21	.23	.24	.24
4700	1433	.04	.08	.12	.16	.19	.21	.23	.24	.24
4800	1463	.04	.09	.13	.16	.19	.22	.24	.25	.25
4900	1494	.04	.09	.13	.16	.20	.22	.24	.25	.26
5000	1524	.05	.09	.13	.17	.20	.23	.24	.26	.26
5100	1554	.05	.09	.13	.17	.20	.23	.25	.26	.27
5200	1585	.05	.09	.14	.17	.21	.23	.25	.27	.27
5300	1615	.05	.09	.14	.18	.21	.24	.26	.27	.28
5400	1646	.05	.10	.14	.18	.22	.24	.26	.28	.28
5500	1676	.05	.10	.14	.18	.22	.25	.27	.28	.29

Reduction of latitude to sea level—Continued.

h	φ	5° 85°	10° 80°	15° 75°	20° 70°	25° 65°	30° 60°	35° 55°	40° 50°	45°
Feet	*Meters*	″	″	″	″	″	″	″	″	″
5600	1707	0.05	0.10	0.15	0.19	0.22	0.25	0.27	0.29	0.29
5700	1737	.05	.10	.15	.19	.23	.26	.28	.29	.30
5800	1768	.05	.10	.15	.19	.23	.26	.28	.30	.30
5900	1798	.05	.11	.15	.20	.24	.27	.29	.30	.31
6000	1829	.05	.11	.16	.20	.24	.27	.29	.31	.31
6100	1859	.06	.11	.16	.20	.24	.28	.30	.31	.32
6200	1890	.06	.11	.16	.21	.25	.28	.30	.32	.32
6300	1920	.06	.11	.16	.21	.25	.28	.31	.32	.33
6400	1951	.06	.11	.17	.21	.26	.29	.31	.33	.33
6500	1981	.06	.12	.17	.22	.26	.29	.32	.33	.34
6600	2012	.06	.12	.17	.22	.26	.30	.32	.34	.34
6700	2042	.06	.12	.17	.22	.27	.30	.33	.34	.35
6800	2073	.06	.12	.18	.23	.27	.31	.33	.35	.35
6900	2103	.06	.12	.18	.23	.28	.31	.34	.35	.36
7000	2134	.06	.12	.18	.23	.28	.32	.34	.36	.36
7100	2164	.06	.13	.19	.24	.28	.32	.35	.36	.37
7200	2195	.07	.13	.19	.24	.29	.33	.35	.37	.38
7300	2225	.07	.13	.19	.24	.29	.33	.36	.37	.38
7400	2256	.07	.13	.19	.25	.30	.33	.36	.38	.39
7500	2286	.07	.13	.20	.25	.30	.34	.37	.38	.39
7600	2316	.07	.14	.20	.25	.30	.34	.37	.39	.40
7700	2347	.07	.14	.20	.26	.31	.35	.38	.40	.40
7800	2377	.07	.14	.20	.26	.31	.35	.38	.40	.41
7900	2408	.07	.14	.21	.26	.32	.36	.39	.41	.41
8000	2438	.07	.14	.21	.27	.32	.36	.39	.41	.42
8100	2469	.07	.14	.21	.27	.32	.37	.40	.42	.42
8200	2499	.07	.15	.21	.27	.33	.37	.40	.42	.43
8300	2530	.08	.15	.22	.28	.33	.37	.41	.43	.43
8400	2560	.08	.15	.22	.28	.34	.38	.41	.43	.44
8500	2591	.08	.15	.22	.28	.34	.38	.42	.44	.44
8600	2621	.08	.15	.22	.29	.34	.39	.42	.44	.45
8700	2652	.08	.16	.23	.29	.35	.39	.43	.45	.45
8800	2682	.08	.16	.23	.29	.35	.40	.43	.45	.46
8900	2713	.08	.16	.23	.30	.36	.40	.44	.46	.46
9000	2743	.08	.16	.23	.30	.36	.41	.44	.46	.47
9100	2774	.08	.16	.24	.30	.36	.41	.45	.47	.47
9200	2804	.08	.16	.24	.31	.37	.42	.45	.47	.48
9300	2835	.08	.17	.24	.31	.37	.42	.46	.48	.48
9400	2865	.09	.17	.24	.31	.38	.42	.46	.48	.49
9500	2896	.09	.17	.25	.32	.38	.43	.47	.49	.50
9600	2926	.09	.17	.25	.32	.38	.43	.47	.49	.50
9700	2957	.09	.17	.25	.32	.39	.44	.48	.50	.51
9800	2987	.09	.17	.26	.33	.39	.44	.48	.50	.51
9900	3018	.09	.18	.26	.33	.40	.45	.48	.51	.52
10000	3048	.09	.18	.26	.33	.40	.45	.49	.51	.52

CORRECTION FOR VARIATION OF POLE.

The reduction to the mean position of the pole is derived from the provisional results published by the Latitude Service of the International Geodetic Association. (See p. 85.)

DISCUSSION OF ERRORS.

In discussing the errors of zenith telescope observations it is desirable to consider separately, as on page 48, the external errors, observer's errors and instrumental errors.

The principal *external errors* are those arising from errors in the adopted declinations and those due to abnormal refraction.

The adopted declinations used in the computation necessarily have probable errors which are sufficiently large to furnish much, often a half, of the error of the computed latitude. This arises from the fact that a good zenith telescope gives results but little, if any, inferior in accuracy to those obtained with the large instruments of the fixed observatories which were used in determining the declinations.

Of the stars observed at thirty-six latitude stations, nearly on the thirty-ninth parallel, between 1880 and 1898, the average value of $e_{\frac{\delta+\delta'}{2}}$ derived from the mean place computations was $\pm 0''.16$ and the extreme values were $\pm 0''.12$ and $\pm 0''.23$. The average probable error of the declination of a star in 1900 as given for the 6188 stars in the Boss catalogue is about $\pm 0''.18$, and hence the average value of $e_{\frac{\delta+\delta'}{2}}$ from the Boss stars would be about $\pm 0''.13$. These figures furnish a good estimate of the accidental errors to be expected from the adopted declinations. To estimate the constant errors to be expected from this source is a rather difficult matter. The principal constant error in declination to be feared is that arising from errors in the adopted systematic corrections applied to the separate catalogues of observed places. The three principal researches in regard to these systematic corrections have been made by Profs. Lewis Boss, A. Auwers, and Simon Newcomb. Judging by the differences between the results of these three researches, the constant error in the mean declinations based upon Professor Boss's researches, may possibly be as great as $0''.3$, but is probably much smaller than that.

In regard to errors arising from abnormal refraction it should be noted that only the *difference* of refraction of the two stars of a pair enters the computed result. The errors in the computed differential refractions are probably very small when all zenith distances are less than 45° and when care is taken to avoid local refraction arising from the temperature inside the observatory being much above that outside, or from masses of heated air from chimneys or other powerful sources of heat near the observatory. If there were a sensible tendency, as has been claimed, for all stars to be seen too far north (or south) on certain nights, because of the existence of a barometric gradient. for example, it should be detected by a comparison of the mean results on different nights at the same station. The conclusion from many such comparisons made by Prof. John F. Hayford is that the variation in the mean results from zenith telescope measurements from night to night is about what should be expected from the known accidental errors of observation and declination; or, in other words, that if there are errors peculiar to each night they are exceedingly small.[1]

The *observer's errors* are those made in bisecting the star and in reading the level and micrometer. Errors due to unnecessary longitudinal pressure on the head of the micrometer may also be placed in this class.

Indirect evidence indicates that the error of bisection of the star is one of the largest errors concerned in the measurement. The bisections should be made with corresponding care. The probable error of a bisection must be but a fraction of the apparent width of the micrometer line if the observations are to be ranked as first class. It is possible to substitute three or more bisections for the one careful bisection recommended in the directions for observing (p. 110), but it is not advisable to do so. On account of the comparative haste with which such bisections must be made, it is doubtful whether the mean of them is much, if any, more accurate than a single careful and deliberate bisection, while the continual handling of the micrometer head, which is necessary when several bisections are made, tends to produce errors.

With care in estimating tenths of divisions on the micrometer head and on the level graduation, each of these readings may be made with a probable error of ± 0.1 division. If one turn of the micrometer screw represents about $60''$ and one division of the level about $1''$, such reading would produce probable errors of $\pm 0''.04$ and $\pm 0''.05$, respectively, in the latitude from a single observation. These errors are small, but not negligible, for the whole probable error of a single observation arising from all sources is often less than $\pm 0''.30$ and sometimes less than $\pm 0''.20$.

[1] See Report of the Boundary Commission upon the Survey and Re-marking of the Boundary between the United States and Mexico West of the Rio Grande, 1891 to 1896 (Washington, 1898), pp. 107–109, for one such comparison.

While reading the level the observer should keep in mind that a very slight unequal or unnecessary heating of the level tube may cause errors several times as large as the mere reading error indicated above, and that if the bubble is found to be moving, a reading taken after allowing it to come to rest deliberately may not be pertinent to the purpose for which it was taken. The level readings are intended to fix the position of the telescope at the instant when the star was bisected.

It requires great care in turning the micrometer head to insure that so little longitudinal force is applied to the screw that the bisection of the star is not affected by it. Such a displacement of 1–4000 of an inch in the position of the micrometer line relative to the objective produces an apparent change of more than 1″ in the position of a star if the focal length of the telescope is less than 50 inches. The whole instrument being elastic, the force required to produce such a displacement is small. An experienced observer has found that in a series of his latitude observations, during which the level was read both before and after the bisections of the star, the former readings continually differed from the latter, from 0″.1 to 0″.9, nearly always in one direction.[1]

Among the *instrumental errors* may be mentioned those due (1) to an inclination of the micrometer line to the horizon; (2) to error in the adopted value of one division of the level; (3) to inclination of the horizontal axis; (4) to erroneous placing of the azimuth stops; (5) to error of collimation; (6) to the instability of the relative positions of different parts of the instrument; (7) to the irregularity of the micrometer screw; (8) to the error of the adopted mean value of one turn of the micrometer screw.

The first of these sources of error must be carefully guarded against, as indicated on page 106, as it tends to introduce a *constant* error into the computed latitudes. The observer, even if he attempts to make the bisection in the middle of the field (horizontally), is apt to make it on one side or the other, according to a fixed habit. If the line is inclined, his micrometer readings are too great on all north stars and too small on all south stars, or *vice versa*.

The error arising from an erroneous level value is smaller the smaller are the level corrections and the more nearly the plus and minus corrections balance each other. If the observer makes it his rule whenever the record shows a level correction of more than one division to correct the inclination of the vertical axis between pairs, this error will be negligible. Little time is needed for this if the observer avoids all reversals by simply manipulating a foot-screw so as to move the bubble as much to the northward (or the southward) as the record indicates the required correction to be.

The errors from the third, fourth, and fifth sources may easily be kept within such limits as to be negligible. An inclination of 1 minute in the horizontal axis, or an error of that amount in either collimation or azimuth, produces only about 0″.01 error in the latitude. All three of these adjustments may easily be kept well within this limit.

The errors arising from instability may be small upon an average, but they undoubtedly become large at times and produce some of the largest residuals. One of the most important functions of the observer is to guard against them by protecting the instrument from sudden temperature changes and from shocks and careless or unnecessary handling, and by avoiding long waits between the two stars of a pair. The closer the agreement in temperature between the observing room and the outer air the more secure is the instrument against sudden and unequal changes of temperature.

Most micrometer screws now used are so regular that the uneliminated error in the mean result for a station arising from the seventh source named above is usually negligible. Irregularities of sufficient size to produce a sensible error in the mean result may be readily detected by inspection of the computation of micrometer value if that computation is made as indicated on pages 126–128. The two forms of irregularity most frequently detected in modern screws on our latitude instruments are those with a period of one turn and those of such a form that the value of one turn increases continuously from one end of the screw to the other. The periodic irregularity operates mainly to increase the computed probable error of observation and must

---

[1] U. S. Coast and Geodetic Survey Report, 1892, part 2, p. 58.

be quite large to have any sensible effect upon the computed mean value of the latitude. If the value of the screw increases continuously and uniformly from one end to the other, the computed results will be free from any error arising from this source, provided all settings are made so that the mean of the two micrometer readings upon a pair falls at the middle of the screw. If this condition is fulfilled within one turn for each pair, the error in the mean result will usually be negligible. If the settings are not so made, it may be necessary to compute and apply a correction for the irregularity.

Evidence has already been presented on pages 126–130 to show that it is difficult to obtain the actual mean micrometer value. It is important, therefore, to guard against errors arising from the eighth source by selecting such pairs that the plus and minus micrometer differences actually observed at a station shall balance as nearly as possible. The final result will be free from error from this source if the weighted mean of the micrometer differences, the signs being preserved, is zero. The only effect of the error in the mean micrometer value in that case is to slightly increase the computed probable errors. The weights are not, however, usually known during the progress of the observations. If the indiscriminate mean of the micrometer differences for each pair, taken with respect to the signs, is made less than one turn at a station, the error of the mean result from this source will usually be less than its computed probable error.

THE ECONOMICS OF LATITUDE OBSERVATIONS.

Two questions imperatively demand an answer under this heading. What ratio of number of observations to number of pairs will give the maximum accuracy for a given expenditure of money and time? What degree of accuracy in the mean result for the station is it desirable and justifiable to strive for?

The answer to the first question depends upon the relative magnitude of the accidental errors of declination and of observation. At 36 stations nearly on the thirty-ninth parallel, at which latitude observations have been made since the beginning of 1880, the average value of $e_{\frac{\delta}{2}}$, the probable error of the mean of two declinations (derived from the mean place computations), is $\pm 0''.16$ and the extreme values were $\pm 0''.12$ and $\pm 0''.23$. At 37 stations occupied with zenith telescopes along the thirty-ninth parallel the extreme values of e, the probable error of a single observation, were $\pm 0''.16$ and $\pm 0''.98$, and at about one-half of the stations it was less than $\pm 0''.42$.[1] Similarly, at 43 stations along that parallel occupied with meridian telescopes e was less than $\pm 0''.45$ at one-half the stations, and the extreme values were $\pm 0''.21$ and $\pm 1''.27$. In the light of these figures one may use the following table to determine the most economical ratio of number of observations to number of pairs:

Weight to be assigned to mean latitude from a single pair.

$$w_n = \left(e^2_{\frac{\delta}{2}} + \frac{e^2}{n_n}\right)^{-1}$$

$e_{\frac{\delta}{2}}$ being assumed to be $\pm 0''.16$.

e	Number of observations on the pair					
	1	2	3	4	5	6
$''$						
±0. 16	20	26	29	31. 2	32. 6	33. 4
±0. 20	15	22	26	28. 1	29. 8	31. 0
±0. 30	9	14	18	20. 8	22. 9	24. 6
±0. 40	5. 4	9. 5	12. 7	15. 2	17. 4	19. 1
±0. 60	2. 6	4. 9	6. 9	8. 7	10. 2	11. 7
±0. 80	1. 5	2. 9	4. 2	5. 4	6. 5	7. 6
±1. 00	1. 0	1. 9	2. 8	3. 6	4. 4	5. 2

[1] One thousand two hundred and seventy-seven observations for variation of latitude at San Francisco in two series gave $e = \pm 0''.19$ and $e = \pm 0''.28$. A similar series at the Hawaiian Islands in 1891–92, 2434 observations, gave $e = \pm 0''.16$. On the Mexican boundary in 1892–93, 1362 observations at fifteen stations gave $e = \pm 0''.19$ to $\pm 0''.38$. All these observations were made with zenith telescopes. (See Coast and Geodetic Survey Reports, 1893, Part 2, p. 494; 1892, Part 2, pp. 54 and 158; 1892, Part 2, p. 50, and Mexican Boundary Report, 1891–1896, p. 101.)

The measure of efficiency of the first observation is the weight shown in the first column, and of each succeeding observation is the resulting *increment* of weight. Thus, if $e = \pm 0''.16$, the first observation gives a weight of 20, while the second observation is less than one-third as efficient, the increment of weight being only 6, and the fifth and sixth observations combined are about one-ninth as efficient as the first observation. Stated otherwise, the probable error of a single observation being in this case the same as the probable error of the mean of two declinations, little is gained by reducing the observation error while the declination error is allowed to remain. If $e = \pm 0''.60$, the table shows that the second and third observations are each nearly as efficient as the first. The larger is e the less difference there is between the first and succeeding observations, but in every case the first observation is more efficient than any later observation.

If each observation after the first involved the same amount of time spent in preparation, observation, and computation as the first, it is evident that to secure a maximum of accuracy for a given expenditure each pair should be observed *but once*. Additional observations on new pairs require appreciably more time than the same number of observations on pairs already observed only in the following items: Preparing the observing list, computing mean places, and computing apparent places. Several observations per pair save an appreciable amount of time in the apparent place computation only when the successive nights of observation follow each other so closely that the apparent places on certain nights may be obtained by interpolation. (The interval over which a straight-line interpolation may be carried with sufficient accuracy is three days.)

After balancing this slight increase in labor against the greater efficiency of the first observation upon a pair over any succeeding observation, it is believed that if e is not greater than $0''.40$, each pair should be observed but once. If e is much greater than $0''.40$, two or possibly even three observations per pair may be advisable.

It is true that if but a single observation is made upon each pair the observer in the field will not be able to determine his error of observation accurately (he may do so approximately by assuming $e_{\frac{m}{2}} = \pm 0''.16$), but the field computation will still perform its essential function of detecting omissions and deficiencies if they exist.

What degree of accuracy in the mean result for a station is it desirable and justifiable to strive for? Omitting from consideration stations occupied to determine the variation of latitude, and stations occupied upon a boundary at which one purpose of the latitude observations is to furnish a means of recovering the same point again, the ordinary purpose of latitude observations in connection with a geodetic survey is to determine the station error in latitude, or, in other words, to determine the deflection of the vertical, measured in the plane of the meridian, from the normal to the spheroid of reference at the station. Broadly stated, the purpose of astronomic observations of latitude and longitude (and to a large extent of azimuth also) in connection with a geodetic survey is to determine the relation between the actual figure of the earth as defined by the lines of action of gravity and the assumed mean figure upon which the geodetic computations are based. In determining this relation three classes of errors are encountered: The errors of the geodetic observations, the errors of the astronomic observations, and the errors arising from the fact that only a few scattered astronomic stations can be occupied in the large area to be covered, and that the station errors as measured at these few *points* must be assumed to represent the facts for the whole *area*. It suffices here in regard to errors of the first class, which are not within the province of this appendix, to state that they are in general of about the same order of magnitude as those of the second class.

The average value of the station error in latitude, without regard to sign, at 381 stations used in the Supplementary Investigation of the Figure of the Earth and Isostasy, is $3''.8$. An examination of these station errors shows that although there is a slight tendency for their values for a given region to be of one sign and magnitude the values at adjacent stations are nevertheless so nearly independent that the nonpredictable rate of change of the station error per mile is frequently more than $0''.1$. Six stations within the District of Columbia show an irregular variation of station error in latitude with a total range of $1''.8$. Stating the result of the examination in another form, if the station error at a point is assumed to represent the average value of the station error for an area, and if the error of that assumption is to be not greater than $\pm 0''.10$, the area adjacent to the station to which the assumption is applied must not be greater than 10 square miles. If one bears in mind that financial considerations so limit the number of latitude stations that in general the above assumption must be extended over hundreds of square miles, it becomes evident that a probable error of $\pm 0''.10$ in the latitude determination is all that it is desirable or justifiable to strive for.[1] One observation upon each of from 15 to 25 pairs will nearly always secure that degree of accuracy, and the observations may be completed in a single night.

As indicated in the General Instructions for Latitude Work, page 104, paragraphs 3 and 4, this Survey has adopted the plan of using such a number of pairs, observed but once, as will make it reasonably certain that the final computation will give a probable error not greater than $\pm 0''.10$ in the resulting latitude.

Between 1905 and 1908, Assistant W. H. Burger determined the latitude at 63 stations in the United States, making only one observation on a pair (unless it was found that some mistake was made on a pair, in which case a second observation was made on it if observations were made on a second night). The average number of pairs observed per station was 16.7, with a maximum of 34 pairs and a minimum of 9 pairs. The average e_p was $\pm 0''.38$ and the average e_ϕ was $\pm 0''.10$. The average number of nights on which observations were made at a station was 1.9.

Assistant Wm. Bowie occupied 7 stations in 1908. The average number of pairs observed per station was 15, with a maximum of 16 and a minimum of 15 pairs. The average e_p was $\pm 0''.31$ and the average e_ϕ was $\pm 0''.08$. Observations were made on only 8 nights for the 7 stations. At only one station were observations made on more than one night.

COST OF ESTABLISHING A LATITUDE STATION.

It is difficult to give accurately the cost per station for recent latitude work as usually the parties were also making observations for azimuth. However, a fair estimate of the cost, including salary of the observer, for latitude stations by this Survey in any except mountainous country is about $200 per station. In a rough area where pack animals would be used extensively the cost might double this estimate. Where transportation is easy and the stations not distant from each other the stations should cost much less than $200 each if the party remains in the field for long seasons.

---

[1] The above discussion also applies, though with less force, to longitude and azimuth observations. In both these cases the errors of observation are necessarily much larger than in latitude observations.

PART IV.

DETERMINATION OF THE ASTRONOMIC AZIMUTH OF A DIRECTION.

GENERAL REMARKS.

Various methods are employed in the Coast and Geodetic Survey for determining astronomically the azimuth of a triangulation line, or what is the same thing, the direction of that line with respect to the meridian, and there are, perhaps, no other geodetic operations in which the choice of the method, the perfection of the instrument, and the skill of the observer enter so directly into the value of the result. It is intended to give here in a concise form an account of several methods now in use, and to present the formulæ as well as specimens of record and examples of computation. If it is proposed to measure a primary or subordinate azimuth, the observer will generally have the choice of the method most suitable and adequate for the purpose, and accordingly provide himself with the proper instrument; yet frequently he may find himself already provided with an instrument, in which case that method will have to be selected which is compatible with the mechanical means at hand and at the same time insures the greatest accuracy.

The astronomic azimuth, or the angle which the plane of the meridian makes with the vertical plane passing through the object whose direction is to be determined, is generally reckoned from the south and in the direction southwest, etc. However, when circumpolar stars are observed it will be found more convenient to reckon from the *north* meridian and eastward—that is, in the same direction as before.

The geodetic azimuth differs from the astronomic azimuth. The former is supposed free from local deflections of the plumb line or vertical, it being the mean of several astronomic azimuths, all referred geodetically to one station, and it may be supposed that in this normal azimuth the several local deflections will have neutralized each other. The astronomic azimuth is, of course, subject to any displacement of the zenith due to local attraction or deflection.

We may distinguish between primary and secondary azimuths—the one fixing the direction of a side in primary triangulation, the other having reference to sides of secondary or tertiary triangulations or to directions in connection with the measure of the magnetic declination. For the determination of a primary azimuth the local time (sidereal) must either be known—as, for instance, when a telegraphic longitude is at the same time determined—or special observations must be made for it. For subordinate azimuths, time and azimuth observations may sometimes be made together, as with the alt-azimuth instrument for magnetic purposes, in which case the sun's limbs are usually observed. In refined work in high latitudes, and for certain rare cases in low latitudes, the transit instrument is needed to furnish the chronometer correction. For primary azimuths, in latitudes not greater than those in the United States, the local time may be found with sufficient accuracy by means of an especially constructed vertical circle, used in the Coast and Geodetic Survey, and shown in illustration No. 8. For secondary azimuths, local time may be found by means of sextants or alt-azimuth instruments.

PRIMARY AZIMUTH.

The requirements for primary azimuth are that the astronomic azimuth observations and the necessary time observations should be made using such methods, instruments, and number of observations as to make it reasonably certain that the probable error of the astronomic azimuth does not exceed $\pm 0''.50$. It is not desirable to spend much time or money in reducing

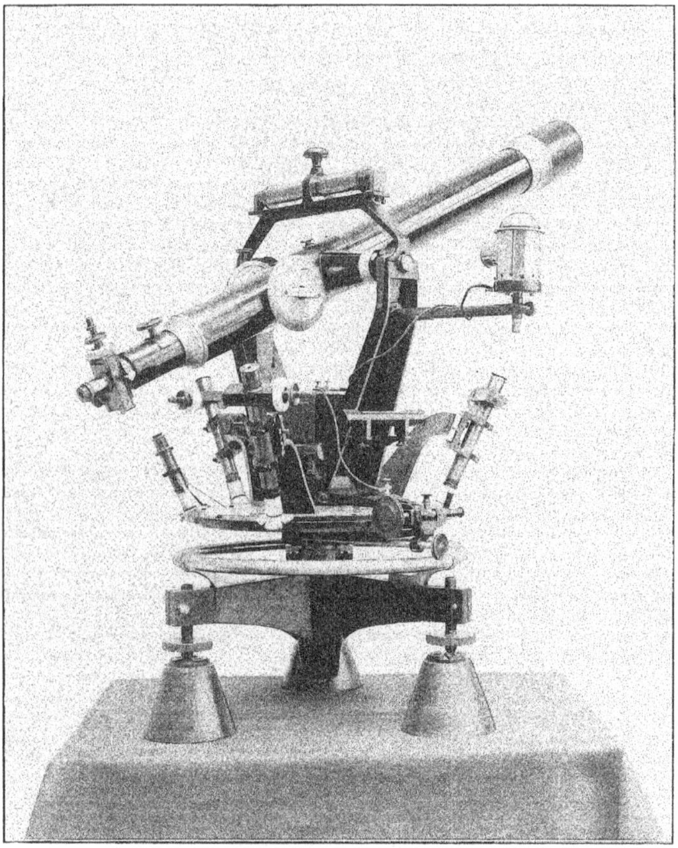

TWELVE-INCH DIRECTION THEODOLITE.

No. 19.

SEVEN-INCH REPEATING THEODOLITE.

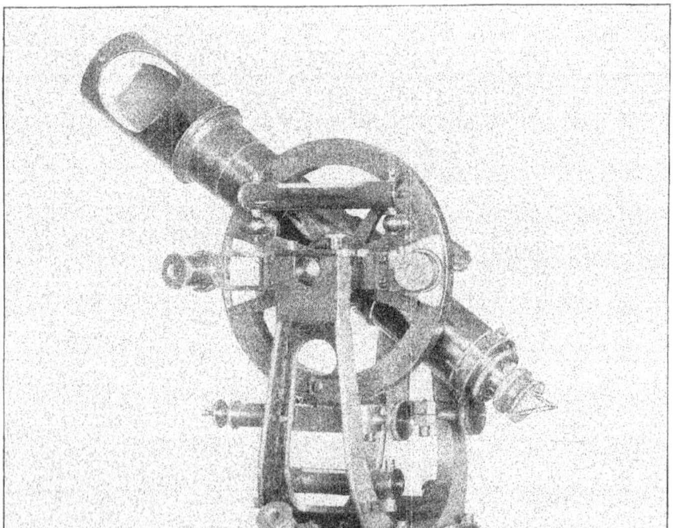

FOUR-INCH THEODOLITE.

the probable error below this amount. At Laplace stations (coincident triangulation, longitude, and azimuth stations), however, the astronomic azimuth should be determined with a probable error not greater than $\pm 0''.30$ and the observations should be made on at least two nights. When observations are made to determine the astronomic azimuth of a line of the primary triangulation, the azimuth station should coincide with a station of the triangulation and the mark used should be some other station of the scheme. In this way the azimuth is referred directly to one of the lines of the triangulation. The probable error of the azimuth of a line obtained from an observed astronomic azimuth on a mark separate from the triangulation is greater than the probable error of the observed azimuth.

The practice in the United States Coast and Geodetic Survey is for the party on primary triangulation to observe all necessary astronomic azimuths during the progress of the triangulation. Where a direction instrument is used, the star is often observed upon in the regular series of observations upon the triangulation stations. In such cases the last object observed upon in any one series is the star, and the instrument is reversed immediately after the first pointing upon it. Where the star is observed upon in connection with two or more triangulation stations, the station next preceding it is the one to which the astronomic azimuth is referred.

INSTRUMENTS.

So great a variety of instruments is used for azimuth determinations that it is of little avail to describe any particular instrument in detail. Illustration No. 18 shows a 12-inch[1] direction theodolite (No. 146) made at this office and now in use for the measurement of horizontal angles and azimuths in primary triangulation. It carries a very accurate graduation, which is read to seconds directly and to tenths by estimation by three microscopes.[2] A glass-hard steel center also contributes toward making this theodolite and others of identical construction furnish results of a very high degree of accuracy. The graduation of the horizontal circle on this instrument is to 5′ spaces. An 8-inch repeating theodolite reading to five seconds by two opposite verniers is shown in illustration No. 19. For observations on the sun for azimuth in connection with magnetic determinations a small 4-inch theodolite is often used. (See illustration No. 20.) This instrument reads to minutes on each of two opposite verniers. The transit instruments and meridian telescopes described in connection with time observations on pages 7–8 are also frequently used for azimuth either in the meridian (p. 160) or in the vertical plane of a circumpolar star at or near elongation (p. 157).

When the azimuth is observed during the progress of the primary triangulation the regular triangulation signal lamps shown in illustrations Nos. 21 and 22 are used. The smaller lamp can be seen under average conditions to a distance of about 30 miles. The larger lamp has been observed in the southwestern portion of the United States, where the atmosphere is very clear, up to distances of 120 miles. Where the mark is only a short distance from the station, an ordinary lantern, a bull's eye lantern, or an electric hand lamp may be used. In connection with a triangulation along the coast the lantern of a lighthouse can be used as the mark.

INSTRUMENT SUPPORTS.

While making observations for a secondary azimuth the instrument used is usually supported upon its own tripod, mounted upon stakes driven firmly into the ground. In primary triangulation the theodolite is frequently mounted upon a tripod which may be as much as 25 or more meters above the ground. Where the instrument is not elevated it is mounted upon a specially constructed wooden tripod or stand which has its legs firmly set into the ground and well braced. On the top of the legs is fitted a wooden cap usually 2 inches thick. On this cap are fastened the plates which receive the foot screws of the theodolite.

The structure shown in illustration No. 23 is used to elevate the instrument in triangulation and azimuth work. It consists of a tripod on which the instrument rests and a four-sided

[1] Following the usual practice, the size of the theodolite is here designated by giving the diameter of the graduated horizontal circle.
[2] For a more complete description of this instrument see Report for 1894, pp. 265–274.

scaffold on which the observer stands. The tripod and scaffold do not touch each other at any point. The top floor of the scaffold is not needed on azimuth work and is only used on primary triangulation when there are two observing parties working in conjunction. A complete description of this type of signal is given on pages 829 to 842 of Appendix 4, Report for 1903. Most of the azimuth stations are in places where it is difficult to carry lumber, and as a result it is usual to have no platform around the stand when the instrument is only elevated above the ground to the height of the observer's eye. Where no platform is used, the observer should be careful not to step close to a leg of the stand while making the observations on the star. Such precautions are not necessary to the same extent while making the observations on the mark (or triangulation station), assuming, of course, that the mark is not far from being in the horizon of the station. As a result of not using an observing platform it may be necessary to make more observations to get the desired degree of accuracy than if a platform had been used. The errors resulting from not having a platform are mainly of the accidental class and their effect on the final azimuth is small.

Where both azimuth and latitude are to be observed at a station, but not at the same time as the triangulation observations, a wooden pier similar to that shown in illustration No. 24 has been found satisfactory in every way. It was used to a great extent by former Assistant W. H. Burger and to a limited extent by Assistant W. Bowie. It will be seen that the spread and slope of the legs of the stand make it possible to mount on it each of the instruments in turn, the top section of the pier being removed when used for latitude. The pier is made as if for the azimuth work, and then the top is sawed off at such point as will make the base of the pier of the required height for the latitude instrument.

AZIMUTH MARK.

When it is necessary to elevate a signal lamp over a triangulation station used as a mark a number of devices may be used. A simple pole well guyed is frequently used, but this is not very satisfactory, for it is difficult to keep the support of the lamp accurately centered over the station mark. A device like that shown in illustration No. 25 may be used, and this has the advantage that the light keeper does not have to climb the pole when posting and inspecting the lamp. A very satisfactory and inexpensive structure frequently used in the United States Coast and Geodetic Survey is shown in illustration No. 26. The legs, of lumber 2 by 4 inches in cross section, are anchored securely in the ground and at intervals the structure is guyed by wire. The light keeper goes up the inside of this signal, and near its top there is an opening leading out to a seat. Such a signal may be built to a height of 140 feet or more. An acetylene lamp, like one of those shown in illustrations Nos. 21 and 22, should be posted at the distant triangulation station used as the mark.

When the azimuth of a line of the triangulation is not measured directly, a special azimuth mark is erected, which is afterwards referred to the triangulation by means of horizontal angles. There has been considerable variety in the azimuth marks so used, each chief of party adapting the mark to the special conditions in which he finds himself and to his own convenience. A box with open top having in its front face a round hole or a slit of suitable size, through which the light of a bull's eye or common lantern can be shown, makes a satisfactory mark. See illustration No. 27. A white or black stripe of paint or signal muslin can be placed on the box, centered over the opening, upon which to make observations during the day in order to refer the astronomic azimuth of the mark to a line of the triangulation.

The location of the mark is generally determined, in part at least, by the configuration of the ground surrounding the station, but it should not be placed any nearer than about one statute mile in order that the sidereal focus of the telescope may not require changing between pointings upon the star and upon the mark, since any such change is likely to change the error of collimation. Should the mark be closer to the station than one mile and no change be made in the sidereal focus when pointing upon the mark, there would probably be errors caused by parallax. If practicable, the mark should be placed nearly in the horizon of the station occupied, in order that small errors of inclination of the horizontal axis of the instrument may not affect the point-

SMALL ACETYLENE SIGNAL LAMP.

No. 22.

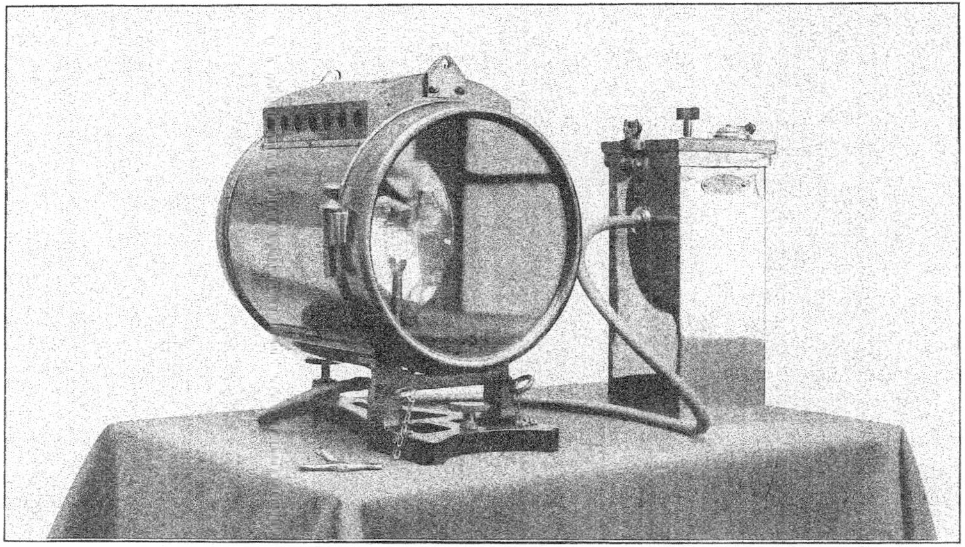

LARGE ACETYLENE SIGNAL LAMP.

ings upon the mark, and corresponding readings of the striding level will be unnecessary. In choosing the position of the mark it should be kept in mind that the higher the line of sight to it above the intervening ground the more steady the light may be expected to show and the smaller the errors to be expected from lateral refraction.

SHELTER FOR THE INSTRUMENT.

An especially designed tent should be used to shield the instrument from the wind. Illustrations 16 and 17 show two tents which have proved satisfactory. The tent should be only as heavy as is necessary to withstand strong winds and protect the instruments from rain. When not in actual use the instruments used for azimuth observations should be dismounted and placed in their packing cases. Owing to the short time during which an azimuth station is occupied for observations it is usually not necessary or desirable to erect a wooden observatory to protect the instruments.

ARTIFICIAL HORIZON.

Instead of determining the inclination of the horizontal axis by readings of a striding level, observations are sometimes taken upon the image of the star as seen reflected from the free surface of mercury (an artificial horizon) in addition to the direct observations upon the star. The error in azimuth produced by the inclination of the horizontal axis is of the same numerical value for the reflected observations as for the direct observations, but is reversed in sign, and the mean result is free from error from this source, *provided* the cross-section of each pivot is circular, or at least that the two pivots have similar cross-sections similarly placed. Considerable care and ingenuity is necessary to protect the mercury effectually against tremors and against wind, either of which will by disturbing the mercury surface make the reflected star image so unsteady as to make accurate pointing upon it difficult or impossible. A glass roof over the mercury to protect it from the wind should never be employed in connection with azimuth observations, since reversal of it does not sufficiently correct for errors arising from refraction at the glass. Large boxes, or tubes of considerable size, with their openings covered with mosquito netting, have proved the most satisfactory protection of the mercury against the wind.

It is believed that the lateral refraction of the direct and reflected ray, when the mercury is set on the ground, may introduce uncertain and possibly large errors into the azimuth. This trouble can be avoided by placing the artificial horizon on a stand nearly as high as the theodolite. This, however, can not be done with the direction theodolite (except in very low latitudes). The artificial horizon can not be used in high latitudes when making observations on Polaris, as the horizontal circle of the theodolite would intercept the reflected ray.

POINTING LINES.

The pointings in azimuth observations are usually taken by using either a single vertical line in a reticle (or attached to a micrometer) or a pair of parallel vertical lines about 20″ (of arc) apart. The first has the advantage over the second that it does not involve the necessity of bisecting a space by eye, as the observation consists simply of noting when the star image appears symmetrical with respect to the line. On the other hand, it has the disadvantage that frequently when a very bright star (or light) is observed the line appears to be "burned off" near the star image; that is, it becomes invisible because of its comparative faintness, and the pointing is correspondingly uncertain. So also if a very faint star (or light) is observed its image may nearly or completely disappear behind the line and so make accurate pointing difficult. For many stars of intermediate degrees of brightness one or the other of these difficulties exists to a greater or less degree. If two vertical lines are used and the distance between them is properly chosen these two difficulties will be avoided and both star (or mark) and lines will always be distinctly visible at the same instant. The observation now consists in noting when the image of the star (or mark) bisects the space between the two lines. This process is probably but slightly less accurate under any conditions of brightness than the direct bisection of a star image under the most favorable conditions as to brightness. In measuring horizontal angles and azimuths in Colorado, Utah, and Nevada, along the thirth-ninth parallel, and on all primary triangulation on the ninety-eighth meridian since 1901, and on the Texas-California arc of primary triangulation, two vertical lines about 20″ apart were used.

During the progress of the triangulation along the western part of the thirty-ninth parallel, observations were made at times upon Polaris in daylight to determine the astronomic azimuth. This is a satisfactory method and occasionally is convenient for the observer.

GENERAL CONSIDERATIONS.

Let the hour angle (t), declination (δ), and latitude (φ) be slightly in error by the quantities dt, $d\delta$, and $d\varphi$, and let dA equal their effect upon the azimuth (A); then, in general, it will be seen that, all other circumstances being equal, dA increases as the zenith distance (ζ) decreases; for a star near the pole and for a latitude not too high a small error in time and in latitude has but a slight effect upon the azimuth, and in the case of a circumpolar star at elongation (when the parallactic angle is 90°) a small error in time, dt, will not affect the azimuth; but small errors in declination, $d\delta$, and in latitude, $d\varphi$, then attain nearly their maximum effect upon the azimuth. If observations are made upon a circumpolar star ($\delta > \varphi$) at the eastern and at the western elongation, effects of $d\delta$ and $d\varphi$ will disappear in the combination of the two results; this, therefore, is the most favorable condition for observing. In general, effects of $d\delta$ and $d\varphi$ disappear in mean results of observations of equal and opposite azimuths. In observations on a circumpolar star in the meridian the effect of a small error in time and in right ascension may be eliminated by a combination of results from upper and lower culminations; for a star in the meridian the quantities $d\delta$ and $d\varphi$ do not enter in the azimuth. If the object to be observed, star (or sun), is of great polar distance (also $\delta < \varphi$), and if δ is positive, the best time for observing is before the eastern transit, or after the western transit over the prime vertical, when the change in azimuth with respect to time is a minimum, but the star (or sun) should not be too near the zenith nor be so low as to be affected by changes of refraction; if δ is negative, the star (or sun) should be observed some distance from the meridian.[1]

These considerations have led to the plan of making first-class azimuth observations almost exclusively upon the close circumpolars α, δ, and λ Ursæ Minoris and 51 Cephei. The apparent places of these four stars are given in the American Ephemeris for every day of the year. Illustration No. 28 will assist in readily finding the two fainter stars λ Ursæ Minoris and 51 Cephei, which barely become visible to the naked eye under the most favorable circumstances; it also shows that when δ Ursæ Minoris and 51 Cephei culminate on either side of the pole, Polaris is not far from its elongation; and, likewise when the pole star culminates, the other two are on opposite sides of the meridian, near their elongations. A similar approximate relation exists between α and λ Ursæ Minoris. Polaris offers the advantage of being observable in daytime with portable instruments; hence it may be observed at eastern and western elongations, or at upper and lower culminations, provided the sun be not too high; λ Ursæ Minoris, from its greater proximity to the pole and its smaller size, presents to the larger instruments a finer and steadier object for bisection than Polaris; 51 Cephei is also advantageously used on account of its small size. The star B. A. C. No. 4165, shown on the diagram, was proposed and used for azimuth work by Assistant G. Davidson. The apparent precessional motion of the pole in 100 years is indicated by the direction and length of the arrow. The sun is employed only to determine azimuths of inferior accuracy, generally in connection with the determination of the magnetic declination.

---

[1] The statements made in a general and somewhat indefinite form in this paragraph may be stated in accurate mathematical form by deriving dA in terms of dt, $d\varphi$, $d\delta$, respectively, from the formula

$$\tan A = \frac{-\sin t}{\cos \varphi \tan \delta - \sin \varphi \cos t}$$

(see p. 143), or from the formulæ used in its derivation.

EIGHTY-FOOT SIGNAL.

No. 24.

WOODEN PIER USED FOR THEODOLITE AND ZENITH TELESCOPE.

GENERAL FORMULÆ.

Four methods of determining azimuth will be treated in detail in this publication, namely, (1) the method in which a direction theodolite is used, as in the measurement of horizontal directions; (2) the method of repetitions with a repeating theodolite; (3) the micrometric method, using an eyepiece micrometer; (4) the determination of azimuth from time observations with a transit or meridian telescope approximately in the meridian.[1] Certain formulæ which are common to the first three of these methods will be stated here for convenient reference.

The computation of the azimuth of a terrestrial line of sight from a set of azimuth observations consists essentially of a computation of the azimuth of the star at the instant of observation, a computation of the horizontal angle between the star and the mark, and the combination of these two results by addition or subtraction.

In the spherical triangle defined by the pole, the zenith, and a star, the side zenith-pole is the co-latitude, the side star-pole is the polar distance of the star, and the angle at the pole is the hour angle[2] or its explement. Starting from these three as known parts, the spherical triangle may be solved by the ordinary formulæ of spherical trigonometry. The solution to obtain the azimuth of the star, which is the angle of this triangle at the zenith, may, *without any approximations*, be put in the form

$$\tan A = -\frac{\sin t}{\cos \varphi \tan \delta - \sin \varphi \cos t}$$

in which A is the azimuth of the star counted from the north in a clockwise direction,[3] and the hour angle t is counted westward from upper culmination continuously to 24^h, or 360°, at the next upper culmination. This is the most convenient formula for use with either of the first three methods. The first term of the denominator changes very slowly and may be tabulated for slightly different values of δ during the period of observation. The second term, for a close circumpolar star, may be computed with sufficient accuracy by five-place logarithms.

The computation of the azimuth from this formula may be considerably shortened by transforming it as indicated below and using the table given on pages 165–173:[4]

$$\tan A = -\frac{\sin t}{\cos \varphi \tan \delta - \sin \varphi \cos t}$$

$$= -\frac{\cot \delta \sec \varphi \sin t}{1 - \cot \delta \tan \varphi \cos t}$$

$$= -\cot \delta \sec \varphi \sin t \left(\frac{1}{1-a}\right)$$

in which $a = \cot \delta \tan \varphi \cos t$.

The second form of this formula is about as convenient as the first. It involves the same number of logarithms as the first and one less reduction from logarithms to numbers.

The third form in connection with the tables given on pages 165–173 gives a much quicker computation process than either of the other two. In using this form and the tables, log cot δ sec φ sin t must be carried to six places and log cot δ tan φ cos t to five places. The most convenient arrangement of the computation is shown on page 148. The formula and tables involve no approximations, and the only errors resulting from their use are those arising from the cast-off decimal places (logarithms limited to *six* places). These errors are of the accidental class, and

[1] The method of determining azimuth by observations upon the sun at any hour angle is not treated in this publication, because it is used mainly in making observations for magnetic declinations and a description of it, with tables for making the parallax and refraction corrections, is given in "Principal Facts of the Earth's Magnetism" published in 1909, and also in "Directions for Magnetic Measurements" published in 1911, both issued by the Coast and Geodetic Survey.

[2] In this publication the hour angle will be reckoned westward from zero at upper culmination (increasing with the lapse of time) to 360° or 24^h.

[3] In astronomic computations it is more convenient to count the azimuth from the north instead of from the south, as in geodetic computations. If the direction of the count is clockwise, as here stated, to change from one reckoning to the other it is only necessary to add or subtract 180°.

[4] The formula and the table are both copied from *Formeln und Hülfstafeln für Geographische Ortsbestimmungen* von Prof. Dr. Th. Albrecht, Leipzig, 1894. The range of the table has, however, been considerably extended.

will seldom exceed 0''.04 for any case covered by the table, and for most observations made below latitude 50° the error will not exceed 0''.01. These quantities are so small in comparison with the errors of observation as to be negligible. A few observations made in Alaska may be beyond the range of the tables on pages 165–173, and when that is found to be the case, one may easily substitute the second formula on page 143 for the third.[1]

To compute the azimuth of a star at the time of *each* pointing made upon it during a set of observations is an unnecessarily laborious process. If for the hour angle, t, of the azimuth formula is taken the mean of the hour angles of the set, the computed azimuth is that corresponding *to the mean hour angle*, but is not the required *mean of the azimuths corresponding to the separate hour angles*, since the rate of change of the azimuth is continually varying because of the curvature of the apparent path of the star. The difference between the two quantities indicated by the italics is small, though not usually negligible, for the interval of time covered by a set of observations. The most convenient way of making the computation for a set of observations is to use the mean hour angle in the azimuth formula and apply to the result a

$$\text{Curvature Correction} = \tan A \, \frac{1}{n} \Sigma \frac{2 \sin^2 \frac{1}{2} \tau}{\sin 1''}$$

in which n is the number of pointings upon the star in the set and τ for each observation is the difference[2] between the time of that observation and the mean of the times for the set. The sign of this curvature correction is always such as to decrease numerically the azimuth reckoned from the north, or in other words, if azimuths are counted clockwise its algebraic sign will be + when the star is west of north and − when the star is east of north. If the star crosses the meridian during the progress of a set the curvature correction will ordinarily be zero. The formula is approximate, but for circumpolars and for the interval of time usually covered by a set of observations its errors are negligible. The value of the term $\frac{2 \sin^2 \frac{1}{2} \tau}{\sin 1''}$ may be found on pages 151–152 of this publication.[3]

If the star observed is Polaris, a convenient rough check on the computation may be obtained from Table V of the American Ephemeris and Nautical Almanac, entitled Azimuth of Polaris at all Hour Angles.

Because of the rapid motion of the observer, due to the rotation of the earth on its axis, a star is seen slightly displaced from its real position. The required

$$\text{Correction for Diurnal Aberration} = 0''.32 \, \frac{\cos A \cos \phi}{\cos h}$$

The sign of the correction is always positive when applied to azimuths counted clockwise. The greatest variation of the correction from its mean value, 0''.32, for the four circumpolars ordinarily observed and for latitudes not greater than 50°, is 0''.02. The correction for diurnal aberration need not be applied to the separate sets but simply to the mean result for a station.

If the horizontal axis is inclined when the pointings are made upon either the star or the mark the corrections indicated below must be applied.

$$\text{Level Correction} = \frac{d}{4}\left\{(w + w') - (e + e')\right\} \tan h$$

if the striding level carries a graduation numbered in both directions from the middle. d is the value of one division of the level and w, e and w', e' are the west and east readings of the

[1] Various other formulæ for computing the azimuth of circumpolar stars have been proposed and used. Each of them requires either the same or a greater time for the computation than that here given, when the whole computation, including the preparation of the auxiliary tables required with some of them, is taken into account. As uniformity of practice is conducive to rapid computation, it is considered desirable that all should use the formulæ given, and therefore no others are here stated. It should be noted that the formula given is accurate and general; that is, it applies to any of the close circumpolars at any hour angle.

[2] If a mean time chronometer is used, the value $\Sigma \frac{2 \sin^2 \frac{1}{2} \tau}{\sin 1''}$ should be increased by its one hundred and eightieth part.

[3] This table was copied from pages 634–637 of Doolittle's Practical Astronomy. These tabular values may be found in various other places.

STRUCTURE FOR ELEVATING SIGNAL LAMP OVER TRIANGULATION STATION USED AS MARK.

No. 26.

STRUCTURE FOR ELEVATING SIGNAL LAMP OVER TRIANGULATION STATION USED AS MARK.

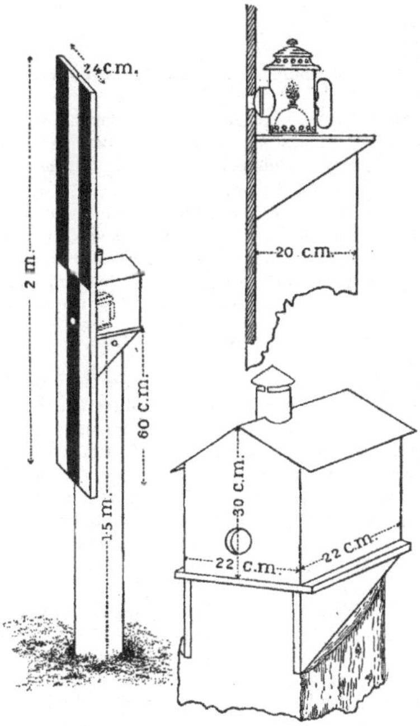

AZIMUTH MARK.

level before and after reversing it. h is the altitude of the star. It is only necessary to know h approximately—an occasional reading of the setting circle will give it with abundant accuracy.

If the graduation on the striding level is numbered continuously in one direction the

$$\text{Level Correction} = \frac{d}{4}\left\{(w - w') + (e - e')\right\}\tan h$$

in which the primed letters refer to readings taken in the position in which the numbering increases toward the east.[1]

If the mark is not in the horizon of the instrument a similar correction, if appreciable, must be applied to readings upon the mark, h now being the altitude of the mark. Ordinarily the mark is so nearly in the horizon of the instrument that tan h is nearly zero and the corrections required to pointings upon the mark are negligible.

The formula as written gives the sign of the correction to be applied to the readings of a horizontal circle of which the numbering increases in a clockwise direction. This is also the sign of the correction to the computed azimuth (counted clockwise) for level readings in connection with pointings upon the mark, but in connection with pointings upon the star the sign must be reversed to give corrections to the computed azimuth of the mark.

DIRECTION METHOD—ADJUSTMENTS.

The measurement of an azimuth by this method is essentially similar to the process of measuring a difference of two horizontal directions with a direction theodolite. The quantity measured in this case is the difference of azimuth of a circumpolar star and a mark instead of a difference of azimuth of two triangulation signals. The fact that the azimuth of the star is continually changing adds new features to the computation, and makes it necessary to know the time of each pointing upon the star. The fact that the star is at a considerable altitude makes readings of the striding level a necessity and decreases the accuracy of the measurement because errors of inclination of the horizontal axis have a marked influence as contrasted with their comparatively unimportant effects upon the measurements of horizontal angles in a triangulation.

The adjustments required are identical with those which are necessary when the instrument is to be used for the measurement of horizontal directions. The adjustments of the focus of the telescope, of the line of collimation, for bringing the vertical lines of the reticle into vertical planes, of the setting circle (if used), and of the striding level may be made as described in connection with a transit on pages 14–16. The vertical axis of the instrument must be made to point as nearly as is feasible to the zenith by bringing the striding level to the proper reading in each of two positions at right angles to each other.

The microscopes with which the horizontal circle is read must be kept in adjustment. Ordinarily it will only be found necessary to adjust the eyepiece by pushing it in or pulling it out until the most distinct vision is obtained of the micrometer lines and of the circle graduation. If the micrometer lines are not apparently parallel to the graduation upon which the pointing is to be made, they should be made so by rotating the micrometer box about the axis of figure of the microscope. If to do this it is necessary to loosen the microscope in its supporting clamp, great caution is necessary to insure that the distance of the objective from the circle of graduation is not changed. The error of run of the reading micrometers should be kept small. In other words, the value of one turn of the micrometer in terms of the circle graduation should not be allowed to differ much from its nominal value. The value of the micrometer may be adjusted by changing the distance of the objective from the graduation. The nearer the objective is to the graduation the smaller is the value of one turn. A change in this distance also necessitates a change in the distance from the objective to the micrometer lines, these lines and the graduation being necessarily at conjugate foci of the

[1] See footnote on p. 23.

objective. This adjustment of the micrometer value is a difficult one to make, but when once well made it usually remains sufficiently good for a long period.

As stated on page 139, primary azimuths are nearly always observed during the progress of the primary triangulation, and the same instrument is used to make the observations on the azimuth star that is used to determine the horizontal directions of the lines of the triangulation. For a number of years past only the 12-inch (30 cm.) direction theodolites (described in Appendix 8, Coast and Geodetic Survey Report for 1894) have been used on primary triangulation. (See illustration No. 18.) Practically all the observations for primary azimuth are made on Polaris. In recent years the azimuth observations have been made at the same time that horizontal observations are being made—that is, Polaris is observed at a setting of the instrument in connection with one or more of the triangulation stations. The observations on Polaris are made at the end of the position in order that the direct and reversed observations on the star may come close together. Instead of determining the astronomic azimuth of the line used as the initial direction for the horizontal angle work it is considered that the azimuth has been determined of the line observed over just previous to the observations on Polaris. If at any station it is necessary to make the observations for azimuth in connection with two lines of the triangulation, then the probable error of the angle between the two lines must be taken into account in deriving the probable error of the azimuth. When a quadrilateral system is used in the triangulation and both diagonal lines are observed, then at each station there will be five primary directions to observe.

Illustration No. 29 shows the lines radiating from such a station. The station A, the first to the east of Polaris, is chosen as the initial and the other stations are observed in turn from left to right, and after observations have been made on E they are made on Polaris. If, for any reason, the line to E is not observed with the other stations during observations for any one position, then Polaris also should not be observed. Later on the instrument should be set for the missing position, and Polaris should be observed in connection with station E.

The observer is instructed to secure an accuracy represented by a probable error of $\pm 0''.50$ for the greater portion of the primary azimuths, and the observations may all be made during one night. This accuracy can usually be secured by observing one set in each of from 12 to 16 positions of the instrument. In no case must an azimuth depend upon less than 10 positions.

At some of the triangulation stations where the accumulated twist of the triangulation is to be determined by a coincident longitude and azimuth station the azimuth is determined with an accuracy represented by a probable error of $\pm 0''.30$, and the observations are made on at least two nights.

DIRECTION METHOD—EXAMPLE OF RECORD AND COMPUTATION.

There are shown below samples of records of azimuth observations on Polaris and the computations. The observations were carried on at the same time that observations of horizontal directions were made at the primary triangulation station, Sears, in Texas. The chronometer correction and rate were determined from observations with a vertical circle on stars approximately on the prime vertical. Examples of the time observations and computations made at Sears for use in the azimuth observations are shown on pages 54 and 55 of this publication.

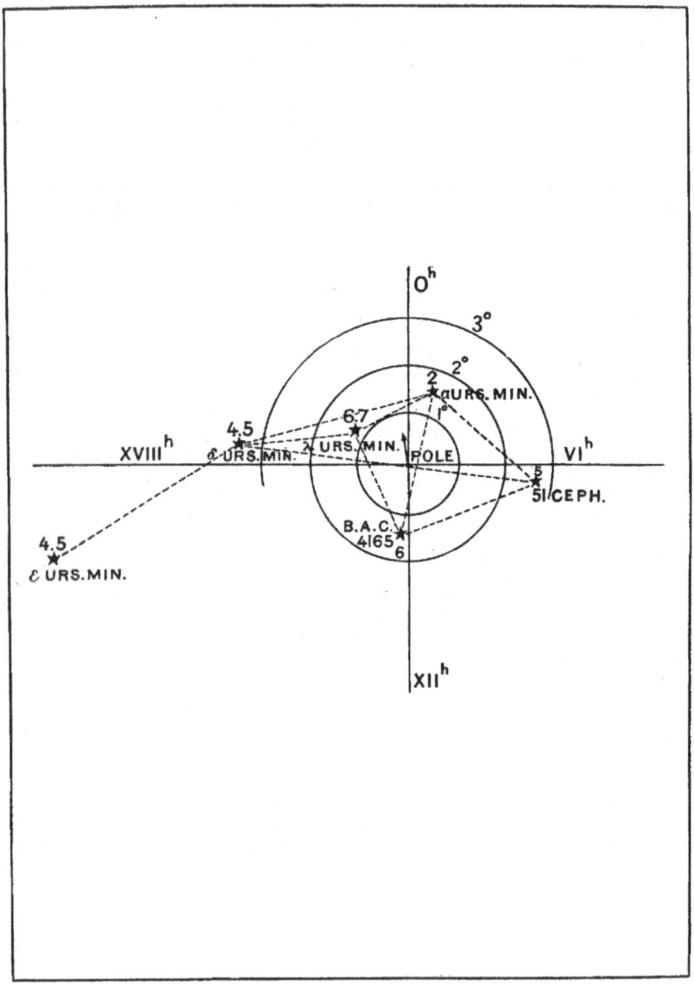

CIRCUMPOLAR STARS.

No. 29.

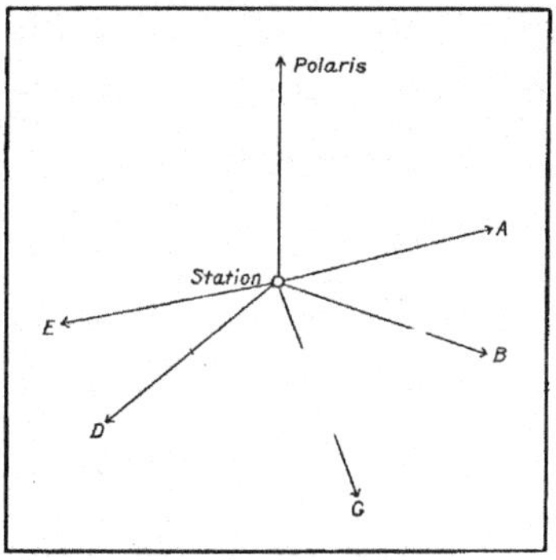

DIAGRAM SHOWING DIRECTIONS TO TRIANGULATION STATIONS AND POLARIS.

Form 251

Horizontal directions.

[Station, Sears, Tex. (Triangulation Station). Observer, W. Bowie. Instrument, Theodolite 168. Date, Dec. 22, 1908.]

Position	Objects observed	Time	Tel. D or R	Mic.	Backward			Forward	Mean	Mean D and R	Direction	Remarks
		h *m*			°	′	″	″			″	
1	Morrison	8 19	D	A	0	0	35	35				1 division of the striding level= 4″.194
				B			41	41				
				C			36	34	37.0			
			R	A	180	00	36	35				
				B			32	31				
				C			35	34	33.8	35.4	00.0	
	Buzzard		D	A	53	30	43	42				
				B			41	42				
				C			34	33	39.2			
			R	A	233	30	39	37				
				B			34	32				
				C			38	38	36.3	37.8	02.4	
	Allen		D	A	170	14	61	62				
				B			57	55				
				C			61	59	59.2			
			R	A	350	14	50	49				
				B			63	60				
				C			53	53	54.7	57.0	21.6	
	Polaris		D	A	252	01	54	53				W E
	h *m* *s*			B			54	53				9.3 28.0
	1 48 35.5			C			51	51	52.7			27.7 9.1
	1 51 06.0											18.4 − 0.5 18.9
	1 49 50.8											
			R	A	72	01	09	09				24.9 6.3
				B			02	01				13.0 31.7
				C			10	08	06.5	29.6		11.9 −13.5 25.4
												− 7.0

Computation of azimuth, direction method.

Form 380..

[Station, Sears, Tex. Chronometer, sidereal 1769. ϕ=32° 33′ 31″. Instrument, theodolite 168. Observer, W. Bowie.]

Date, 1908, position	Dec. 22, 1	2	3	4
Chronometer reading	1 49 50. 8	2 01 33. 0	2 16 31. 0	2 43 28. 8
Chronometer correction	− 4 37. 5	− 4 37. 5	− 4 37. 4	− 4 37. 3
Sidereal time	1 45 13. 3	1 56 55. 5	2 11 53. 6	2 38 51. 5
α of Polaris	1 26 41. 9	1 26 41. 9	1 26 41. 8	1 26 41. 8
t of Polaris (time)	0 18 31. 4	0 30 13. 6	0 45 11. 8	1 12 09. 7
t of Polaris (arc)	4° 37′ 51″. 0	7° 33′ 24″. 0	11° 17′ 57″. 0	18° 02′ 25″. 5
δ of Polaris	88 49 27. 4			
log cot δ	8. 31224	8. 31224	8. 31224	8. 31224
log tan ϕ	9. 80517	9. 80517	9. 80517	9. 80517
log cos t	9. 99858	9. 99621	9. 99150	9. 97811
log a (to five places)	8. 11590	8. 11362	8. 10891	8. 09552
log cot δ	8. 312243	8. 312243	8. 312243	8. 312243
log sec ϕ	0. 074254	0. 074254	0. 074254	0. 074254
log sin t	8. 907064	9. 118948	9. 292105	9. 490924
log $\frac{1}{1-a}$	0. 005710	0. 005679	0. 005618	0. 005445
log ($-$tan A) (to 6 places)	7. 299271	7. 511124	7. 684220	7. 882866
A=Azimuth of Polaris, from north*	0 06 50. 8	0 11 09. 2	0 16 36. 9	0 26 15. 0
Difference in time between D. and R.	*m s* 2 30	*m s* 2 00	*m s* 3 18	*m s* 1 38
Curvature correction	0	0	0	0
	° ′ ″	° ′ ″	° ′ ″	° ′ ″
Altitude of Polaris=h	33 46	33 46	33 46	33 46
$\frac{d}{4}$ tan h=level factor	0. 701	0. 701	0. 701	0. 701
Inclination †	−7. 0	−7. 2	−7. 0	−1. 8
Level correction	−4. 9	−5. 0	−4. 9	−1. 3
Circle reads on Polaris	252 01 29. 6	86 58 11. 2	281 54 27. 0	116 45 48. 6
Corrected reading on Polaris	252 01 24. 7	86 58 06. 2	281 54 22. 1	116 45 47. 3
Circle reads on mark	170 14 57. 0	5 15 58. 2	200 17 42. 4	35 18 45. 4
Difference, mark—Polaris	278 13 32. 3	278 17 52. 0	278 23 20. 3	278 32 58. 1
Corrected azimuth of Polaris, from north*	0 06 50. 8	0 11 09. 2	0 16 36. 9	0 26 15. 0
	180 00 00. 0	180 00 00. 0	180 00 00. 0	180 00 00. 0
Azimuth of Allen (Clockwise from south)	98 06 41. 5	98 06 42. 8	98 06 43. 4	98 06 43. 1

To the mean result from the above computation must be applied corrections for diurnal aberration and eccentricity (if any) of Mark. Carry times and angles to tenths of seconds only.

* Minus, if west of north.

† The values shown in this line are actually four times the inclination of the horizontal axis in terms of level divisions.

Summary of azimuth results.

[Sears, Tex., Dec. 22, 1908.]

Position	Azimuth of Allen	v	v^2
	° ′ ″		
1	98 06 41.5	+0.8	.64
2	42.8	−0.5	.25
3	43.4	−1.1	1.21
4	43.1	−0.8	.64
5	39.7	+2.6	6.76
6	42.7	−0.4	.16
7	41.6	+0.7	.49
8	43.3	−1.0	1.00
9	40.0	+2.3	5.29
10	45.0	−2.7	7.29
11	43.3	−1.0	1.00
12	40.7	+1.6	2.56
			$\Sigma v^2=27.29$

$$e=\pm0.6745\sqrt{\frac{\Sigma v^2}{n(n-1)}}=\pm0''.31$$

The mean observed azimuth	$=98°\ 06'\ 42''.26\pm0''.31.$
Diurnal aberration	$=+0.32.$
Correction for eccentric light	$=+0.04.$
Correction for elevation of mark	$=-0.01.$
Reduction to mean position of pole [1]	$=-0.29.$
Azimuth of the line from Sears to Allen [2]	$=98\ \ 06\ \ 42.32\pm0.31.$

DIRECTION METHOD—EXPLANATION OF RECORD AND COMPUTATION.

The triangulation stations and Polaris which were observed at one setting of the instrument (in this case position No. 1) are placed in the record in the order of their azimuths (left to right) from the initial station, "Morrison." The telescope in its direct position is pointed upon each station in turn and finally upon Polaris. The telescope is then reversed, and the first pointing after reversal is upon Polaris; then pointings are made upon the triangulation stations in the reverse order of azimuth (from right to left). The readings in the reversed position of the telescope are placed directly under the direct reading. The mean of the readings in the direct and in the reversed positions of the telescope is used in computing the direction of a line with reference to the initial line. There are three microscope micrometers on the instrument used in making the observations at Sears, and at each pointing a backward and forward reading of each micrometer was made on the two graduations of the circle nearest the center of the comb.

The mean run of the micrometers was kept very small and as the micrometer was placed upon a different portion of the five-minute space between successive graduations, the resultant effect of the micrometer run was negligible. The initial positions (minutes and seconds) of the micrometer wire on the circle for the first four positions were 00′ 40″, 01′ 50″, 03′ 10″, and 04′ 20″. In general, 12 or 16 positions of the circle are used for the initial settings and these readings of the minutes and seconds on the initial are repeated in each group of four positions; that is, in positions 5 to 8, 9 to 12, and 13 to 16. It can be shown that on any object the error due to run is practically zero in each set of four positions of the circle, if the mean run of the three micrometers with regard to sign is less than 1″.0 and the run of no one micrometer is larger than 3″.0. Special observations are made in primary triangulation to determine whether the run of the micrometers is within these limits.

[1] See Astronomische Nachrichten No. 4414.

[2] Sears and Allen are triangulation stations.

The chronometer time of the observations on Polaris and also the level readings are shown in the record. The time of making an observation may be noted by the observer who picks up and carries the beat of the chronometer, or an assistant may note the clock time upon a signal from the observer. When the latter method is used the observer calls "Mark" when the star is bisected.

The chronometer corrections shown in the computations resulted from a special series of time observations with the vertical circle at the station (see pp. 54 and 55).

The formula used in making the computation is the third form of the azimuth formula shown on page 143. The tables on pages 165 to 173 which give the logarithm of $\frac{1}{1-a}$ were used in the computations. Much time is saved in such computations as the above by carrying along all the different sets at one time and thus working along the horizontal lines of the form shown instead of down each column. Also tan ϕ and sec ϕ are constants for the station, cos t and sin t may be taken out at one opening of the logarithm table, etc. A comparison of corresponding parts of different columns furnishes rough checks which serve to locate any large errors quickly. The value of one division of the striding level is 4''.194. In general, one set like the above, in each of 12 to 16 positions of one of the 12-inch theodolites, will give a probable error of the result less than ±0''.50. Even where the observations for azimuth are made coincidently with those for horizontal directions in a triangulation there is no difficulty in completing the azimuth observations at a station in one evening. For special stations a probable error of the result of ±0''.30 or less must be gotten and observations must be made on more than one night. The general practice now in the Coast and Geodetic Survey is to make only one pointing on the star in each of the positions of the telescope and therefore the correction for curvature of the path of the star between the two pointings is usually negligible. When there is a delay in making the second pointing the curvature correction should be computed by the formula shown on page 144.

Tabular values of $\frac{2 \sin^2 \frac{1}{2}\tau}{\sin 1''}$ are given on pages 151–152. The small table shown below gives the values of the curvature correction direct for values of the interval, 2τ, between the two pointings on the star, from 2 to 7 minutes, and azimuths of Polaris less than 2° 30′, for use with the direction method, when only two observations are made on Polaris for one setting of the instrument.

Curvature correction.

Azimuth of Polaris. \ 2τ	2^m	3^m	4^m	5^m	6^m	7^m
° ′	″	″	″	″	″	″
0 10	.0	.0	.0	.0	.1	.1
0 20	.0	.0	.0	.1	.1	.1
0 30	.0	.0	.1	.1	.2	.2
0 40	.0	.1	.1	.1	.2	.3
0 50	.0	.1	.1	.2	.3	.3
1 00	.0	.1	.1	.2	.3	.4
1 10	.0	.1	.2	.2	.4	.5
1 20	.0	.1	.2	.3	.4	.6
1 30	.0	.1	.2	.3	.5	.6
1 40	.1	.1	.2	.4	.5	.7
1 50	.1	.1	.3	.4	.6	.8
2 00	.1	.2	.3	.4	.6	.8
2 10	.1	.2	.3	.5	.7	.9
2 20	.1	.2	.3	.5	.7	1.0
2 30	.1	.2	.3	.5	.8	1.1

$$\frac{2 \sin^2 \frac{1}{2} \tau}{\sin 1''}$$

τ	0m	1m	2m	3m	4m	5m	6m	7m	8m
s	"	"	"	"	"	"	"	"	"
0	0.00	1.96	7.85	17.67	31.42	49.09	70.68	96.20	125.65
1	0.00	2.03	7.98	17.87	31.68	49.41	71.07	96.66	126.17
2	0.00	2.10	8.12	18.07	31.94	49.74	71.47	97.12	126.70
3	0.00	2.16	8.25	18.27	32.20	50.07	71.86	97.58	127.22
4	0.01	2.23	8.39	18.47	32.47	50.40	72.26	98.04	127.75
5	0.01	2.31	8.52	18.67	32.74	50.73	72.66	98.50	128.28
6	0.02	2.38	8.66	18.87	33.01	51.07	73.06	98.97	128.81
7	0.02	2.45	8.80	19.07	33.27	51.40	73.46	99.43	129.34
8	0.03	2.52	8.94	19.28	33.54	51.74	73.86	99.90	129.87
9	0.04	2.60	9.08	19.48	33.81	52.07	74.26	100.37	130.40
10	0.05	2.67	9.22	19.69	34.09	52.41	74.66	100.84	130.94
11	0.06	2.75	9.36	19.90	34.36	52.75	75.06	101.31	131.47
12	0.08	2.83	9.50	20.11	34.64	53.09	75.47	101.78	132.01
13	0.09	2.91	9.64	20.32	34.91	53.43	75.88	102.25	132.55
14	0.11	2.99	9.79	20.53	35.19	53.77	76.29	102.72	133.09
15	0.12	3.07	9.94	20.74	35.46	54.11	76.69	103.20	133.63
16	0.14	3.15	10.09	20.95	35.74	54.46	77.10	103.67	134.17
17	0.16	3.23	10.24	21.16	36.02	54.80	77.51	104.15	134.71
18	0.18	3.32	10.39	21.38	36.30	55.15	77.93	104.63	135.25
19	0.20	3.40	10.54	21.60	36.58	55.50	78.34	105.10	135.80
20	0.22	3.49	10.69	21.82	36.87	55.84	78.75	105.58	136.34
21	0.24	3.58	10.84	22.03	37.15	56.19	79.16	106.06	136.88
22	0.26	3.67	11.00	22.25	37.44	56.55	79.58	106.55	137.43
23	0.28	3.76	11.15	22.47	37.72	56.90	80.00	107.03	137.98
24	0.31	3.85	11.31	22.70	38.01	57.25	80.42	107.51	138.53
25	0.34	3.94	11.47	22.92	38.30	57.60	80.84	107.99	139.08
26	0.37	4.03	11.63	23.14	38.59	57.96	81.26	108.48	139.63
27	0.40	4.12	11.79	23.37	38.88	58.32	81.68	108.97	140.18
28	0.43	4.22	11.95	23.60	39.17	58.68	82.10	109.46	140.74
29	0.46	4.32	12.11	23.82	39.46	59.03	82.52	109.95	141.29
30	0.49	4.42	12.27	24.05	39.76	59.40	82.95	110.44	141.85
31	0.52	4.52	12.43	24.28	40.05	59.75	83.38	110.93	142.40
32	0.56	4.62	12.60	24.51	40.35	60.11	83.81	111.43	142.96
33	0.59	4.72	12.76	24.74	40.65	60.47	84.23	111.92	143.52
34	0.63	4.82	12.93	24.98	40.95	60.84	84.66	112.41	144.08
35	0.67	4.92	13.10	25.21	41.25	61.20	85.09	112.90	144.64
36	0.71	5.03	13.27	25.45	41.55	61.57	85.52	113.40	145.20
37	0.75	5.13	13.44	25.68	41.85	61.94	85.95	113.90	145.76
38	0.79	5.24	13.62	25.92	42.15	62.31	86.39	114.40	146.33
39	0.83	5.34	13.79	26.16	42.45	62.68	86.82	114.90	146.89
40	0.87	5.45	13.96	26.40	42.76	63.05	87.26	115.40	147.46
41	0.91	5.56	14.13	26.64	43.06	63.42	87.70	115.90	148.03
42	0.96	5.67	14.31	26.88	43.37	63.79	88.14	116.40	148.60
43	1.01	5.78	14.49	27.12	43.68	64.16	88.57	116.90	149.17
44	1.06	5.90	14.67	27.37	43.99	64.54	89.01	117.41	149.74
45	1.10	6.01	14.85	27.61	44.30	64.91	89.45	117.92	150.31
46	1.15	6.13	15.03	27.86	44.61	65.29	89.89	118.43	150.88
47	1.20	6.24	15.21	28.10	44.92	65.67	90.33	118.94	151.45
48	1.26	6.36	15.39	28.35	45.24	66.05	90.78	119.45	152.03
49	1.31	6.48	15.57	28.60	45.55	66.43	91.23	119.96	152.61
50	1.36	6.60	15.76	28.85	45.87	66.81	91.68	120.47	153.19
51	1.42	6.72	15.95	29.10	46.18	67.19	92.12	120.98	153.77
52	1.48	6.84	16.14	29.36	46.50	67.58	92.57	121.49	154.35
53	1.53	6.96	16.32	29.61	46.82	67.96	93.02	122.01	154.93
54	1.59	7.09	16.51	29.86	47.14	68.35	93.47	122.53	155.51
55	1.65	7.21	16.70	30.12	47.46	68.73	93.92	123.05	156.09
56	1.71	7.34	16.89	30.38	47.79	69.12	94.38	123.57	156.67
57	1.77	7.46	17.08	30.64	48.11	69.51	94.83	124.09	157.25
58	1.83	7.60	17.28	30.90	48.43	69.90	95.29	124.61	157.84
59	1.89	7.72	17.47	31.16	48.76	70.29	95.74	125.13	158.43

$$\frac{2 \sin^2 \frac{1}{2} \tau}{\sin 1''}$$

τ	9m	10m	11m	12m	13m	14m	15m	16m
s	"	"	"	"	"	"	"	"
0	159.02	196.32	237.54	282.68	331.74	384.74	441.63	502.46
1	159.61	196.97	238.26	283.47	332.59	385.65	442.62	503.50
2	160.20	197.63	238.98	284.26	333.44	386.56	443.60	504.55
3	160.80	198.28	239.70	285.04	334.29	387.48	444.58	505.60
4	161.39	198.94	240.42	285.83	335.15	388.40	445.56	506.65
5	161.98	199.60	241.14	286.62	336.00	389.32	446.55	507.70
6	162.58	200.26	241.87	287.41	336.86	390.24	447.54	508.76
7	163.17	200.92	242.60	288.20	337.72	391.16	448.53	509.81
8	163.77	201.59	243.33	289.00	338.58	392.09	449.51	510.86
9	164.37	202.25	244.06	289.79	339.44	393.01	450.50	511.92
10	164.97	202.92	244.79	290.58	340.30	393.94	451.50	512.98
11	165.57	203.58	245.52	291.38	341.16	394.86	452.49	514.08
12	166.17	204.25	246.25	292.18	342.02	395.79	453.48	515.09
13	166.77	204.92	246.98	292.98	342.88	396.72	454.48	516.15
14	167.37	205.59	247.72	293.78	343.75	397.65	455.47	517.21
15	167.97	206.26	248.45	294.58	344.62	398.58	456.47	518.27
16	168.58	206.93	249.19	295.38	345.49	399.52	457.47	519.34
17	169.19	207.60	249.93	296.18	346.36	400.45	458.47	520.40
18	169.80	208.27	250.67	296.99	347.23	401.38	459.47	521.47
19	170.41	208.94	251.41	297.79	348.10	402.32	460.47	522.53
20	171.02	209.62	252.15	298.60	348.97	403.26	461.47	523.60
21	171.63	210.30	252.89	299.40	349.84	404.20	462.48	524.67
22	172.24	210.98	253.63	300.21	350.71	405.14	463.48	525.74
23	172.85	211.66	254.37	301.02	351.58	406.08	464.48	526.81
24	173.47	212.34	255.12	301.83	352.46	407.02	465.49	527.89
25	174.08	213.02	255.87	302.64	353.34	407.96	466.50	528.96
26	174.70	213.70	256.62	303.46	354.22	408.90	467.51	530.03
27	175.32	214.38	257.37	304.27	355.10	409.84	468.52	531.11
28	175.94	215.07	258.12	305.09	355.98	410.79	469.53	532.18
29	176.56	215.75	258.87	305.90	356.86	411.73	470.54	533.26
30	177.18	216.44	259.62	306.72	357.74	412.68	471.55	534.33
31	177.80	217.12	260.37	307.54	358.62	413.63	472.57	535.41
32	178.43	217.81	261.12	308.36	359.51	414.59	473.58	536.50
33	179.05	218.50	261.88	309.18	360.39	415.54	474.60	537.58
34	179.68	219.19	262.64	310.00	361.28	416.49	475.62	538.67
35	180.30	219.88	263.39	310.82	362.17	417.44	476.64	539.75
36	180.93	220.58	264.15	311.65	363.07	418.40	477.65	540.83
37	181.56	221.27	264.91	312.47	363.96	419.35	478.67	541.91
38	182.19	221.97	265.68	313.30	364.85	420.31	479.70	543.00
39	182.82	222.66	266.44	314.12	365.75	421.27	480.72	544.09
40	183.46	223.36	267.20	314.95	366.64	422.23	481.74	545.18
41	184.09	224.06	267.96	315.78	367.53	423.19	482.77	546.27
42	184.72	224.76	268.73	316.61	368.42	424.15	483.79	547.36
43	185.35	225.46	269.49	317.44	369.31	425.11	484.82	548.45
44	185.99	226.16	270.26	318.27	370.21	426.07	485.85	549.55
45	186.63	226.86	271.02	319.10	371.11	427.04	486.88	550.64
46	187.27	227.57	271.79	319.94	372.01	428.01	487.91	551.73
47	187.91	228.27	272.56	320.78	372.91	428.97	488.94	552.83
48	188.55	228.98	273.34	321.62	373.82	429.93	489.97	553.93
49	189.19	229.68	274.11	322.45	374.72	430.90	491.01	555.03
50	189.83	230.39	274.88	323.29	375.62	431.87	492.05	556.13
51	190.47	231.10	275.66	324.13	376.52	432.84	493.08	557.24
52	191.12	231.81	276.43	324.97	377.43	433.82	494.12	558.34
53	191.76	232.52	277.20	325.81	378.34	434.79	495.15	559.44
54	192.41	233.24	277.98	326.66	379.26	435.76	496.19	560.55
55	193.06	233.95	278.76	327.50	380.17	436.73	497.23	561.65
56	193.71	234.67	279.55	328.35	381.08	437.71	498.28	562.76
57	194.36	235.38	280.33	329.19	381.99	438.69	499.32	563.87
58	195.01	236.10	281.12	330.04	382.90	439.67	500.37	564.98
59	195.66	236.82	281.90	330.89	383.82	440.65	501.41	566.08

METHOD OF REPETITIONS—EXAMPLE OF RECORD AND COMPUTATION.

Remarks similar to those appearing on page 145 apply here also. The observations required to determine the azimuth of a mark by the method of repetitions are the same as those required to measure a horizontal angle in a triangulation with the same repeating theodolite, with the addition of level readings, and readings of the chronometer at the instants of the pointings upon the star.

The adjustments required are those mentioned on page 145, with the exception that a repeating theodolite is ordinarily read by verniers instead of microscopes.

Record—Azimuth by repetitions.

[Station, Kahatchee △. State, Alabama. Date, June 6, 1898. Observer, O. B. F. Instrument, 10-inch Gambey No. 63. Star, Polaris.]

[One division striding level=2″.67.]

Objects	Chr. time on star	Pos. of tel.	Repetitions	Level readings W	Level readings E	Circle readings °	′	A ″	B ″	Mean	Angle
Mark		D	0			178	03	22.5	20	21.2	
Star	14h 46m 30s		1	4.5 9.2	10.7 5.9						
	49 08		2								
	52 51	D	3	9.6 5.2	5.6 10.0						
	56 10	R	4	11.3 7.8	4.0 7.4						
Set No. 5	14 59 12		5								
	15 01 55	R	6	8.7 11.9	6.6 3.4	100	16	20	20	20	72° 57′ 50″.2
	14 54 17.7			68.2 +14.6	53.6						
Star	15 04 44	R	1	11.9 8.5	3.4 6.8						
	07 18		2								
	09 54	R	3	7.9 11.2	7.3 4.1						
Set No. 6	14 15	D	4	9.0 5.9	6.1 9.6						
	16 14		5								
	15 18 24		6	5.9 9.1	9.6 6.2						
Mark		D				177	27	00	00	00	72° 51′ 46″.7
	15 11 48.2			69.4 +16.3	53.1						

Computation—Azimuth by repetitions.

[Kahatchee, Ala. $\phi=33°\ 13'\ 40''.33$.]

Date, 1898, set	June 6 5	June 6 6
Chronometer reading	14 54 17.7	15 11 48.2
Chronometer correction	−31.1	−31.1
Sidereal time	14 53 46.6	15 11 17.1
α of Polaris	1 21 20.3	1 21 20.3
t of Polaris (time)	13 32 26.3	13 49 56.8
t of Polaris (arc)	203° 06′ 34″.5	207° 29′ 12″.0
δ of Polaris	88 45 46.9	
log cot δ	8.33430	8.33430
log tan ϕ	9.81629	9.81629
log cos t	9.96367n	9.94798n
log a (to five places)	8.11426n	8.09857n
log cot δ	8.334305	8.334305
log sec ϕ	0.077535	0.077535
log sin t	9.593830n	9.664211n
$\log \frac{1}{1-a}$	9.994387	9.994584
log (−tan A) (to 6 places)	8.000057n	8.070635n
A=Azimuth of Polaris, from north *	0° 34′ 22″.8	0° 40′ 26″.8
	m s ″	m s ″
τ and $\frac{2 \sin^2 \frac{1}{2} \tau}{\sin 1''}$	7 47.7 119.3	7 04.2 98.1
	5 09.7 52.3	4 30.2 39.8
	1 26.7 4.1	1 54.2 7.1
	1 52.3 6.9	2 26.8 11.8
	4 54.3 47.2	4 25.8 38.5
	7 37.3 114.0	6 35.8 85.4
Sum	343.8	280.7
Mean	57.3	46.8
$\log \frac{1}{n} \Sigma \frac{2 \sin^2 \frac{1}{2} \tau}{\sin 1''}$	1.758	1.670
log (curvature corr.)	9.758	9.741
Curvature correction	−0.6	−0.6
Altitude of Polaris=h	32° 07′	
$\frac{d}{4}$ tan h=level factor	.419	.419
Inclination †	+3.6	+4.1
Level correction	−1″.5	−1″.7
Angle, star−mark	72 57 50.2	72 51 46.7
Corrected angle	72 57 48.7	72 51 45.0
Corrected azimuth of star *	0 34 22.2	0 40 26.2
Azimuth of mark E of N	73 32 10.9	73 32 11.2
	180 00 00.0	180 00 00.0
Azimuth of mark (Clockwise from south)	253 32 10.9	253 32 11.2

To the mean result from the above computation must be applied corrections for diurnal aberration and eccentricity (if any) of Mark. Carry times and angles to tenths of seconds only.

* Minus if west of north.

† See footnote on p. 148.

METHOD OF REPETITIONS—EXPLANATION OF RECORD AND COMPUTATION.

Throughout the observations the instrument was always turned in a *clockwise direction* about its vertical axis. In set No. 5 the swing from the mark to the star was made with the upper motion loose and lower motion clamped, and therefore with the circle reading changing, and in set No. 6 the reverse was the case. In set No. 5 the explement of the small angle between the star and the mark was really measured, while in No. 6 the angle itself was measured. Both results may be computed directly in terms of the angle by making the subtractions thus, in set No. 5.

$$\text{angle}=\frac{(360°+178°\ 03'\ 21''.2)-100°\ 16'\ 20''.0}{6}=72°\ 57'\ 50''.2$$

in set No. 6,

$$\text{angle}=\frac{(360°+177°\ 27'\ 00''.0)-100°\ 16'\ 20''.0}{6}=72°\ 51'\ 46''.7^{1}.$$

If the clamp on the horizontal circle produces a constant error, either by dragging or overrunning, these two results will be equally in error with opposite signs, and their mean will be free from the constant part of the clamp error. Hence, it is desirable to observe the sets alternately in the order Mark-Star, Star-Mark, as here indicated.

The summary of results for this station shows 37 sets of observations were made on four nights. From the 18 sets observed in the order Star-Mark the mean azimuth was 73° 32′ 12″.07, and from the 19 sets observed in the order Mark-Star the mean was 73° 32′ 12″.89, showing that the clamp error was very small. The adopted indiscriminate mean of all the 37 sets was 73° 32′ 12″.49. The correction for diurnal aberration (+0″.31) being applied, the resulting azimuth of the mark, E. of N. equals 73° 32′ 12″.80 ± 0″.16. The probable error of a single set $=\sqrt{\frac{0.455\ \Sigma v^2}{(n-1)}}=\pm 0''.98$.

During these observations the instrument was supported upon its tripod, the legs of which were set upon large stakes driven solidly into the ground.

The level readings were taken with the first, third, fourth, and sixth pointings upon the star, that is, at the beginning and end of the set and just before and just after the reversal of the telescope. In each case the level was read in one position just before perfecting the pointing upon the star, and in the other position immediately after the pointing upon the star. The value of one division of the level was 2″.67.

The computation needs no further explanation. The formula

$$\tan A=-\cot\delta\sec\varphi\sin t\left(\frac{1}{1-a}\right)$$

was used.

The correction for elevation of mark, when appreciable, is applied in the final summary of results, just as in the case of the direction method. The reduction to the mean position of the pole is also applied to the final result, but for observations previous to the year 1900 no such reduction can now be made. (See p. 85.)

MICROMETRIC METHOD—EXAMPLE OF RECORD AND COMPUTATION.

In the micrometric method[2] the small difference of azimuth of the star and the mark is measured with an eyepiece micrometer, independently of the graduated horizontal circle of the instrument, even if it has one. The mark must therefore be placed nearly in the vertical of the star at the time at which it is to be observed. The method may be used with the star at any hour-angle, but unless the star is near elongation it will pass beyond the safe range of the micrometer after but two or three sets of observations have been taken, whereas if the mark

[1] The computer should notice the convenient fact that in dividing an angle by six the remainder, when the degrees are divided, is the tens figure in the minutes, and the remainder in the minutes is the tens figure in the seconds.

[2] For an account of this method, together with some historical notes, see Appendix No. 2 of the Report for 1891.

is placed nearly under the star at elongation (preferably one or two minutes of arc inside), the observations may be continued for two hours or more and the results will also be nearly independent of the chronometer error. The instrument used may be a theodolite, a meridian telescope, a transit, or, in fact, any instrument having a stable horizontal axis and furnished with an eyepiece micrometer capable of measuring angles in the plane defined by the telescope and its horizontal axis.

Record and computation—Azimuth by micrometric method.

[Station No. 10, Mexican Boundary. Date, October 13, 1892. Observer, J. F. H. Instrument, Fauth Repeating Theodolite, No. 725 (10 in.). Star, Polaris near eastern elongation.]

Circle	Level readings W	Level readings E	Chronometer time	τ	$\frac{2 \sin^2 \frac{1}{2}\tau}{\sin 1''}$	Micrometer readings— On star	Micrometer readings— On mark	
			h m s	m s				
E	8.0	9.9	9 06 38.0	3 58.6	31.05	18ᵗ.379	18ᵗ.310	$\lambda=2^h\ 12^m$ W. of Washington
	10.0	7.3	07 32.0	3 04.6	18.59	.388	.315	$\phi=31°\ 19'\ 35''$.
	+18.0	−17.2	08 05.5	2 31.1	12.45	.400	.315	1 div. of level $=3''.68$
		+0.8	09 13.0	1 23.6	3.82	.424	.311	1 turn of mic. $=123''.73$
E			09 48.0	0 48.6	1.29	.430	.316	
						18.4042	18.3134	Means
W	9.0	9.0	9 12 01.8	1 25.2	3.96	18.100	18.290	
	7.0	10.9	12 24.7	1 48.1	6.37	.100	.275	
	+16.0	−19.9	12 48.3	2 11.7	9.46	.090	.279	
		−3.9	13 36.3	2 59.7	17.61	.086	.281	
W	Mean	1ᵈ.55	13 58.1	3 21.5	22.14	.080	.279	
			9 10 36.6		12.67	18.0912	18.2808	Means

ζ of star at middle of first half of set$=58°\ 48'$. cosec $\zeta=1.1691$. cot $58°\ 47'=0.606$.
ζ of star at middle of second half of set$=58°\ 46'$. cosec $\zeta=1.1695$.
$\alpha=1^h\ 20^m\ 07^s.4$. $\delta=88°\ 44'\ 10''.4$.

Collimation axis reads ½ (18.3134+18.2808)[1] $=18^t.2971$
Mark east of collimation axis 18.3134−18.2971 $= 0.0163 = 02''.02$
Circle E., star E. of collimation axis (18.4042−18.2971) (1.1691) $= 0.1252$
Circle W., star E. of collimation axis (18.2971−18.0912) (1.1695) $= 0.2408$

Mean, star E. of collimation axis $= 0.1830 = 22''.64$

Mark west of star $= 20''.62$
Level correction (1.55) (0.92) (0.606) $= -0''.86$

Mark west of star, corrected $= 19''.76$

Mean chronometer time of observation $= 21^h\ 10^m\ 36^s.6$
Chronometer correction $= -2\ \ 11\ \ 28.2$
Sidereal time $= 18\ \ 59\ \ 08.4$
α $= 1\ \ 20\ \ 07.4$

Hour-angle, t, in time $17\ \ 39\ \ 01.0$
" in arc $264°\ 45'\ 15''.0$

log. cot δ = 8.34362
log. tan ϕ = 9.78436
log. cos t = 8.96108 n

log. a = 7.08906 n

[1] In this instrument increased readings of the micrometer correspond to a movement of the line of sight toward the east when the vertical circle is to the east, and toward the west when the vertical circle is to the west.

log. cot δ	= 8.343618
log. sec ϕ	= 0.068431
log. sin t	= 9.998177 n
log. $\frac{1}{1-a}$	= 9.999467
log. $(-\tan A)$	= 8.409693 n
A	$=+1° 28' 16''.91$
log. 12.67	= 1.10278
log. curvature corr.	= 9.51247
Curvature corr.	= −0.33
Diur. Aber. corr.	= +0.32
Mean azimuth of star	$=+1° 28' 16''.90$
Mark west of star	19 .76
Azimuth of mark, E. of N.	$=+1° 27' 57''.14$

The correction for elevation of mark and the reduction to the mean position of the pole are applied to the final result of the separate measures at a station. In the case of this particular station the necessary information is not yet available for reduction to the mean position of the pole. (See p. 85.)

MICROMETRIC METHOD—EXPLANATION OF RECORD AND COMPUTATION.

The compact form of record shown above does not indicate the order in which the observations were taken. The micrometer line is placed nearly in the collimation axis of the telescope, a pointing made upon the mark by turning the horizontal circle, and the instrument is then clamped in azimuth. The program is then to take five pointings upon the mark; direct the telescope to the star; place the striding level in position; take three pointings upon the star with chronometer times; read and reverse the striding level; take two more pointings upon the star, noting the times; read the striding level. This completes a half-set. The horizontal axis of the telescope is then reversed in its Y's; the telescope pointed approximately to the star; the striding level placed in position; three pointings taken upon the star with observed chronometer times; the striding level is read and reversed; two more pointings are taken upon the star, with observed times; the striding level is read, and finally five pointings upon the mark are taken.

Three such complete sets may be observed in from thirty to fifty minutes. The effect of a uniform twisting of the instrument in azimuth is eliminated from the result. The bubble of the striding level has plenty of time to settle without delaying the observer an instant for that purpose.

The zenith distance of the star should be read occasionally, once during each set, say, as it is needed in making the computation. If it is read with one of the star pointings in each set, its value at any other time may be obtained with sufficient accuracy by interpolation.

It should be borne in mind in making the computation that the micrometer measures angles in the plane defined by the telescope and its horizontal axis. To reduce the measured angle between the collimation axis and the star to a horizontal angle, it must be multiplied by cosec ζ, as indicated in the computation. To avoid all approximation in the computation it would be necessary to reduce each pointing upon the star separately, as the zenith distance is constantly changing. It is sufficiently accurate, however, to reduce the mean of the pointings of a half-set with the mean zenith distance of that half-set, as indicated in the computation. To use a single zenith distance for the *whole* set will sometimes introduce errors which are rather too large to be neglected. The factor cosec ζ will not, in general, be necessary in connection with pointings upon the mark, because the mark will usually be nearly in the horizon of the instrument, and cosec ζ therefore nearly unity, and because the collimation axis is purposely placed as nearly as possible upon the mark and the angle concerned is therefore very small.

The micrometer value may be determined by observations upon a star near culmination by the process outlined on page 124. If the striding level is read in connection with such observations, the correction to be applied to each observed time to reduce it to what it would have been with the transverse axis horizontal is

$$\left\{(w+w')-(e+e')\right\}\frac{d\cos\zeta\sec\delta}{60}$$

for upper culmination and for a level of which the graduation is numbered both ways from the middle. For lower culmination the sign of the correction must be reversed.

Another convenient way of determining the micrometer value, all in daylight, is to measure a small horizontal angle at the instrument between two terrestrial objects, both with the micrometer and the horizontal circle of the theodolite. This method is less liable to constant errors than the circumpolar method.

If the azimuth mark is placed to the *southward* of the station, the program of observing and the computation are but slightly modified.

DISCUSSION OF ERRORS.

It is convenient and conducive to conciseness to discuss separately the external errors, observer's errors, and instrumental errors, which together comprise the errors of observation.

The *external errors* affecting azimuth determinations are those due to lateral refraction of the rays of light from the star or mark to the instrument, to errors in the adopted right ascension and declination of the star observed, and to error in the adopted latitude of the station of observation.

Examination of many series of azimuth observations indicates that, in general, they are subject to some error which is peculiar to each night of observation, and constant for that night, but changes from night to night. For example, from 144 sets of micrometric observations of azimuth, made on 36 different nights at 15 stations on the Mexican boundary in 1892–93, it was found that the error peculiar to each night was represented by the probable error $\pm 0''.38$, and that the probable error of each set, exclusive of this error, was $\pm 0''.54$.[1] In other words, in this series of observations the error peculiar to each night, which could not have been eliminated by increasing the number of observations on that night, was two-thirds as large, on an average, as the error of observation in the result from a single set. Similarly, from the published results of 418 sets of micrometric observations on 8 nights at a European station,[2] it follows that the error peculiar to each night was $\pm 0''.55$, while the probable error of a single set was $\pm 0''.80$. The micrometric observations are peculiarly adapted to exhibiting this error, because of their great accuracy and the rapidity with which they may be taken. Azimuth was observed at 73 stations on the transcontinental triangulation along the thirty-ninth parallel. Most of these observations were taken by the direction method, and although they are, for various reasons, but poorly adapted, as a rule, to exhibiting the errors peculiar to the separate nights, there are no less than 16 cases out of the 73 in which a mere inspection indicates that there were errors of that character.

The most plausible explanation of the above facts seems to be that there is lateral refraction between the mark and the instrument, dependent upon the peculiar atmospheric conditions of each night. Whether that explanation be true or not, the fact remains that an increase of accuracy in an azimuth determination at a given station may be attained much more readily by increasing the number of nights of observation than by increasing the number of sets on each night. If one series of observations is made early in the evening and another series just before dawn on the same night, these series may be considered, in so far as the preceding sentence is concerned, to be on different nights, as the atmospheric conditions will have been given an opportunity to change.

The line from the station to the mark should not pass close to any objects, such as a smokestack, building, clump of trees, or a hill. Even when the line is close to the ground which has

[1] See Report of International Boundary Commission, United States and Mexico, 1891–96 (Washington, 1898), pp. 69–72.

[2] Station Kampenwand. See pp. 68–92, Veröffentlichung der Königl. Bayerischen Commission für die Internationale Erdmessung, Astronomische-Geodätische Arbeiten, Heft 2 (München, 1897).

a decided slope normal to the line, there may be decided lateral refraction. During the primary triangulation in the city of Greater New York the errors on the lines which were close to stacks and buildings were much greater than on the clear lines. There was a line in the Texas-California arc of primary triangulation which at one point was very close to the side of a steep hill. The line was observed from the end nearest the hill on several days and nights, with a total range in the means for the several observing periods of 7·7 seconds of arc. It was found that the observations made when the wind was blowing across the line toward the hill gave the more reliable results. (See p. 62 of Special Publication No. 11 of the U. S. Coast and Geodetic Survey.)

The positions of the four principal close circumpolars have been determined by so many observations at the fixed observatories under such favorable conditions that it is difficult to believe that the errors in their adopted right ascensions and declinations are sufficiently large to produce errors in the computed azimuths that are otherwise than small in comparison with the other errors involved in the azimuth observations. On the other hand, when Polaris (or some other circumpolar) has been observed at both culminations or both elongations, at a given station, the observations at one culmination (or elongation) have often shown a tendency to differ by a constant from those at the other culmination (or elongation), as if the adopted right ascension (or declination) were in error. It should be borne in mind in such cases that the atmospheric conditions have been reversed, so to speak, between the culminations (or elongations); for in one case the temperature will be rising and in the other falling, in general, the two cases occurring at the extreme ends of darkness or of light, or one in the darkness and the other in the light. Hence only a long and careful investigation will determine whether such constant differences are due to defective star places or to changed atmospheric conditions. It is important from a practical point of view to note that if the azimuth observations at a station are all made upon one star and are equally distributed between two hour-angles, differing by about twelve hours, the mean result will be sensibly independent of the errors of the adopted right ascension and declination, and the conditions will be decidedly favorable to eliminating the effects of lateral refraction from the mean result.

An error in the adopted latitude of the station produces the maximum effect when the star is observed at elongation and is without effect if the star is observed at culmination. For Polaris at elongation, to produce an error of 0″.01 in the computed azimuth the adopted latitude must be in error by 0″.70 for a station in latitude 30°, and by 0″.14 for a station in latitude 60°. The error in the computed azimuth from this source will be larger for a star farther from the pole. The astronomic latitude (defined by the actual line of gravity at the station) is required for the computation, and not the geodetic latitude. This error, which will in general be very small, will also be eliminated by observing the star at two positions about twelve hours apart.

The *observer's errors* are his errors of pointing upon the mark and star, errors of pointing upon the circle graduation if reading microscopes are used, errors of vernier reading if verniers are used, errors of reading the micrometer heads, errors in reading the striding level, and errors in estimating the times of the star pointings. There is such a large range of difference in the designs of the various instruments used for azimuth observations that little can be said of the relative and absolute magnitude of these errors that will be of general application. Each observer should investigate these errors for himself with the particular instrument in hand. It will be found in general that the observer's errors play a minor part in furnishing the final errors of the results, except perhaps in the micrometric method.

The effect of errors in time, either errors in estimating the times of the star pointings, the personal equation of the observer, or errors in the adopted chronometer correction, may be estimated by noting the rate at which the star was moving in azimuth when the observations were made. Such errors are usually small, but not insensible except near elongation, and will tend to be eliminated by observations of the same star at two hour-angles differing by about twelve hours.

Of the magnitude of the *instrumental errors* arising from imperfect adjustment and imperfect construction and imperfect stability little of general application can be said, because of the great variety of the instruments.

With the larger and more powerful instruments the errors due to instability become relatively great and should be guarded against by careful manipulation and rapid observing, by using a carefully planned program of observations, and by protecting the instrument against temperature changes as far as possible. In this connection it should be noted that each of the programs of observation given on the preceding pages is especially adapted to elimination of the effect of any continuous twisting of the instrument in azimuth, and is so planned that the observer will not ordinarily lose time in waiting for the bubble of the striding level to come to rest. That observer of azimuth will be most successful in avoiding errors due to instability who keeps it most clearly and continuously in mind that the instrument and its support are made of elastic material of such a large coefficient of thermal expansion that no part remains of fixed dimensions or shape. He will be especially careful about the thermal conditions and the stresses to which his instrument is subjected and will observe with the greatest rapidity consistent with allowable observer's errors.

The errors due to the striding level become more serious the farther north is the station, as may be seen by inspection of the formula for the level correction (p. 144).

The errors of graduation of the horizontal circle have the same effect in azimuth observations as in observations of horizontal angles. The number of positions in which the circle must be used in the direction method may therefore be decided upon the same basis as in the angle measurements.

The micrometric method gives a higher degree of accuracy than either the method of repetitions or the method of directions. This is probably due largely to the great rapidity with which the observations may be made, a condition which is very favorable to the elimination of errors due to instability of the instrument and its support. The error, in the final result for a station by this method, due to an error in the adopted value of one turn of the micrometer may be made very small by so placing the azimuth mark (or marks) and so timing the observations that the sum of the angles measured eastward from the mark (or marks) to the star shall be nearly equal to the sum of such angles measured westward.

STATEMENT OF COSTS.

When azimuths are observed with a theodolite during the progress of a triangulation the cost is very small. This is the method now employed in the primary triangulation by the Coast and Geodetic Survey. It is probable that the observations and field computations for an azimuth do not involve an additional cost of more than $50 per azimuth station.

If, however, the azimuths are observed by a separate party some time later than the triangulation, and when there is more or less building of signals at the stations at each end of the line for which the azimuth is determined, the cost per station will vary during a season's operations from $200 to $500. When an observer must go out in the field to determine a single azimuth at a distant point the expense may be more than $500. A season's work should be planned so that the cost and time of traveling between stations will be a minimum. If practicable, the work in any locality should be done at that time of the year when the most favorable weather conditions may be expected.

AZIMUTH FROM TIME OBSERVATIONS.

For a number of years azimuths of a secondary degree of accuracy for use in connection with tertiary triangulation in Alaska have been obtained directly from time observations with a transit or meridian telescope, with little additional labor of observing and computing. With the adoption of the transit micrometer the accuracy of the results was greatly increased, approaching primary in character. This method of determining azimuths has proved of great value in high latitudes where slow-moving stars are high in altitude, and, consequently, satisfactory azimuths from observations with a theodolite are difficult to obtain.

Observations on a mark which is set closely in the meridian are made during each half set of observations for time. See page 80 for description of method of observing time in high latitudes. The azimuth correction, computed from the time observations, is combined with the reading on the mark to get the azimuth.

It is necessary, of course, to have the mark near enough to the meridian of the instrument to fall within the field that can be measured by means of the reticle or with the micrometer wire. It is best, in the case of the transit micrometer, to place the mark so nearly in the meridian that its image will fall inside the range of the comb, so that the number of turns of the micrometer screw may be readily counted between the pointings in the direct and reversed positions. The mark may be placed either to the north or south and should, if practicable, be at least a mile from the instrument.

The method of observing is as follows: Just before beginning time observations with the telescope band east, say, a number of observations are taken on the mark; the telescope is reversed to the position band west, and an equal number of observations is made on the mark. The stars of the first half set are then observed, followed by observations on the mark. The telescope is next reversed to the position band east, the mark observed, and then the stars of the second half set are taken. Finally, observations are taken on the mark, the telescope is reversed to position band west, and the same number of observations is made on the mark. This completes the first set of azimuth observations, and the observations on the stars for a full time set.

The mean of all the readings on the mark band east, is adopted as the final value in this position of the axis and, similarly, the mean is taken for all readings with band west. The mean of these two adopted values for band east and band west gives the reading of the collimation axis, and the difference between either of the two values and the mean is the angle between the mark and the collimation axis of the telescope. This angle, combined with the azimuth constant of the time set, gives the azimuth of the mark. The angle is observed as so many turns of the micrometer head or screw, or spaces of the reticle. This angle is considered to be positive when the mark is east of the collimation axis, when pointing south, or west of that axis when pointing north. To this angle (reduced to seconds of time) is added algebraically the azimuth constant, a (see p. 25), derived from the computation of the time set. This azimuth constant is the angle between the meridian and the collimation axis. It is considered to be positive if the collimation axis is east of the meridian, with the telescope pointing south, or if the axis is west of the meridian with the telescope north.

If the mark is much out of the horizon of the instrument, readings of the striding level should be made while observing on the mark, and its elevation should be measured roughly with the finder circle. The correction for inclination of axis is applied as on page 145 and the reduction to the horizon, of the angle between mark and collimation axis, is made as on page 157.

If readings on the mark are obtained in only one position of the telescope axis, it will be necessary to take into consideration the collimation constant of the time set and the equatorial interval [1] of the assumed zero as well as the azimuth constant. The reading on the mark made with the micrometer screw, or estimated on the reticle, is referred to some assumed zero of the screw or diaphragm. Combining the angle between the mark and this zero with the equatorial interval of the zero gives the angle between the mark and the line of collimation. This latter angle, combined with the collimation constant of the time set, gives the angle between the mark and the collimation axis. This last angle, the angle between the mark and the collimation axis, combined with the azimuth constant of the time set, gives the desired angle between the mark and the meridian. That part of the azimuth angle which lies between the collimation axis of the telescope and the mark must be reduced to the horizon if the mark is not in the horizontal plane of the instrument. Any inclination of the horizontal axis must be corrected for, as explained on page 145.

[1] This is the angle between the mean position of the micrometer wire or the mean lines of the reticle and the assumed zero. See p. 32.

The following examples with explanations will show this method of determining azimuth:

Example of record—Readings on azimuth mark.

TRANSIT MICROMETER.

[Station, Fairbanks, Alaska. Date, Aug. 9, 1910. Observer, E. Smith. Instrument: Transit No. 18, with transit micrometer. Mark to northward.]

Before observations for time on first half-set		Between the two half-sets		After observations for time on second half-set	
Band East	West	West	East	East	West
T	*T*	*T*	*T*	*T*	*T*
+5. 050	+0. 952	+0. 890	+5. 050	+5. 120	+1. 000
5. 070	0. 915	0. 960	5. 070	5. 090	0. 946
5. 110	0. 940	0. 950	5. 093	5. 121	0. 985
5. 110	0. 990	0. 965	5. 082	5. 120	0. 930
5. 040	0. 920	0. 938	5. 060	5. 068	0. 985
5. 020	0. 990	0. 910	5. 049	5. 140	0. 982
5. 055	0. 930	0. 970	5. 023	5. 140	0. 960
5. 110	0. 930	0. 959	5. 100	5. 110	0. 930
5. 090	0. 950	0. 960	5. 110	5. 080	0. 959
5. 120	0. 985	0. 958	5. 098	5. 090	0. 967
Means: +5. 078	+0. 947	+0. 946	+5. 074	+5. 108	+0. 946

Computation of azimuth from time observations.

TRANSIT MICROMETER.

[Fairbanks, Alaska, 1910. Transit No. 18. Equatorial interval of one turn of micrometer, $2^s.826$. Mark to northward.]

Date	August 8		August 8		August 9	
Band	East	West	East	West	East	West
	T *s*	*T* *s*	*T* *s*	*T* *s*	*T* *s*	*T* *s*
Mean reading on mark	5. 074	1. 023	5. 067	0. 996	5. 087	0. 946
Mean reading of E. and W. (reading of collimation axis)	3. 048	3. 048	3. 032	3. 032	3. 016	3. 016
Angle, mark to collimation axis	−2. 026 = −5. 73	−2. 025 = −5. 72	−2. 035 = −5. 75	−2. 036 = −5. 75	−2. 071 = −5. 85	−2. 070 = −5. 85
a (from time set)	−0. 16	−0. 36	−0. 21	−0. 25	−0. 04	−0. 12
Angle, mark to meridian	−5. 89	−6. 08	−5. 96	−6. 00	−5. 89	−5. 97
Mean for set (in time)	$-5^s.98$		$-5^s.98$		$-5^s.93$	
Mean for set (in arc)	−89″.7		−89″.7		−89″.0	

Mean azimuth of mark east of north, 1′ 29″.5.
Correction for elevation of mark, 0.0.
Reduction to mean position of pole,[1] +0.8.
Azimuth of mark, 180° 01′ 30″.3.

The comb should be considered as being numbered from one side to the other and in such a way that the numbers increase with increasing numbers on the micrometer head as the wire is moved across the field. For convenience the first tooth may be given the number 1 rather than zero. The observer in the field must note in the record for one position of the telescope (band west or east) whether the line of sight points farther east or west with increasing readings on the micrometer head.

In the example above, with band east, the readings increase on the micrometer head as the line of sight moves toward the east. That is, for the reading of five turns, band east, the line of sight is about two turns east of the collimation axis. With band west increasing readings correspond to a motion of the line of sight toward the west, a reading of one turn, band west, corresponding to a postion of the line of sight of about two turns east of the collimation axis.

A set of azimuth observations was made with each of two time sets on August 8.

[1] See Astronomische Nachrichten No. 4504.

Computation of azimuth from time observations.

DIAPHRAGM.

[St. Michael, Alaska, 1898. Meridian telescope No. 13. Equatorial interval of one space of reticle, 3ˢ.455. Mark to southward.]

Date	July 13		July 14		July 15	
Clamp	East	West	East	West	East	West
	Spaces s	*Spaces* s	*Spaces* s	*Spaces* s	*Spaces* s	*Spaces* s
Angle, mark to center line	−0.20 = −0.69	0.00 = 0.00	−0.175 = −0.60	−0.025 = −0.09	−0.75 = −2.59	−0.15 = −0.52
Mean of E and W (Angle mark to collimation axis)	−0.34	−0.34	−0.34	−0.34	−1.56	−1.56
a (from time set)	+0.39	+0.86	+0.40	+0.72	+1.75	+1.63
Angle, mark to meridian	+0.05	+0.52	+0.06	+0.38	+0.19	+0.07
Mean for set (in time)	+0ˢ.28		+0ˢ.22		+0ˢ.13	
Mean for set (in arc)	+4″.2		+3″.3		+2″.0	

Date	July 18		Sept. 13		Sept. 17	
Clamp	East	West	East	West	East	West
	Spaces s	*Spaces* s	*Spaces* s	*Spaces* s	*Spaces* s	*Spaces* s
Angle, mark to center line	−0.975 = −3.37	−0.05 = −0.17	0.00 = 0.00	0.00 = 0.00	+0.25 = +0.86	+0.825 = +2.85
Mean of E and W (Angle mark to collimation axis)	−1.77	−1.77	0.00	0.00	+1.86	+1.86
a (from time set)	+2.78	+2.64	+0.41	+0.06	−2.01	−1.42
Angle, mark to meridian	+1.01	+0.87	+0.41	+0.06	−0.15	+0.44
Mean for set (in time)	+0ˢ.94		+0ˢ.24		+0ˢ.14	
Mean for set (in arc)	+14″.1		+3″.6		+2″.1	

Final mean, mark east of south,	0° 00′ 04″.9
Correction for elevation of mark	0.0
Azimuth of mark	359° 59′ 55″.1

There is no essential difference between the above method and that with the transit micrometer. The angle between the mark and the center line of the diaphragm is estimated in spaces of the reticle. The accuracy of the resulting azimuth in this case as well as in that of the transit micrometer depends largely on the accuracy with which the azimuth constant is determined from the time observations. The effect of errors of pointing and reading on the mark may be made relatively small by repeated observations.

The work of the Latitude Service of the International Geodetic Association began in 1899, so it is only for observations made after that year that a satisfactory reduction can now be made to the mean position of the pole. It is probable that in a few years a reliable value of this reduction can be had, based on theoretical grounds.

Computation of azimuth from time observations.

DIAPHRAGM.

[St. Michael, Alaska, 1898. Meridian telescope No. 13. Readings on mark in only one position of telescope axis. Equatorial interval of one space of reticle, 3ˢ.455. Mark to southward.]

Date	July 13	July 14
Clamp	East	East
	Spaces s	*Spaces* s
Mark east of center line	−0.20 = −0.69	−0.175 = −0.60
Eq. interval of center line	0.00	0.00
c	+0.12	+0.18
a	+0.39	+0.40
Mark east of south	−0.18	−0.02
Mark east of south	−2″.7	−0″.3

The above is taken from the example already given for observations in both positions of the telescope. In this case of deriving the azimuth from observations on the mark in only one position of the axis, the equatorial interval of the assumed zero and the collimation constant of the time set must be applied to the reading on the mark. The collimation constant is applied with the same sign as derived from the computation of the time set when the observations on the mark are made with band west, mark south, and with the opposite sign when made with band east, mark south. The equatorial interval, i, of the assumed zero of the reticle or micrometer is considered positive when west of the mean line or position, band west. It follows, then, that when i and c are combined in the azimuth angle they are applied with opposite signs. Defining the measured angle between the mark and the assumed zero as positive when the mark is east of the zero, pointing south, and using a, c, and i, with their conventional signs, the following general expressions cover all cases:

$$\text{Mark south}\begin{cases}\text{Band W} \ldots & \alpha=360^\circ-\{a_W+(M+c-i)\sec h\}15\\ \text{Band E} \ldots & \alpha=360^\circ-\{a_E+(M-c+i)\sec h\}15\end{cases}$$

$$\text{Mark north}\begin{cases}\text{Band W} \ldots & \alpha=180^\circ-\{a_W+(M-c+i)\sec h\}15\\ \text{Band E} \ldots & \alpha=180^\circ-\{a_E+(M+c-i)\sec h\}15\end{cases}$$

a_W and a_E are the azimuth constants from the time set. M is the angle (in seconds of time) between the mark and the assumed zero of the micrometer or diaphragm. It is assumed to be positive when the mark is east of the zero when pointing south. It is also positive when the mark is west, pointing north. c is the collimation constant of the time set. i is the equatorial interval, in seconds of time, between the mean position of the micrometer wire and the assumed zero of the micrometer, or between the mean line of the reticle and the assumed zero. h is the angle of elevation or depression of the mark. The quantity to be subtracted from 360° or 180° is in seconds of arc.

CORRECTION FOR ELEVATION OF MARK.

When the object used as an azimuth mark is at a considerable elevation, it is necessary to apply a correction to obtain the astronomic azimuth of the projection of the mark on the spheroidal surface of reference. This correction, in seconds, is:

$$+\frac{e^2\,h}{2a\sin 1''}\cos^2\phi\sin 2\,\alpha,$$

in which e^2 is the square of the eccentricity and a the semi-major axis of the spheroid of reference; ϕ is the latitude of the observing station; α is the azimuth of the line to the mark; and h is the elevation of the mark. For h in meters, and Clarke's 1866 dimensions of the spheroid, as stated in meters, this expression becomes:

$$+0''.000109\,h\cos^2\phi\sin 2\alpha\text{, or}$$
$$+[\,\bar{6}.0392]\,h\cos^2\phi\sin 2\alpha,$$

where the number in brackets is a logarithm, the dash over the characteristic indicating that 10 is to be substracted from it. The sign of the expression shows that when the mark is either southwest or northeast of the observing station the observed azimuth of the mark must be increased to obtain the correct azimuth, while for mark northwest or southeast, the observed azimuth must be decreased.

CORRECTION FOR VARIATION OF THE POLE.

A correction is necessary to reduce the observed astronomic azimuth to the mean position of the pole. This correction may amount to a half-second or more for points in the northern part of the United States. The secant of the latiude is a factor of the correction, so the value becomes larger for the higher latitudes. (See p. 85.)

$$Log\ \frac{1}{1-a}$$

Log a	0	1	2	3	4	5	6	7	8	9
9.00	0.045758	5869	5980	6092	6204	6317	6429	6542	6656	6769
8.99	0.044660	4769	4878	4987	5096	5205	5315	5425	5536	5647
98	3591	3697	3803	3909	4016	4122	4229	4337	4444	4552
97	2549	2652	2755	2858	2962	3066	3171	3275	3380	3486
96	1532	1633	1733	1834	1936	2037	2139	2241	2343	2446
95	0.040541	0639	0737	0836	0935	1034	1133	1232	1332	1432
94	0.039575	9670	9766	9862	9959	$\bar{0}$055	$\bar{0}$152	$\bar{0}$249	$\bar{0}$346	$\bar{0}$443
93	8633	8726	8819	8913	9007	9101	9195	9290	9385	9480
92	7714	7805	7896	7987	8079	8171	8263	8355	8447	8540
91	6818	6907	6996	7085	7174	7263	7353	7443	7533	7624
8.90	0.035944	6031	6118	6204	6291	6379	6466	6554	6642	6730
89	5092	5177	5261	5346	5431	5516	5601	5687	5772	5858
88	4261	4343	4426	4508	4591	4674	4757	4841	4924	5008
87	3451	3531	3611	3692	3772	3853	3934	4016	4097	4179
86	2660	2738	2816	2895	2974	3053	3132	3211	3291	3371
85	0.031888	1965	2041	2118	2195	2272	2349	2426	2504	2582
84	1136	1210	1285	1360	1435	1510	1585	1660	1736	1812
83	0402	0474	0547	0620	0693	0766	0840	0914	0987	1061
82	0.029685	9756	9827	9898	9970	$\bar{0}$041	$\bar{0}$113	$\bar{0}$185	$\bar{0}$257	$\bar{0}$329
81	8987	9056	9125	9194	9264	9334	9404	9474	9544	9615
8.80	0.028305	8372	8440	8508	8576	8644	8712	8780	8849	8918
79	7640	7705	7771	7838	7904	7970	8037	8103	8170	8237
78	6990	7055	7119	7183	7248	7313	7378	7443	7509	7574
77	6357	6420	6482	6545	6608	6672	6735	6799	6862	6926
76	5739	5800	5861	5923	5984	6046	6108	6170	6232	6294
75	0.025136	5195	5255	5315	5375	5435	5496	5556	5617	5678
74	4547	4605	4664	4722	4781	4840	4899	4958	5017	5076
73	3973	4029	4086	4143	4201	4258	4316	4373	4431	4489
72	3412	3467	3523	3579	3635	3691	3747	3803	3859	3916
71	2865	2919	2973	3027	3082	3137	3191	3246	3301	3357
8.70	0.022331	2383	2436	2489	2543	2596	2649	2703	2757	2811
69	1809	1861	1913	1964	2016	2068	2121	2173	2225	2278
68	1301	1351	1401	1452	1503	1553	1604	1655	1707	1758
67	0804	0853	0902	0952	1001	1051	1100	1150	1200	1250
66	0319	0367	0415	0463	0512	0560	0609	0657	0706	0755
65	0.019846	9893	9940	9987	$\bar{0}$034	$\bar{0}$081	$\bar{0}$128	$\bar{0}$176	$\bar{0}$223	$\bar{0}$271
64	9384	9430	9475	9521	9567	9613	9660	9706	9752	9799
63	8933	8978	9022	9067	9112	9157	9202	9247	9293	9338
62	8493	8536	8580	8624	8667	8711	8755	8800	8844	8888
61	8063	8105	8148	8191	8233	8276	8319	8363	8406	8449
8.60	0.017643	7685	7726	7768	7810	7852	7894	7936	7978	8020
59	7233	7274	7315	7355	7396	7437	7478	7519	7560	7602
58	6833	6873	6913	6952	6992	7032	7072	7112	7153	7193
57	6443	6482	6520	6559	6598	6637	6676	6715	6755	6794
56	6062	6099	6137	6175	6213	6251	6289	6328	6366	6404
55	0.015689	5726	5763	5800	5837	5874	5912	5949	5986	6024
54	5326	5362	5398	5434	5470	5507	5543	5579	5616	5653
53	4971	5006	5041	5077	5112	5147	5183	5218	5254	5290
52	4624	4659	4693	4727	4762	4797	4831	4866	4901	4936
51	4286	4319	4353	4387	4420	4454	4488	4522	4556	4590
8.50	0.013955	3988	4021	4054	4087	4120	4153	4186	4219	4253

Proportional parts

	111	108	105	102	99
1	11.1	10.8	10.5	10.2	9.9
2	22.2	21.6	21.0	20.4	19.8
3	33.3	32.4	31.5	30.6	29.7
4	44.4	43.2	42.0	40.8	39.6
5	55.5	54.0	52.5	51.0	49.5
6	66.6	64.8	63.0	61.2	59.4
7	77.7	75.6	73.5	71.4	69.3
8	88.8	86.4	84.0	81.6	79.2
9	99.9	97.2	94.5	91.8	89.1
	96	**93**	**90**	**87**	**84**
1	9.6	9.3	9.0	8.7	8.4
[illegible]	[illegible]	[illegible]	18.0	[illegible]	[illegible].8
	[illegible]	.	27.	[illegible]	[illegible] .2
		.	36.	.	.6
	48	.	45.	.	42.0
	57	.	54.	.	50.4
	[illegible]	.	63.	60.	[illegible].8
		.	72.	69.	.2
	.	.	81.	78.	.6
	81	**78**	**75**	**72**	**69**
[illegible]	8.1	7.8	7.5	7.2	6.9
	16.[illegible]	[illegible]	[illegible]	14.[illegible]	13.[illegible]
	24.	[illegible]	[illegible]	21.	20.
	32.	.	.	28.	27.
	40.	.	.	36.	[illegible]4.
	48.	.	45.	43.	1.
	56.	.	52.	50.	8.
	[illegible]4.	.	60.	57.	5.
	2.	.	67.	64.	2.
	66	**63**	**60**	**57**	**55**
[illegible]	6.6	6.3	6.0	5.7	5.5
	[illegible]	[illegible]	12.0	1[illegible].4	11.[illegible]
	[illegible]	[illegible]	[illegible]8.	1 .1	16.
	.	.	4.	2 .8	22.
	.	.	0.	2 .5	27.
	.	.	6.	3 .2	33.
	46.	44.	42.	3 .9	38.
	52.	50.	48.	45.6	44.
	59.	56.	54.	51.3	49.
	53	**51**	**49**	**47**	**45**
[illegible]	5.[illegible]	5.1	4.9	4.[illegible]	4.5
	10.[illegible]	10.[illegible]	[illegible].8	9.	[illegible]
	15.	[illegible]5.	1 .7	14.	1 .
	21.	0.	1 .6	18.	[illegible] .
	26.	5.	24.5	23.	.
	31.	0.	29.4	28.	.
	37.	5.	[illegible].3	32.	.
	42.	40.	.2	37.	.
	47.	45.	44.1	42.	40.
	43	**41**	**39**	**37**	**35**
1	4.3	4.1	3.9	3.7	3.5
2	8.6	8.2	7.8	7.4	7.0
3	12.9	12.3	11.7	11.1	10.5
4	17.2	16.4	15.6	14.8	14.0
5	21.5	20.5	19.5	18.5	17.5
6	25.8	24.6	23.4	22.2	21.0
7	30.1	28.7	27.3	25.9	24.5
8	34.4	32.8	31.2	29.6	28.0
9	38.7	36.9	35.1	33.3	31.5

$$Log\ \frac{1}{1-a}$$

Log a	0	1	2	3	4	5	6	7	8	9
8.50	0.013955	3988	4021	4054	4087	4120	4153	4186	4219	4253
49	3633	3665	3697	3729	3761	3793	3825	3858	3890	3923
48	3318	3349	3380	3411	3443	3474	3506	3537	3569	3601
47	3010	3040	3071	3101	3132	3163	3194	3225	3256	3287
46	2709	2739	2769	2799	2829	2859	2889	2919	2949	2979
45	0.012416	2445	2474	2503	2532	2562	2591	2621	2650	2680
44	2129	2158	2186	2215	2243	2272	2300	2329	2358	2387
43	1849	1877	1905	1933	1961	1989	2017	2045	2073	2101
42	1576	1603	1630	1657	1685	1712	1739	1767	1794	1822
41	1309	1335	1362	1388	1415	1442	1468	1495	1522	1549
8.40	0.011048	1074	1100	1126	1152	1178	1204	1230	1256	1283
39	0794	0819	0844	0869	0895	0920	0946	0971	0997	1023
38	0545	0570	0594	0619	0644	0669	0694	0718	0743	0769
37	0302	0326	0350	0374	0399	0423	0447	0472	0496	0520
36	0065	0088	0112	0135	0159	0183	0207	0230	0254	0278
35	0.009833	9856	9879	9902	9925	9948	9972	9995	0018	0041
34	9607	9629	9652	9674	9697	9719	9742	9765	9787	9810
33	9386	9408	9430	9452	9474	9496	9518	9540	9562	9584
32	9170	9191	9213	9234	9256	9277	9299	9320	9342	9364
31	8959	8980	9001	9022	9043	9064	9085	9106	9127	9149
8.30	0.008753	8773	8794	8814	8835	8855	8876	8897	8917	8938
29	8552	8572	8592	8612	8632	8652	8672	8692	8712	8733
28	8355	8375	8394	8414	8433	8453	8473	8492	8512	8532
27	8163	8182	8201	8220	8239	8259	8278	8297	8316	8336
26	7976	7994	8013	8031	8050	8069	8088	8106	8125	8144
25	0.007792	7811	7829	7847	7865	7884	7902	7920	7939	7957
24	7614	7631	7649	7667	7685	7702	7720	7738	7756	7774
23	7439	7456	7473	7491	7508	7526	7543	7561	7578	7596
22	7268	7285	7302	7319	7336	7353	7370	7387	7404	7421
21	7101	7118	7134	7151	7167	7184	7201	7218	7234	7251
8.20	0.006938	6954	6971	6987	7003	7019	7036	7052	7068	7085
19	6779	6795	6811	6826	6842	6858	6874	6890	6906	6922
18	6624	6639	6654	6670	6685	6701	6716	6732	6748	6763
17	6472	6487	6502	6517	6532	6547	6562	6578	6593	6608
16	6323	6338	6353	6367	6382	6397	6412	6427	6442	6457
15	0.006178	6193	6207	6221	6236	6250	6265	6279	6294	6309
14	6037	6051	6065	6079	6093	6107	6121	6135	6150	6164
13	5898	5912	5926	5940	5953	5967	5981	5995	6009	6023
12	5763	5777	5790	5803	5817	5830	5844	5857	5871	5885
11	5631	5644	5657	5670	5684	5697	5710	5723	5737	5750
8.10	0.005502	5515	5528	5541	5553	5566	5579	5592	5605	5618
09	5376	5389	5401	5414	5426	5439	5451	5464	5477	5489
08	5253	5265	5277	5290	5302'	5314	5327	5339	5351	5364
07	5133	5145	5157	5169	5181	5193	5205	5217	5229	5241
06	5015	5027	5038	5050	5062	5074	5085	5097	5109	5121
05	4900	4912	4923	4935	4946	4957	4969	4980	4992	5004
04	4788	4799	4810	4822	4833	4844	4855	4866	4878	4889
03	4679	4690	4700	4711	4722	4733	4744	4755	4766	4777
02	4572	4582	4593	4603	4614	4625	4636	4646	4657	4668
01	4467	4477	4488	4498	4509	4519	4529	4540	4550	4561
8.00	0.004365	4375	4385	4395	4405	4416	4426	4436	4446	4457

Proportional parts

	34	33	32	31	30
[illegible]	3.4	3.3	3.2	3.1	3.0
	6.[illegible]	6.[illegible]	6.[illegible]	6.[illegible]	6.0
	10.	9.	9.	9.	9.
	13.	13.	12.	12.	12.
	17.	16.	16.	15.	15.
	20.	19.	19.	18.	18.
	23.	23.	22.	21.	21.
	27.	26.	25.	24.	24.
	30.	29.	28.	27.	2 .
	29	**28**	**27**	**26**	**25**
[illegible]	2.9	2.8	2.7	2.6	2.5
	5.[illegible]	[illegible]	5.[illegible]	5.[illegible]	5.0
	8.	.	28.	7.	7.
	11.	1 .	0.	10.	10.
	14.	2 .	[illegible].	13.	12.
	17.	.	.	15.	15.
	20.	.	.	18.	17.
	23.	.	.	20.	20.
	26.	25.	.	23.	22.
	24	**23**	**22**	**21**	**20**
[illegible]	2.4	2.3	2.2	2.1	2.0
	[illegible]	[illegible]	[illegible]	[illegible] 2	[illegible].0
	.	.		3	.
	.	.		4	.
	1 .	1 .	1	1 5	1 .
	1 .	1 .	1	1 6	1 .
	1 .	1 .		7	.
	1 .	1 .		8	.
	2 .	20.		9	.
	19	**18**	**17**	**16**	**15**
[illegible]	1.9	1.8	1.7	1.6	1.5
	[illegible]	[illegible]	[illegible].4	[illegible]	[illegible]
	.	.	.1.	.	
	.	.	.8	.	
	.	.	.5	.	
	1 .	1 .	1 .2	.	
	1 .	1 .	1 .9	1 .	1
	.	.	1 .6	1 .	1
	.	.	1 .3	.	.
	14	**13**	**12**	**11**	**10**
[illegible]	1.4	1.3	1.2	1.1	1.0
	[illegible]	[illegible]	2.4	[illegible]	2.0
	.	.	3.6	.	3.0
	.	.	4.8	.	4.0
	.	.	6.0	.	5.0
	.	.	7.2	.	6.0
	.	.	8.4	.	7.0
	1 .	1 .	9.6	.	8.0
	1 .	1 .	10.8	.	9.0

$$Log \frac{1}{1-a}$$

Log a	0	1	2	3	4	5	6	7	8	9
8.00	0.004365	4375	4385	4395	4405	4416	4426	4436	4446	4457
7.99	4265	4275	4285	4295	4305	4315	4325	4335	4345	4355
98	4167	4177	4187	4196	4206	4216	4226	4235	4245	4255
97	4072	4082	4091	4100	4110	4119	4129	4139	4148	4158
96	3979	3988	3997	4007	4016	4025	4035	4044	4053	4063
95	0.003888	3897	3906	3915	3924	3933	3942	3951	3961	3970
94	3799	3808	3817	3826	3834	3843	3852	3861	3870	3879
93	3712	3721	3729	3738	3747	3755	3764	3773	3782	3790
92	3627	3636	3644	3653	3661	3670	3678	3687	3695	3704
91	3545	3553	3561	3569	3577	3586	3594	3602	3611	3619
7.90	0.003463	3472	3480	3488	3496	3504	3512	3520	3528	3536
89	3384	3392	3400	3408	3416	3424	3432	3440	3448	3456
88	3307	3315	3322	3330	3338	3345	3353	3361	3369	3377
87	3231	3239	3246	3254	3261	3269	3277	3284	3292	3299
86	3158	3165	3172	3180	3187	3194	3202	3209	3217	3224
85	0.003086	3093	3100	3107	3114	3121	3129	3136	3143	3150
84	3015	3022	3029	3036	3043	3050	3057	3064	3071	3078
83	2946	2953	2960	2967	2974	2980	2987	2994	3001	3008
82	2879	2886	2892	2899	2906	2912	2919	2926	2933	2939
81	2813	2820	2826	2833	2839	2846	2852	2859	2866	2872
7.80	0.002749	2755	2762	2768	2774	2781	2787	2794	2800	2807
79	2686	2692	2699	2705	2711	2717	2724	2730	2736	2743
78	2625	2631	2637	2643	2649	2655	2661	2668	2674	2680
77	2565	2571	2577	2583	2589	2595	2601	2607	2613	2619
76	2506	2512	2518	2524	2530	2535	2541	2547	2553	2559
75	0.002449	2455	2460	2466	2472	2478	2483	2489	2495	2501
74	2393	2399	2404	2410	2415	2421	2427	2432	2438	2443
73	2339	2344	2349	2355	2360	2366	2371	2377	2382	2388
72	2285	2290	2296	2301	2306	2312	2317	2322	2328	2.33
71	2233	2238	2243	2249	2254	2259	2264	2269	2275	2280
7.70	0.002182	2187	2192	2197	2202	2207	2213	2218	2223	2228
69	2132	2137	2142	2147	2152	2157	2162	2167	2172	2177
68	2084	2088	2093	2098	2103	2108	2113	2118	2122	2127
67	2036	2041	2046	2050	2055	2060	2065	2069	2074	2079
66	1990	1994	1999	2003	2008	2013	2017	2022	2027	2031
65	0.001944	1949	1953	1958	1962	1967	1971	1976	1980	1985
64	1900	1904	1909	1913	1918	1922	1926	1931	1935	1940
63	1857	1861	1865	1869	1874	1878	1882	1887	1891	1896
62	1814	1818	1823	1827	1831	1835	1840	1844	1848	1852
61	1773	1777	1781	1785	1789	1793	1798	1802	1806	1810
7.60	0.001732	1736	1740	1744	1748	1753	1757	1761	1765	1769
59	1693	1697	1701	1705	1709	1713	1716	1720	1724	1728
58	1654	1658	1662	1666	1670	1673	1677	1681	1685	1689
57	1617	1620	1624	1628	1632	1635	1639	1643	1647	1650
56	1580	1583	1587	1591	1594	1598	1602	1605	1609	1613
55	0.001544	1547	1551	1554	1558	1562	1565	1569	1572	1576
54	1508	1512	1515	1519	1522	1526	1529	1533	1537	1540
53	1474	1477	1481	1484	1488	1491	1495	1498	1502	1505
52	1440	1444	1447	1450	1454	1457	1461	1464	1467	1471
51	1408	1411	1414	1417	1421	1424	1427	1431	1434	1437
7.50	0.001376	1379	1382	1385	1388	1391	1395	1398	1401	1404

Proportional parts

	11	10
1	1.1	1.0
2	2.2	2.0
3	3.3	3.0
4	4.4	4.0
5	5.5	5.0
6	6.6	6.0
7	7.7	7.0
8	8.8	8.0
9	9.9	9.0

	9	8
1	0.9	0.8
2	1.8	1.6
3	2.7	2.4
4	3.6	3.2
5	4.5	4.0
6	5.4	4.8
7	6.3	5.6
8	7.2	6.4
9	8.1	7.2

	7	6
1	0.7	0.6
2	1.4	1.2
3	2.1	1.8
4	2.8	2.4
5	3.5	3.0
6	4.2	3.6
7	4.9	4.2
8	5.6	4.8
9	6.3	5.4

	5	4
1	0.5	0.4
2	1.0	0.8
3	1.5	1.2
4	2.0	1.6
5	2.5	2.0
6	3.0	2.4
7	3.5	2.8
8	4.0	3.2
9	4.5	3.6

$$Log \frac{1}{1-a}$$

Log a	0	1	2	3	4	5	6	7	8	9
7.50	0.001376	1379	1382	1385	1388	1391	1395	1398	1401	1404
49	1344	1347	1350	1354	1357	1360	1363	1366	1369	1372
48	1314	1317	1320	1323	1326	1329	1332	1335	1338	1341
47	1284	1287	1290	1292	1295	1298	1301	1304	1307	1311
46	1254	1257	1260	1263	1266	1269	1272	1275	1278	1281
45	0.001226	1229	1231	1234	1237	1240	1243	1246	1249	1251
44	1198	1201	1203	1206	1209	1212	1214	1217	1220	1223
43	1170	1173	1176	1179	1181	1184	1187	1190	1192	1195
42	1144	1146	1149	1152	1154	1157	1160	1162	1165	1168
41	1118	1120	1123	1126	1128	1131	1133	1136	1139	1141
7.40	0.001092	1095	1097	1100	1102	1105	1107	1110	1113	1115
39	1067	1070	1072	1075	1077	1080	1082	1085	1087	1090
38	1043	1045	1048	1050	1053	1055	1058	1060	1062	1065
37	1019	1022	1024	1026	1029	1031	1033	1036	1038	1041
36	0.000996	998	1001	1003	1005	1008	1010	1012	1015	1017
35	0.000973	976	978	980	982	985	987	989	991	994
34	951	953	956	958	960	962	964	967	969	971
33	929	932	934	936	938	940	942	945	947	949
32	908	910	913	915	917	919	921	923	925	927
31	888	890	892	894	896	898	900	902	904	906
7.30	0.000867	869	871	873	875	877	879	882	884	886
29	848	850	852	854	855	857	859	861	863	865
28	828	830	832	834	836	838	840	842	844	846
27	809	811	813	815	817	819	821	823	825	826
26	791	793	795	796	798	800	802	804	806	808
25	0.000773	775	777	778	780	782	784	786	787	789
24	755	757	759	761	762	764	766	768	769	771
23	738	740	742	743	745	747	748	750	752	754
22	721	723	725	726	728	730	731	733	735	736
21	705	707	708	710	711	713	715	716	718	720
7.20	0.000689	690	692	694	695	697	698	700	702	703
19	673	675	676	678	679	681	683	684	686	687
18	658	659	661	662	664	665	667	669	670	672
17	643	644	646	647	649	650	652	653	655	656
16	628	630	631	633	634	635	637	638	640	641
15	0.000614	615	617	618	620	621	622	624	625	627
14	600	601	603	604	605	607	608	610	611	612
13	586	588	589	590	592	593	594	596	597	599
12	573	574	576	577	578	580	581	582	584	585
11	560	561	562	564	565	566	568	569	570	572
7.10	0.000547	548	550	551	552	553	555	556	557	559
09	535	536	537	538	540	541	542	543	545	546
08	522	524	525	526	527	529	530	531	532	533
07	511	512	513	514	515	516	518	519	520	521
06	499	500	501	502	504	505	506	507	508	509
05	0.000488	489	490	491	492	493	494	495	497	498
04	476	478	479	480	481	482	483	484	485	486
03	466	467	468	469	470	471	472	473	474	475
02	455	456	457	458	459	460	461	462	463	465
01	445	446	447	448	449	450	451	452	453	454
7.00	0.000435	436	437	438	439	440	441	442	443	444

Proportional parts

	4	3
1	0.4	0.3
2	0.8	0.6
3	1.2	0.9
4	1.6	1.2
5	2.0	1.5
6	2.4	1.8
7	2.8	2.1
8	3.2	2.4
9	3.6	2.7

	2	1
1	0.2	0.1
2	0.4	0.2
3	0.6	0.3
4	0.8	0.4
5	1.0	0.5
6	1.2	0.6
7	1.4	0.7
8	1.6	0.8
9	1.8	0.9

$$Log \frac{1}{1-a}$$

Log a	0	1	2	3	4	5	6	7	8	9
7.00	0.000435	436	437	438	439	440	441	442	443	444
6.9	345	353	361	370	378	387	396	405	415	425
8	274	280	287	294	301	308	315	322	330	337
7	218	223	228	233	239	244	250	256	262	268
6	173	177	181	185	190	194	199	203	208	213
5	0.000137	141	144	147	151	154	158	161	165	169
4	109	112	114	117	120	122	125	128	131	134
3	87	89	91	93	95	97	100	102	104	107
2	69	70	72	74	75	77	79	81	83	85
1	55	56	57	59	60	61	63	64	66	67
6.0	0.000043	44	45	47	48	49	50	51	52	53
5.9	34	35	36	37	38	39	40	41	41	42
8	27	28	29	29	30	31	31	32	33	34
7	22	22	23	23	24	24	25	26	26	27
6	17	18	18	19	19	19	20	20	21	21
5	0.000014	14	14	15	15	15	16	16	17	17
4	11	11	11	12	12	12	13	13	13	13
3	9	9	9	9	10	10	10	10	10	11
2	7	7	7	7	8	8	8	8	8	8
1	5	6	6	6	6	6	6	6	7	7
5.0	0.000004	4	5	5	5	5	5	5	5	5
4	0.000000	1	1	1	1	1	2	2	3	3
4 n	1.000000	9999	9999	9999	9999	9999	9998	9998	9997	9997
5.0 n	9.999996	96	95	95	95	95	95	95	95	95
1 n	95	94	94	94	94	94	94	94	93	93
2 n	93	93	93	93	92	92	92	92	92	92
3 n	91	91	91	91	90	90	90	90	90	89
4 n	89	89	89	88	88	88	87	87	87	87
5 n	9.999986	86	86	85	85	85	84	84	83	83
6 n	83	82	82	81	81	81	80	80	79	79
7 n	78	78	77	77	76	76	75	74	74	73
8 n	73	72	71	71	70	69	69	68	67	66
9 n	66	65	64	63	62	61	60	59	59	58
6.0 n	9.999957	56	55	53	52	51	50	49	48	47
1 n	45	44	43	41	40	39	37	36	34	33
2 n	31	30	28	26	25	23	21	19	17	15
3 n	13	11	09	07	05	03	01	898	896	893
4 n	9.999891	888	886	883	880	878	875	872	869	866
5 n	9.999863	859	856	853	849	846	842	839	835	831
6 n	827	823	819	815	810	806	802	797	792	787
7 n	782	777	772	767	761	756	750	744	738	732
8 n	726	720	713	706	700	693	685	678	671	663
9 n	655	647	639	631	622	613	604	595	585	576
7.00 n	9.999566	565	564	563	562	561	560	559	558	557

Proportional parts

	10	9	8	7	6	5	4	3	2	1
1	1.0	0.9	0.8	0.7	0.6	0.5	0.4	0.3	0.2	0.1
2	2.0	1.8	1.6	1.4	1.2	1.0	0.8	0.6	0.4	0.2
3	3.0	2.7	2.4	2.1	1.8	1.5	1.2	0.9	0.6	0.3
4	4.0	3.6	3.2	2.8	2.4	2.0	1.6	1.2	0.8	0.4
5	5.0	4.5	4.0	3.5	3.0	2.5	2.0	1.5	1.0	0.5
6	6.0	5.4	4.8	4.2	3.6	3.0	2.4	1.8	1.2	0.6
7	7.0	6.3	5.6	4.9	4.2	3.5	2.8	2.1	1.4	0.7
8	8.0	7.2	6.4	5.6	4.8	4.0	3.2	2.4	1.6	0.8
9	9.0	8.1	7.2	6.3	5.4	4.5	3.6	2.7	1.8	0.9

$$Log \frac{1}{1-a}$$

Log a	0	1	2	3	4	5	6	7	8	9
7.00 n	9.999566	565	564	563	562	561	560	559	558	557
01 n	556	555	554	553	552	551	550	549	548	547
02 n	545	544	543	542	541	540	539	538	537	536
03 n	535	534	533	532	531	530	528	527	526	525
04 n	524	523	522	521	520	519	517	516	515	514
05 n	9.999513	512	511	510	508	507	506	505	504	503
06 n	502	501	499	498	497	496	495	494	492	491
07 n	490	489	488	487	485	484	483	482	481	479
08 n	478	477	476	475	473	472	471	470	469	467
09 n	466	465	464	462	461	460	459	457	456	455
7.10 n	9.999454	452	451	450	449	447	446	445	443	442
11 n	441	440	438	437	436	434	433	432	430	429
12 n	428	427	425	424	423	421	420	419	417	416
13 n	415	413	412	410	409	408	406	405	404	402
14 n	401	400	398	397	395	394	393	391	390	388
15 n	9.999387	386	384	383	381	380	378	377	376	374
16 n	373	371	370	368	367	365	364	363	361	360
17 n	358	357	355	354	352	351	349	348	346	345
18 n	343	342	340	339	337	336	334	333	331	329
19 n	328	326	325	323	322	320	319	317	315	314
7.20 n	9.999312	311	309	307	306	304	303	301	299	298
21 n	296	295	293	291	290	288	286	285	283	282
22 n	280	278	277	275	273	272	270	268	266	265
23 n	263	261	260	258	256	255	253	251	249	247
24 n	246	244	242	241	239	237	235	234	232	230
25 n	9.999228	227	225	223	221	219	218	216	214	212
26 n	210	209	207	205	203	201	199	198	196	194
27 n	192	190	188	186	185	183	181	179	177	175
28 n	173	171	169	168	166	164	162	160	158	156
29 n	154	152	150	148	146	144	142	140	138	136
7.30 n	9.999134	132	130	128	126	124	122	120	118	116
31 n	114	112	110	108	106	104	102	100	098	096
32 n	094	091	089	087	085	083	081	079	077	075
33 n	072	070	068	066	064	062	060	057	055	053
34 n	051	049	047	044	042	040	038	036	033	031
35 n	9.999029	027	024	022	020	018	015	013	011	009
36 n	006	004	002	8999	8997	8995	8992	8990	8988	8985
37 n	9.998983	8981	8978	8976	8974	8971	8909	8967	8964	8962
38 n	8959	8957	8955	8952	8950	8947	8945	8943	8940	8938
39 n	8935	8933	8930	8928	8925	8923	8920	8918	8915	8913
7.40 n	9.998910	8908	8905	8903	8900	8898	8895	8893	8890	8888
41 n	8885	8883	8880	8877	8875	8872	8870	8867	8864	8862
42 n	8859	8857	8854	8851	8849	8846	8843	8841	8838	8835
43 n	8833	8830	8827	8825	8822	8819	8816	8814	8811	8808
44 n	8805	8803	8800	8797	8794	8792	8789	8786	8783	8781
45 n	9.998778	8775	8772	8769	8766	8764	8761	8758	8755	8752
46 n	8749	8746	8744	8741	8738	8735	8732	8729	8726	8723
47 n	8720	8717	8714	8711	8708	8705	8702	8699	8696	8693
48 n	8690	8687	8684	8681	8678	8675	8672	8669	8666	8663
49 n	8660	8657	8654	8651	8648	8644	8641	8638	8635	8632
7.50 n	9.998629	8626	8622	8619	8616	8613	8610	8607	8603	8600

Proportional parts

	1	2
1	0.1	0.2
2	0.2	0.4
3	0.3	0.6
4	0.4	0.8
5	0.5	1.0
6	0.6	1.2
7	0.7	1.4
8	0.8	1.6
9	0.9	1.8

	3	4
1	0.3	0.4
2	0.6	0.8
3	0.9	1.2
4	1.2	1.6
5	1.5	2.0
6	1.8	2.4
7	2.1	2.8
8	2.4	3.2
9	2.7	3.6

$$Log \frac{1}{1-a}$$

Log a	0	1	2	3	4	5	6	7	8	9
7.50 n	9.998629	8626	8622	8619	8616	8613	8610	8607	8603	8600
51 n	8597	8594	8590	8587	8584	8581	8577	8574	8571	8568
52 n	8564	8561	8558	8554	8551	8548	8544	8541	8538	8534
53 n	8531	8528	8524	8521	8517	8514	8511	8507	8504	8500
54 n	8497	8493	8490	8486	8483	8479	8476	8472	8469	8465
55 n	9.998462	8458	8455	8451	8448	8444	8440	8437	8433	8430
56 n	8426	8422	8419	8415	8411	8408	8404	8400	8397	8393
57 n	8389	8386	8382	8378	8375	8371	8367	8363	8360	8356
58 n	8352	8348	8344	8341	8337	8333	8329	8325	8321	8318
59 n	8314	8310	8306	8302	8298	8294	8290	8286	8282	8278
7.60 n	9.998274	8271	8267	8263	8259	8255	8251	8246	8242	8238
61 n	8234	8230	8226	8222	8218	8214	8210	8206	8202	8197
62 n	8193	8189	8185	8181	8177	8172	8168	8164	8160	8156
63 n	8151	8147	8143	8139	8134	8130	8126	8121	8117	8113
64 n	8108	8104	8100	8095	8091	8087	8082	8078	8073	8069
65 n	9.998064	8060	8055	8051	8047	8042	8038	8033	8028	8024
66 n	8019	8015	8010	8006	8001	7997	7992	7987	7983	7978
67 n	7973	7969	7964	7959	7955	7950	7945	7941	7936	7931
68 n	7926	7922	7917	7912	7907	7902	7898	7893	7888	7883
69 n	7878	7873	7868	7863	7859	7854	7849	7844	7839	7834
7.70 n	9.997829	7824	7819	7814	7809	7804	7799	7794	7789	7783
71 n	7778	7773	7768	7763	7758	7753	7748	7742	7737	7732
72 n	7727	7722	7716	7711	7706	7700	7695	7690	7685	7679
73 n	7674	7669	7663	7658	7652	7647	7642	7636	7631	7625
74 n	7620	7614	7609	7603	7598	7592	7587	7581	7576	7570
75 n	9.997565	7559	7553	7548	7542	7537	7531	7525	7519	7514
76 n	7508	7502	7497	7491	7485	7479	7473	7468	7462	7456
77 n	7450	7444	7438	7433	7427	7421	7415	7409	7403	7397
78 n	7391	7385	7379	7373	7367	7361	7355	7349	7343	7337
79 n	7330	7324	7318	7312	7306	7300	7293	7287	7281	7275
7.80 n	9.997268	7262	7256	7250	7243	7237	7231	7224	7218	7211
81 n	7205	7199	7192	7186	7179	7173	7166	7160	7153	7147
82 n	7140	7134	7127	7120	7114	7107	7100	7094	7087	7080
83 n	7074	7067	7060	7053	7047	7040	7033	7026	7019	7013
84 n	7006	6999	6992	6985	6978	6971	6964	6957	6950	6943
85 n	9.996936	6929	6922	6915	6908	6901	6894	6887	6880	6872
86 n	6865	6858	6851	6844	6836	6829	6822	6814	6807	6800
87 n	6792	6785	6778	6770	6763	6755	6748	6740	6733	6725
88 n	6718	6710	6703	6695	6688	6680	6672	6665	6657	6650
89 n	6642	6634	6626	6619	6611	6603	6595	6587	6580	6572
7.90 n	9.996564	6556	6548	6540	6532	6524	6516	6508	6500	6492
91 n	6484	6476	6468	6460	6452	6444	6435	6427	6419	6411
92 n	6403	6394	6386	6378	6369	6361	6353	6344	6336	6328
93 n	6319	6311	6302	6294	6285	6277	6268	6260	6251	6242
94 n	6234	6225	6217	6208	6199	6190	6182	6173	6164	6155
95 n	9.996146	6138	6129	6120	6111	6102	6093	6084	6075	6066
96 n	6057	6048	6039	6030	6021	6012	6003	5993	5984	5975
97 n	5966	5956	5947	5938	5929	5919	5910	5900	5891	5882
98 n	5872	5863	5853	5844	5834	5825	5815	5805	5796	5786
99 n	5777	5767	5757	5747	5738	5728	5718	5708	5698	5689
8.00 n	9.995679	5669	5659	5649	5639	5629	5619	5609	5599	5589

Proportional parts

	4	5
1	0.4	0.5
2	0.8	1.0
3	1.2	1.5
4	1.6	2.0
5	2.0	2.5
6	2.4	3.0
7	2.8	3.5
8	3.2	4.0
9	3.6	4.5

	6	7
1	0.6	0.7
2	1.2	1.4
3	1.8	2.1
4	2.4	2.8
5	3.0	3.5
6	3.6	4.2
7	4.2	4.9
8	4.8	5.6
9	5.4	6.3

	8	9
1	0.8	0.9
2	1.6	1.8
3	2.4	2.7
4	3.2	3.6
5	4.0	4.5
6	4.8	5.4
7	5.6	6.3
8	6.4	7.2
9	7.2	8.1

	10	11
1	1.0	1.1
2	2.0	2.2
3	3.0	3.3
4	4.0	4.4
5	5.0	5.5
6	6.0	6.6
7	7.0	7.7
8	8.0	8.8
9	9.0	9.9

$$Log\ \frac{1}{1-a}$$

Log a	0	1	2	3	4	5	6	7	8	9
8.00 n	9.995679	5669	5659	5649	5639	5629	5619	5609	5599	5589
01 n	5578	5568	5558	5548	5538	5528	5517	5507	5497	5486
02 n	5476	5466	5455	5445	5434	5424	5413	5403	5392	5382
03 n	5371	5361	5350	5339	5329	5318	5307	5296	5286	5275
04 n	5264	5253	5242	5231	5220	5209	5198	5187	5176	5165
05 n	9.995154	5143	5132	5121	5110	5098	5087	5076	5065	5053
06 n	5042	5031	5019	5008	4996	4985	4973	4962	4950	4939
07 n	4927	4916	4904	4892	4881	4869	4857	4845	4833	4822
08 n	4810	4798	4786	4774	4762	4750	4738	4726	4714	4702
09 n	4690	4677	4665	4653	4641	4628	4616	4604	4591	4579
8.10 n	9.994567	4554	4542	4529	4517	4504	4492	4479	4466	4454
11 n	4441	4428	4415	4403	4390	4377	4364	4351	4338	4325
12 n	4312	4299	4286	4273	4260	4247	4234	4220	4207	4194
13 n	4181	4167	4154	4141	4127	4114	4100	4087	4073	4060
14 n	4046	4032	4019	4005	3991	3978	3964	3950	3936	3922
15 n	9.993908	3894	3880	3866	3852	3838	3824	3810	3796	3782
16 n	3767	3753	3739	3725	3710	3696	3681	3667	3652	3638
17 n	3623	3609	3594	3579	3565	3550	3535	3521	3506	3491
18 n	3476	3461	3446	3431	3416	3401	3386	3371	3356	3340
19 n	3325	3310	3295	3279	3264	3248	3233	3218	3202	3186
8.20 n	9.993171	3155	3140	3124	3108	3092	3077	3061	3045	3029
21 n	3013	2997	2981	2965	2949	2933	2917	2900	2884	2868
22 n	2852	2835	2819	2803	2786	2770	2753	2736	2720	2703
23 n	2687	2670	2653	2636	2619	2603	2586	2569	2552	2535
24 n	2518	2501	2483	2466	2449	2432	2414	2397	2380	2362
25 n	9.992345	2327	2310	2292	2275	2257	2239	2222	2204	2186
26 n	2168	2150	2132	2114	2096	2078	2060	2042	2024	2006
27 n	1987	1969	1951	1932	1914	1896	1877	1858	1840	1821
28 n	1803	1784	1765	1746	1727	1709	1690	1671	1652	1633
29 n	1613	1594	1575	1556	1537	1517	1498	1478	1459	1440
8.30 n	9.991420	1400	1381	1361	1341	1322	1302	1282	1262	1242
31 n	1222	1202	1182	1162	1142	1122	1101	1081	1061	1040
32 n	1020	0999	0979	0958	0938	0917	0896	0875	0855	0834
33 n	0813	0792	0771	0750	0729	0708	0686	0665	0644	0622
34 n	0601	0580	0558	0537	0515	0493	0472	0450	0428	0406
35 n	9.990385	0363	0341	0319	0297	0274	0252	0230	0208	0186
36 n	0163	0141	0118	0096	0073	0051	0028	0005	$\bar{9}982$	$\bar{9}960$
37 n	9.989937	9914	9891	9868	9845	9821	9798	9775	9752	9728
38 n	9705	9682	9658	9634	9611	9587	9563	9540	9516	9492
39 n	9468	9444	9420	9396	9372	9348	9323	9299	9275	9250
8.40 n	9.989226	9201	9177	9152	9127	9103	9078	9053	9028	9003
41 n	8978	8953	8928	8903	8877	8852	8827	8801	8776	8750
42 n	8725	8699	8673	8647	8622	8596	8570	8544	8518	8492
43 n	8465	8439	8413	8386	8360	8334	8307	8280	8254	8227
44 n	8200	8173	8147	8120	8093	8066	8038	8011	7984	7957
45 n	9.987929	7902	7874	7847	7819	7791	7764	7736	7708	7680
46 n	7652	7624	7596	7568	7539	7511	7483	7454	7426	7397
47 n	7369	7340	7311	7282	7253	7224	7195	7166	7137	7108
48 n	7079	7049	7020	6990	6961	6931	6902	6872	6842	6812
49 n	6782	6752	6722	6692	6662	6631	6601	6571	6540	6510
8.50 n	9.986479	6448	6418	6387	6356	6325	6294	6263	6232	6200

Proportional parts

	10	11	12	13	14
1	1.0	1.1	1.2	1.3	1.4
2	[illegible]	[illegible]	[illegible]	[illegible]	[illegible]
3	[illegible]	[illegible]	[illegible]	[illegible]	[illegible]
4	[illegible]	[illegible]	[illegible]	[illegible]	[illegible]
5	[illegible]	[illegible]	[illegible]	[illegible]	[illegible]
6	[illegible]	[illegible]	[illegible]	[illegible]	[illegible]
7	[illegible]	[illegible]	[illegible]	[illegible]	[illegible]
8	[illegible]	[illegible]	[illegible]	[illegible]	[illegible]
9	[illegible]	[illegible]	[illegible]	[illegible]	[illegible]

	15	16	17	18	19
1	1.5	1.6	1.7	1.8	1.9
2	[illegible]	[illegible]	[illegible]	[illegible]	[illegible]
3	[illegible]	[illegible]	[illegible]	[illegible]	[illegible]
4	[illegible]	[illegible]	[illegible]	[illegible]	[illegible]
5	[illegible]	[illegible]	[illegible]	[illegible]	[illegible]
6	[illegible]	[illegible]	[illegible]	[illegible]	[illegible]
7	[illegible]	[illegible]	[illegible]	[illegible]	[illegible]
8	[illegible]	[illegible]	[illegible]	[illegible]	[illegible]
9	[illegible]	[illegible]	[illegible]	[illegible]	[illegible]

	20	21	22	23	24
1	2.0	[illegible]	2.2	2.3	2.4
2	[illegible]	[illegible]	[illegible]	[illegible]	[illegible]
3	[illegible]	[illegible]	[illegible]	[illegible]	[illegible]
4	[illegible]	[illegible]	[illegible]	[illegible]	[illegible]
5	[illegible]	[illegible]	[illegible]	[illegible]	[illegible]
6	[illegible]	[illegible]	[illegible]	[illegible]	[illegible]
7	[illegible]	[illegible]	[illegible]	[illegible]	[illegible]
8	[illegible]	[illegible]	[illegible]	[illegible]	[illegible]
9	[illegible]	[illegible]	[illegible]	[illegible]	[illegible]

	25	26	27	28	29
1	2.5	2.6	2.7	2.8	2.9
2	[illegible]	[illegible]	[illegible]	[illegible]	[illegible]
3	[illegible]	[illegible]	[illegible]	[illegible]	[illegible]
4	[illegible]	[illegible]	[illegible]	[illegible]	[illegible]
5	[illegible]	[illegible]	[illegible]	[illegible]	[illegible]
6	[illegible]	[illegible]	[illegible]	[illegible]	[illegible]
7	[illegible]	[illegible]	[illegible]	[illegible]	[illegible]
8	[illegible]	[illegible]	[illegible]	[illegible]	[illegible]
9	[illegible]	[illegible]	[illegible]	[illegible]	[illegible]

	30	31	32
1	3.0	3.1	3.2
2	6.0	6.2	6.4
3	9.0	9.3	9.6
4	12.0	12.4	12.8
5	15.0	15.5	16.0
6	18.0	18.6	19.2
7	21.0	21.7	22.4
8	24.0	24.8	25.6
9	27.0	27.9	28.8

$$Log \frac{1}{1-a}$$

Log a	0	1	2	3	4	5	6	7	8	9
8.50 n	9.986479	6448	6418	6387	6356	6325	6294	6263	6232	6200
51 n	6169	6138	6106	6075	6043	6011	5980	5948	5916	5884
52 n	5852	5820	5788	5756	5723	5691	5659	5626	5593	5561
53 n	5528	5495	5462	5429	5396	5363	5330	5297	5263	5230
54 n	5197	5163	5129	5096	5062	5028	4994	4960	4926	4892
55 n	9.984858	4823	4789	4755	4720	4685	4651	4616	4581	4546
56 n	4511	4476	4441	4406	4370	4335	4300	4264	4228	4193
57 n	4157	4121	4085	4049	4013	3977	3941	3904	3868	3831
58 n	3795	3758	3721	3684	3648	3611	3573	3536	3499	3462
59 n	3424	3387	3349	3312	3274	3236	3198	3160	3122	3084
8.60 n	9.983046	3007	2969	2930	2892	2853	2814	2776	2737	2698
61 n	2658	2619	2580	2541	2501	2462	2422	2382	2343	2303
62 n	2263	2223	2183	2142	2102	2062	2021	1981	1940	1899
63 n	1858	1817	1776	1735	1694	1653	1611	1570	1528	1486
64 n	1444	1403	1361	1319	1276	1234	1192	1149	1107	1064
65 n	9.981022	0979	0936	0893	0850	0807	0763	0720	0677	0633
66 n	0589	0546	0502	0458	0414	0370	0325	0281	0237	0192
67 n	0147	0103	0058	0013	$\bar{9}$968	$\bar{9}$923	$\bar{9}$878	$\bar{9}$832	$\bar{9}$787	$\bar{9}$741
68 n	9.979695	9650	9604	9558	9512	9466	9420	9373	9327	9280
69 n	9234	9187	9140	9093	9046	8999	8952	8904	8857	8809
8.70 n	9.978762	8714	8666	8618	8570	8522	8473	8425	8376	8328
71 n	8279	8230	8181	8132	8083	8034	7985	7935	7885	7836
72 n	7786	7736	7686	7636	7586	7535	7485	7434	7384	7333
73 n	7282	7231	7180	7128	7077	7026	6974	6922	6870	6818
74 n	6766	6714	6662	6610	6557	6505	6452	6399	6346	6293
75 n	9.976240	6187	6133	6080	6026	5972	5918	5864	5810	5756
76 n	5702	5647	5593	5538	5483	5428	5373	5318	5262	5207
77 n	5152	5096	5040	4984	4928	4872	4816	4759	4703	4646
78 n	4589	4532	4475	4418	4361	4304	4246	4188	4131	4073
79 n	4015	3957	3898	3840	3781	3723	3664	3605	3546	3487
8.80 n	9.973428	3368	3309	3249	3189	3129	3069	3009	2949	2888
81 n	2828	2767	2706	2645	2584	2523	2461	2400	2338	2276
82 n	2215	2153	2090	2028	1966	1903	1840	1777	1714	1651
83 n	1588	1525	1461	1398	1334	1270	1206	1141	1077	1013
84 n	0948	0883	0818	0753	0688	0623	0557	0492	0426	0360
85 n	9.970294	0228	0161	0095	0028	$\bar{9}$962	$\bar{9}$895	$\bar{9}$828	$\bar{9}$760	$\bar{9}$693
86 n	9.969626	9558	9490	9422	9354	9286	9218	9149	9081	9012
87 n	8943	8874	8804	8735	8666	8596	8526	8456	8386	8316
88 n	8245	8175	8104	8033	7962	7891	7819	7748	7676	7604
89 n	7532	7460	7388	7316	7243	7170	7097	7024	6951	6878
8.90 n	9.966804	6731	6657	6583	6509	6435	6360	6285	6211	6136
91 n	6061	5985	5910	5834	5759	5683	5607	5531	5454	5378
92 n	5301	5224	5147	5070	4992	4915	4837	4759	4681	4603
93 n	4525	4446	4368	4289	4210	4130	4051	3972	3892	3812
94 n	3732	3652	3571	3491	3410	3329	3248	3167	3086	3004
95 n	9.962922	2840	2758	2676	2594	2511	2428	2845	2262	2179
96 n	2095	2012	1928	1844	1760	1675	1591	1506	1421	1336
97 n	1251	1165	1080	0994	0908	0822	0735	0649	0562	0475
98 n	9.960388	0301	0213	0126	0038	$\bar{9}$950	$\bar{9}$862	$\bar{9}$773	$\bar{9}$685	$\bar{9}$596
99 n	9.959507	9418	9329	9239	9149	9059	8969	8879	8789	8698
9.00 n	8607	8516	8425	8334	8242	8150	8058	7966	7874	7781

Proportional parts

	32	34	36	38	40
1	3.2	3.4	3.6	3.8	4.0
2	6.4	6.8	7.2	7.6	8.0
3	9.6	10.2	10.8	11.4	12.0
4	12.8	13.6	14.4	15.2	16.0
5	16.0	17.0	18.0	19.0	20.0
6	19.2	20.4	21.6	22.8	24.0
7	22.4	23.8	25.2	26.6	28.0
8	25.6	27.2	28.8	30.4	32.0
9	28.8	30.6	32.4	34.2	36.0

	42	44	46	48	50
1	4.2	4.4	4.6	4.8	5.0
2	8.4	8.8	9.2	9.6	10.0
3	12.6	13.2	13.8	14.4	15.0
4	16.8	17.6	18.4	19.2	20.0
5	21.0	22.0	23.0	24.0	25.0
6	25.2	26.4	27.6	28.8	30.0
7	29.4	30.8	32.2	33.6	35.0
8	33.6	35.2	36.8	38.4	40.0
9	37.8	39.6	41.4	43.2	45.0

	52	54	56	58	60
1	5.2	5.4	5.6	5.8	6.0
2	10.[illegible]	10.[illegible]	1[illegible].[illegible]	1[illegible].[illegible]	12.0
3	[illegible]5.	16.	[illegible].	1 .	18.
4	0.	21.	.	2 .	24.
5	6.	27.	.	2 .	30.
6	1.	32.	.	3 .	36.
7	6.	37.	.	40.	42.
8	1.	43.	44.	46.	48.
9	46.	48.	50.	52.	54.

	62	64	66	68	70
1	6.2	6.4	6.6	6.8	7.0
2	12.4	12.8	13.2	13.6	14.0
3	18.6	19.2	19.8	20.4	21.0
4	24.8	25.6	26.4	27.2	28.0
5	31.0	32.0	33.0	34.0	35.0
6	37.2	38.4	39.6	40.8	42.0
7	43.4	44.8	46.2	47.6	49.0
8	49.6	51.2	52.8	54.4	56.0
9	55.8	57.6	59.4	61.2	63.0

	72	74	76	78	80
1	7.2	7.4	7.6	7.8	8.0
2	14.4	14.8	15.2	15.6	16.0
3	21.6	22.2	22.8	23.4	24.0
4	28.8	29.6	30.4	31.2	32.0
5	36.0	37.0	38.0	39.0	40.0
6	43.2	44.4	45.6	46.8	48.0
7	50.4	51.8	53.2	54.6	56.0
8	57.6	59.2	60.8	62.4	64.0
9	64.8	66.6	68.4	70.2	72.0

	82	84	86	88	90
1	8.2	8.4	8.6	8.8	9.0
2	16.4	16.8	17.2	17.6	18.0
3	24.6	25.2	25.8	26.4	27.0
4	32.8	33 6	34.4	35.2	36.0
5	41.0	42.0	43.0	44.0	45.0
6	49.2	50.4	51.6	52.8	54.0
7	57.4	58.8	60.2	61.6	63.0
8	65.6	67.2	68.8	70.4	72.0
9	73.8	75.6	77.4	79.2	81.0

INDEX.

80639°—13——12

Spotted something that isn't quite right?

Let us know and we'll fix it for future reprints. And as a thank you for your diligence, we'll send you a free Sothis book of your choice.

publisher@sothis.press

This page intentionally left blank

Made in the USA
Monee, IL
08 November 2026

67982455R00132